Lecture Notes in Computer Science   2277

Edited by G. Goos, J. Hartmanis, and J. van Leeuwen

Springer
Berlin
Heidelberg
New York
Barcelona
Hong Kong
London
Milan
Paris
Tokyo

Paul Callaghan   Zhaohui Luo
James McKinna   Robert Pollack (Eds.)

Types for
Proofs and Programs

International Workshop, TYPES 2000
Durham, UK, December 8-12, 2000
Selected Papers

 Springer

Series Editors

Gerhard Goos, Karlsruhe University, Germany
Juris Hartmanis, Cornell University, NY, USA
Jan van Leeuwen, Utrecht University, The Netherlands

Volume Editors

Paul Callaghan
Zhaohui Luo
James McKinna
Robert Pollack
University of Durham, Department of Computer Science
South Road, Durham DH1 3LE, United Kingdom
E-mail: {p.c.callaghan,zhaohui.luo,j.h.mckinna}@durham.ac.uk

Robert Pollack
The University of Edinburgh, Division of Informatics
Laboratory for Foundations of Computer Science
James Clerk Maxwell Building, The King's Buildings
Mayfield Road, Edinburgh EH9 3JZ, Scotland, United Kingdom
E-mail: rap@dcs.ed.ac.uk

Cataloging-in-Publication Data applied for

Die Deutsche Bibliothek - CIP-Einheitsaufnahme

Types for proofs and programs : international workshop, TYPES 2000, Durham,
UK, December 8 - 12, 2000 ; selected papers / Paul Callaghan ... (ed.). -
Berlin ; Heidelberg ; New York ; Barcelona ; Hong Kong ; London ; Milan ;
Paris ; Tokyo : Springer, 2002
  (Lecture notes in computer science ; Vol. 2277)
  ISBN 3-540-43287-6

CR Subject Classification (1998):F.3.1, F.4.1, D.3.3, I.2.3

ISSN 0302-9743
ISBN 3-540-43287-6 Springer-Verlag Berlin Heidelberg New York

Springer-Verlag Berlin Heidelberg New York
a member of BertelsmannSpringer Science+Business Media GmbH

http://www.springer.de

© Springer-Verlag Berlin Heidelberg 2002
Printed in Germany

Typesetting: Camera-ready by author, data conversion by Christian Grosche, Hamburg
Printed on acid-free paper   SPIN 10846238   06/3142   5 4 3 2 1 0

Preface

This book contains a selection of papers presented at the first annual workshop of the TYPES Working Group (Computer-Assisted Reasoning Based on Type Theory, EU IST project 29001), which was held 8th – 12th of December, 2000 at the University of Durham, Durham, UK. It was attended by about 80 researchers.

The workshop follows a series of meetings organised in 1993, 1994, 1995, 1996, 1998, and 1999 under the auspices of the Esprit BRA 6435 and the Esprit Working Group 21900 for the previous TYPES projects. Those proceedings were also published in the LNCS series, edited by Henk Barendregt and Tobias Nipkow (Vol. 806, 1993), by Peter Dybjer, Bengt Nordström, and Jan Smith (Vol. 996, 1994), by Stefano Berardi and Mario Coppo (Vol. 1158, 1995), by Christine Paulin-Mohring and Eduardo Gimenez (Vol. 1512, 1996), by Thorsten Altenkirch, Wolfgang Naraschewski, and Bernhard Reus (Vol. 1657, 1998), and by Thierry Coquand, Peter Dybjer, Bengt Nordström, and Jan Smith (Vol. 1956, 1999). The Esprit BRA 6453 was itself a continuation of the former Esprit Action 3245, Logical Frameworks: Design, Implementation, and Experiments. The articles from the annual workshops under that Action were edited by Gerard Huet and Gordon Plotkin in the books Logical Frameworks and Logical Environments, both published by Cambridge University Press.

Acknowledgements

We are very grateful to members of Durham's Computer Assisted Reasoning Group, especially Robert Kießling, for helping to organise the workshop. Robert's contribution was key to the success of the meeting. We also wish to thank Louise Coates for administrative support.

We would also like to thank the invited speakers, Susumu Hayashi and Giuseppe Longo, for accepting our invitation and for contributing to this volume.

The Working Group received funding from the EU IST programme, which has made the annual workshops possible.

December 2001

Paul Callaghan, Zhaohui Luo
James McKinna, Robert Pollack

Referees

We would like to thank the following people for their kind work in reviewing the papers submitted to these proceedings.

Thorsten Altenkirch
Jamie Andrews
Marcin Benke
Stefano Berardi
Stefan Berghofer
Yves Bertot
Marc Bezem
Catarina Coquand
David Delahaye
Ewen Denney
Gilles Dowek
Uwe Egly
Matt Fairtlough

Daniel Fridlender
Pietro Di Gianantonio
Healfdene Goguen
Masami Hagiya
Michael Hedberg
Hugo Herbelin
Martin Hofmann
Paul Jackson
Conor McBride
Michael Mendler
Eugenio Moggi
Milad Niqui
Erik Palmgren

Christine Paulin
Loïc Pottier
David Pym
Christophe Raffalli
Fred Richman
Natarajan Shankar
Alex Simpson
Jan Smith
Sergei Soloviev
Björn von Sydow
Makoto Takeyama
Benjamin Werner
Freek Wiedijk

Table of Contents

Collection Principles in
Dependent Type Theory*

Peter Aczel[1] and Nicola Gambino[2]

[1] Departments of Mathematics and Computer Science
University of Manchester
petera@cs.man.ac.uk
[2] Department of Computer Science, University of Manchester
ngambino@cs.man.ac.uk

Abstract. We introduce logic-enriched intuitionistic type theories, that extend intuitionistic dependent type theories with primitive judgements to express logic. By adding type theoretic rules that correspond to the collection axiom schemes of the constructive set theory **CZF** we obtain a generalisation of the type theoretic interpretation of **CZF**. Suitable logic-enriched type theories allow also the study of reinterpretations of logic. We end the paper with an application to the double-negation interpretation.

Introduction

In [1] the constructive set theory **CZF** was given an interpretation in the dependent type theory $\mathbf{ML_1 V}$. This type theory is a version of Martin-Löf's intuitionistic type theory with one universe of small types, but no W-types except for the special W-type V which is used to interpret the universe of sets of **CZF**. In [2] the interpretation was extended to an interpretation of $\mathbf{CZF^+} = \mathbf{CZF} + \mathbf{REA}$ in $\mathbf{ML_{1w} V}$. Here **REA** is the Regular Extension Axiom and $\mathbf{ML_{1w} V}$ is obtained from $\mathbf{ML_1 V}$ by adding rules to express that the universe of small types is closed under the formation of W-types, although there are no rules for the general formation of W-types, except for the special W-type V.

In intuitionistic type theories such as $\mathbf{ML_1 V}$ logic is usually represented using the propositions-as-types idea. This is indeed how the intuitionistic logic of **CZF** is interpreted in $\mathbf{ML_1 V}$. The propositions-as-types interpretation of logic plays an important role in the type theoretic interpretation of **CZF**. Recent work by I. Moerdijk and E. Palmgren shows however that it is possible to interpret **CZF** in predicative categorical universes in which logic is not interpreted using the propositions-as-types idea [17, 18].

---

* This paper was written while visiting the Mittag-Leffler Institute, The Royal Swedish Academy of Sciences. Both authors wish to express their gratitude for the invitation to visit the Institute. The first author is also grateful to his two departments for supporting his visit. The second author is grateful to his department and to the "Fondazione Ing. A. Gini" for supporting his visit.

P. Callaghan et al. (Eds.): TYPES 2000, LNCS 2277, pp. 1–23, 2002.

One of the aims of this paper is to show how the interpretation of **CZF** can be generalised to an interpretation of **CZF** in a logic-enriched intuitionistic type theory **ML(CZF)**, which itself has a natural interpretation back into **CZF**. By a logic-enriched intuitionistic type theory we mean a pure intuitionistic type theory like $\mathbf{ML_1V}$ that is extended with extra judgement forms to express, relative to a context of variable declarations, being a proposition and assertions that one proposition follows from others.

We expect this generalisation to be fruitful. We will give just one indication of this in the paper. We will show how a logic-enriched type theory can accommodate the j-translation reinterpretations of logic. The idea of a j-translation is really a folklore idea to generalise the double negation translation of classical logic into intuitionistic logic by using any defined modal operator j to generalise the double negation operator provided that j satisfies suitable conditions that correspond to the conditions for a Lawvere-Tierney topology in an elementary topos or for a nucleus on a frame. In general only intuitionistic logic gets reinterpreted. See [5] for a recent treatment.

In order to carry through our interpretation of **CZF** in a logic-enriched type theory it will be necessary to include two rules in the type theory corresponding to the two collection schemes of **CZF**, Strong Collection and Subset Collection. In the original type theoretic interpretation these new type theoretic rules, called Collection rules, were not needed as they are both consequences of the type theoretic axiom of choice that holds in the propositions-as-types interpretation of logic. Fortunately we will see that the type theoretic Collection rules are preserved in a j-translation, although an extra condition on j is required to show that the rule corresponding to Subset Collection is preserved. A set theoretical counterpart of this condition was used in [10] to develop frame-valued semantics for **CZF**.

We have discussed the issue of translating **CZF** into an intuitionistic type theory. It is also natural to consider translations in the opposite direction. For example we may interpret types as sets and objects of a type as elements of the corresponding set. Then to each type forming operation there is the natural set forming operation that corresponds to it. For example corresponding to the Σ and Π forms of type are their set theoretical versions. In this way we get a conceptually very simple set theoretical interpretation of the type theory **ML**, which has no universes or W-types, in **CZF**, and this extends to an interpretation of **MLW** in $\mathbf{CZF^+}$ and $\mathbf{MLW_1}$ in an extension $\mathbf{CZF^+u}$ of $\mathbf{CZF^+}$ expressing the existence of a universe in the sense of [19]. Here **MLW** is **ML** with W-types and $\mathbf{MLW_1}$ is its extension with a universe of small types reflecting all the forms of type of **MLW**. The syntactic details of this kind of translation can be found in [4].

A weakness of these types-as-sets interpretations, when linked with the reverse sets-as-trees interpretations, is that there seems to be a mismatch between the set theories and the type theories. So although we get a translation of **CZF** into $\mathbf{ML_1V}$ we only seem to get a translation of **ML** into **CZF** and to translate $\mathbf{ML_1V}$ into a constructive set theory using types-as-sets we seem to need to go to the set theory $\mathbf{CZF^+u}$, which is much stronger than $\mathbf{CZF}$. This mismatch is overcome in [4] by having axioms for an infinite hierarchy of universes on both the type theory side and the set theory side. This allows for the two sides to catch up with each other.

This brings us to the second main aim of this paper, which is to present another approach to resolving the mismatch problem by replacing the types-as-sets approach by a types-as-classes approach to interpreting type theories in set theories. To carry this approach through it is necessary to restrict the formation of Π-types. This is because in set theory we may only form the class $\Pi_{x \in A} B_x$, where A is a class and B_x is a class for each $x \in A$, when the class A is a set. The corresponding restriction in type theory is to require that $(\Pi x : A)B(x)$ be allowed as a type only when B is a family of types indexed by a small type A. When W-types are also wanted then again it is necessary to put a restriction on their formation. This time the restriction is to only allow the formation of a type $(Wx : A)B(x)$ when each type $B(x)$ is small, although the type A need not be small.

With these restrictions we are led to consider a pure type theory $\mathbf{ML_1^-} + \mathbf{W^-}$ and a logic-enriched extension $\mathbf{ML(CZF)}$ and give a types-as-classes interpretation of $\mathbf{ML(CZF)}$ in $\mathbf{CZF}$. With the reverse sets-as-trees interpretation of $\mathbf{CZF}$ in $\mathbf{ML(CZF)}$ we get a match between a type theory and $\mathbf{CZF}$.

Remark. Due to space constraints, unfortunately this paper does not contain any proofs. Most of the type theoretic rules are also omitted. We hope to present in a future occasion a full version of the paper, including detailed proofs and a complete list of type theoretic rules. A draft of the full version is available from the authors' web pages.

Plan of the Paper. Section 1 recalls pure type theories and introduces logic-enriched type theories. Section 2 is devoted to the propositions-as-types interpretation of logic-enriched type theories into pure type theories. Section 3 contains the types-as-classes interpretation of logic-enriched type theories into $\mathbf{CZF}$. Section 4 presents the Collection rules and their proposition-as-types and types-as-classes interpretation. Section 5 develops a generalised type theoretic interpretation of $\mathbf{CZF}$ into $\mathbf{ML(CZF)}$. Section 6 discusses j-translation reinterpretations of logic in logic-enriched type theories.

1 Logic-Enriched Type Theories

1.1 Standard Pure Type Theories

A **standard pure type theory** has the forms of judgement $(\Gamma)\ \mathcal{B}$ where Γ is a context consisting of a list of declarations $x_1 : A_1, \ldots, x_n : A_n$ of distinct variables $x_1, \ldots, x_n$, and B has one of the forms

$$A : \mathsf{type},$$
$$A = A' : \mathsf{type},$$
$$a : A,$$
$$a = a' : A.$$

For the context Γ to be well-formed it is required that

$$(\;) \; A_1 \quad : \quad \mathsf{type},$$
$$(x_1 : A_1) \; A_2 \quad : \quad \mathsf{type},$$
$$\cdots\cdots$$
$$(x_1 : A_1, \ldots, x_{n-1} : A_{n-1}) \; A_n \quad : \quad \mathsf{type}.$$

The well-formedness of each of these forms of judgement has other presuppositions. Thus, in a well-formed context Γ, the judgement $A = A' : \mathsf{type}$ presupposes that $A : \mathsf{type}$ and $A' : \mathsf{type}$, the judgement $a : A$ presupposes that $A : \mathsf{type}$, and the judgement $a = a' : A$ presupposes that $a : A$ and $a' : A$. In the rest of the paper we will prefer to leave out the empty context whenever possible. So $(\;) \; A_1 : \mathsf{type}$, as above, will be written just $A_1 : \mathsf{type}$.

Any standard type theory will have certain general rules for deriving well-formed judgements, each instance of a rule having the form

$$\frac{J_1 \quad \cdots \quad J_k}{J}$$

where $J_1, \ldots, J_k, J$ are all judgements. In stating a rule of a standard type theory it is very convenient to suppress mention of a context that is common to both the premisses and the conclusion of the rule. For example we will write the reflexivity rule for type equality as just

$$\frac{A : \mathsf{type}}{A = A : \mathsf{type}}.$$

But in applying this rule we are allowed to infer $(\Gamma) \; A = A : \mathsf{type}$ from $(\Gamma) \; A : \mathsf{type}$ for any well-formed context Γ.

It will be convenient in stating certain results to add the following additional form of judgement to a standard type theory

$$(\Gamma) \; B_1, \ldots, B_m \Rightarrow B$$

where $(\Gamma) \; B_i : \mathsf{type}$ for $i = 1, \ldots, m$ and $(\Gamma) \; B : \mathsf{type}$ are well-formed judgements. The only rule involving this form of judgement is

$$\frac{(y_1 : B_1, \ldots, y_m : B_m) \; b : B}{B_1, \ldots, B_m \Rightarrow B}.$$

As there are no other rules involving the new judgement form this extension of a standard type theory is conservative.

1.2 Review of Some Pure Type Theories

We will use **ML** to stand for a variant of Martin-Löf's type theory without universes or W-types. We prefer to avoid having any identity types. Also, rather than have finite types $\mathbb{N}_k$ for all $k = 0, 1, \ldots$ we will just have them for $k = 0, 1, 2$ and use the notation $\mathbb{0}, \mathbb{1}, \mathbb{2}$ for them. As usual we define binary product and function types as follows:

$$A_1 \times A_2 \stackrel{\text{def}}{=} (\Sigma_{\_} : A_1)A_2\,,$$
$$A_1 \to A_2 \stackrel{\text{def}}{=} (\Pi_{\_} : A_1)A_2\,,$$

where the symbol $\_$ indicates an anonymous bound variable. Finally we do not take binary sums as primitive but define them. To do so we will allow dependent types to be defined by cases on $\mathbb{2}$; i.e. given $A_1, A_2 :$ **type** we allow the formation of $\mathbb{R}_2(A_1, A_2, c) :$ **type** whenever $c : \mathbb{2}$ so that for $i = 1, 2$,

$$\mathbb{R}_2(A_1, A_2, i) = A_i : \textsf{type},$$

where $1, 2 : \mathbb{2}$ are the canonical elements of $\mathbb{2}$. Using $\mathbb{R}_2$ we define, for types A_1, A_2,

$$A_1 + A_2 \stackrel{\text{def}}{=} (\Sigma z : \mathbb{2})\mathbb{R}_2(A_1, A_2, z)\,.$$

So the primitive forms of type of **ML** are

$$\mathbb{0}, \mathbb{1}, \mathbb{2}, \mathbb{N}, \mathbb{R}_2(A_1, A_2, e), (\Sigma x : A)B, (\Pi x : A)B\,.$$

The type theory $\mathbf{ML_1}$ is the pure standard theory obtained from **ML** by adding a type universe $\mathbb{U}$ of small types, or rather of representatives for small types as we will use the universe, *à la* Tarski, where for each $a : \mathbb{U}$, $\mathbb{T}(a)$ is the small type represented by a. The rules for $\mathbb{U}$ express that $\mathbb{U}$ reflects all the forms of type of **ML**.

The type theory **MLW** is obtained from **ML** by adding rules for the W-types $(Wx : A)B$ and $\mathbf{MLW_1}$ is like $\mathbf{ML_1}$ except that the rules for the W-types are added and the type universe $\mathbb{U}$ also reflects the W-types[1]. There are two natural subtheories of $\mathbf{MLW_1}$. The first one[2], $\mathbf{ML_1 + W}$, has W-types but they are not reflected in $\mathbb{U}$. The second one, $\mathbf{ML_{1w}}$, only has small W-types.

The pure type theory $\mathbf{ML_1 + W}$ is a stronger theory than the set theory **CZF**. By replacing the Π and W forms of type by restricted versions we will get a pure type theory $\mathbf{ML_1^- + W^-}$ which will have a straightforward translation into **CZF** in which types are interpreted as classes of **CZF** and terms are interpreted as sets of **CZF**. This interpretation is to be discussed in sec. 3. In $\mathbf{MLW_1}$ we

---

[1] In the literature essentially this theory has been written $\mathbf{ML_1 W}$. But this seems a bit misleading as it might suggest that the universe does not reflect the W-types.

[2] It would be natural to name this type theory $\mathbf{ML_1 W}$, but this is in conflict with the notation of the literature.

can define the following restricted versions of the Π and W forms of type. For $a : \mathbb{U}$ and $(x : \mathbb{T}(a))\, B :$ type let

$$(\Pi^- x : a)B \stackrel{\mathrm{def}}{=} (\Pi x : \mathbb{T}(a))B\,,$$

and for $A :$ type and $(x : A)\, b : \mathbb{U}$ let

$$(W^- x : A)b \stackrel{\mathrm{def}}{=} (W x : A)\mathbb{T}(b)\,.$$

The type theory $\mathbf{ML_1^-} + \mathbf{W^-}$ does not have the Π and W forms of type but instead has rules for Π^- and W^- as primitive type forming operators. Note that the type universe $\mathbb{U}$ still reflects the Π^- forms of type, but does not reflect the W^- forms.

It seems necessary to add extra elimination rules for the type 2 and the W^- forms of type. In the case of 2 this is so as to be able to derive the elimination rules for the defined binary sums, $A_1 + A_2$. In the case of the W^- forms of type we will need double recursion on W^--types when we come to define extensional equality on the W^--type $\mathbb{V}$, to be defined in sec. 5. In both cases the Π^--types do not seem to be enough to get what we want, although Π-types are enough. We can no longer define function types $A_1 \to A_2$ for types A_1, A_2. Instead we define $a \to A'$ for $a : \mathbb{U}$ and $A' :$ type as follows:

$$a \to A' \stackrel{\mathrm{def}}{=} (\Pi^-\_ : a)A'\,.$$

1.3   Adding Predicate Logic

Given a standard pure type theory we may consider enriching it with the following two additional forms of judgement, $(\Gamma)\, \mathcal{B}$, where Γ should be a well-formed context as before and $\mathcal{B}$ has one of the following forms:

$$\phi : \mathsf{prop}\,,$$
$$\phi_1, \ldots, \phi_m \Rightarrow \phi\,.$$

In the context Γ the well-formedness of $\phi_1, \ldots, \phi_m \Rightarrow \phi$ presupposes that $\phi_i :$ prop for $i = 1, \ldots, m$ and $\phi :$ prop. Using these new judgement forms it is straightforward to add the standard formation and inference rules for the intuitionistic logical constants; i.e. the canonical true and false propositions $\top, \bot$, the binary connectives $\wedge, \vee, \supset$ and the quantifiers $(\forall x : A), (\exists x : A)$ for each type A. As an example, in Table 1 we give formation and inference rules for disjunction and existential quantification.

As always, in the statement of formation rules we suppress a context that is common to the premises and conclusion. In the inference rules we will also suppress a list of assumptions appearing on the left hand side of $\Rightarrow$ in the logical premises and conclusion of each inference rule. Moreover we will write $(\Gamma)\, \phi$ rather than $(\Gamma) \Rightarrow \phi$ and just ϕ rather than the judgement $\Rightarrow \phi$.

Given a standard pure type theory $\mathbf{T}$ let $\mathbf{T} + \mathbf{IL}$ be the theory obtained from $\mathbf{T}$ by enriching it with intuitionistic logic as just described. Each type determines

Table 1. Formation and inference rules for disjunction and existential quantification

$$\frac{\phi_1 : \text{prop} \qquad \phi_2 : \text{prop}}{\phi_1 \vee \phi_2 : \text{prop}} \qquad\qquad \frac{A : \text{type} \qquad (x : A)\ \phi : \text{prop}}{(\exists x : A)\ \phi : \text{prop}}$$

$$\frac{\phi_1 : \text{prop} \qquad \phi_2 : \text{prop} \qquad \phi_i}{\phi_1 \vee \phi_2}\ (i = 1, 2) \qquad\qquad \frac{\phi_1 \vee \phi_2 \qquad \phi_1 \Rightarrow \theta \qquad \phi_2 \Rightarrow \theta}{\theta}$$

$$\frac{a : A \quad (x : A)\ \phi : \text{prop} \quad \phi[a/x]}{(\exists x : A)\phi} \qquad\qquad \frac{(\exists x : A)\phi \quad \theta : \text{prop} \quad (x : A)\ \phi \Rightarrow \theta}{\theta}$$

the proposition that there is an object of that type; i.e. we can associate with each type expression A of $\mathbf{T}$, the formula $!A \overset{\text{def}}{=} (\exists_- : A)\top$ of $\mathbf{T} + \mathbf{IL}$.

The proof of the direction from left to right in the following result involves simple applications of the inference rules for $\top$ and $\exists$. The result in the other direction is an 'Explicit Definability' result that generalises the Explicit Definability for Numbers result for Heyting Arithmetic. The proof for Heyting Arithmetic, as given in 5.10 of Chapter 3 of [21], carries over here.

Proposition 1. *For any standard pure type theory* $\mathbf{T}$, *if* $\mathbf{T} \vdash (\Gamma)\ B_i : \text{type}$ *for* $i = 1, \ldots, m$, *and* $\mathbf{T} \vdash (\Gamma)\ B : \text{type}$ *then*

$$\mathbf{T} \vdash (\Gamma)\ B_1, \ldots, B_m \Rightarrow B \quad \textit{iff} \quad \mathbf{T} + \mathbf{IL} \vdash (\Gamma)\ !B_1, \ldots, !B_m \Rightarrow !B\,.$$

1.4   Induction Rules

It is natural to extend a standard logic-enriched type theory with additional non-logical rules to express properties of the various forms of type. For example it is natural to add a rule for mathematical induction to the rules concerning the type of natural numbers and there are similar rules for the other inductive forms of type.

So, for each inductive type $C : \text{type}$ of $\mathbf{MLW} + \mathbf{IL}$, if $(z : C)\ \phi : \text{prop}$ and $e : C$ we have the induction rule

$$\frac{\text{Premisses}}{\phi[e/z]}\,,$$

where the inductive types C and the correspondence between the form of C and the premisses is given in Table 2.

Table 2. The Inductive types of **MLW** and the premisses of their induction rules

C	Premisses
0	None,
1	$\phi[0/z]$,
2	$\phi[1/z]$, $\phi[2/z]$,
$\mathbb{N}$	$\phi[0/z]$, $(x : \mathbb{N})\, \phi[x/z] \Rightarrow \phi[\mathsf{succ}(x)/z]$,
$(\Sigma x : A)B$	$(x : A, y : B)\, \phi[\mathsf{pair}(x,y)/z]$
$(W x : A)B$	$(x : A, u : B \to C)\, (\forall y : B)\phi[\mathsf{app}(u,y)/z] \Rightarrow \phi[\mathsf{sup}(x,u)/z]$.

2   Propositions-as-Types

2.1   Propositions-as-Types Interpretation for ML and MLW

We present the familiar propositions-as-types translation, here abbreviated PaT translation, of **ML + IL** into **ML**. The PaT translation has no effect on the pure **ML** part but, relative to any context, associates with each $\phi :$ prop a type $Pr(\phi) :$ type, so that for each derivation of a judgement $(\Gamma)\ \ \phi_1, \ldots, \phi_m \Rightarrow \phi$ in **ML + IL** there is a derivation in **ML** of $(\Gamma)\ \ Pr(\phi_1), \ldots, Pr(\phi_m) \Rightarrow Pr(\phi)$. The PaT translation is defined as follows:

$$Pr(\bot) \overset{\mathrm{def}}{=} 0$$

$$Pr(\top) \overset{\mathrm{def}}{=} 1$$

$$Pr(\phi_1 \wedge \phi_2) \overset{\mathrm{def}}{=} Pr(\phi_1) \times Pr(\phi_2)$$

$$Pr(\phi_1 \vee \phi_2) \overset{\mathrm{def}}{=} Pr(\phi_1) + Pr(\phi_2)$$

$$Pr(\phi_1 \supset \phi_2) \overset{\mathrm{def}}{=} Pr(\phi_1) \to Pr(\phi_2)$$

$$Pr((\forall x : A)\phi_0) \overset{\mathrm{def}}{=} (\Pi x : A)Pr(\phi_0)$$

$$Pr((\exists x : A)\phi_0) \overset{\mathrm{def}}{=} (\Sigma x : A)Pr(\phi_0)$$

We will need the following rule (**PaT**) to state our main result below about the propositions-as-types interpretation

$$\frac{\phi :\ \mathsf{prop}}{\phi \ \equiv\ !\,Pr(\phi)} \qquad\qquad \textbf{(PaT)}$$

where, for $\phi, \psi :$ prop, we define $\phi \equiv \psi \overset{\mathrm{def}}{=} (\phi \supset \psi) \wedge (\psi \supset \phi)$. Recall the familiar fact that the type theoretic axiom of choice holds in the propositions-as-types interpretation. We express this version of the axiom of choice as the rule:

$$\frac{A :\ \mathsf{type} \qquad (x : A)\ B :\ \mathsf{type} \qquad (x : A, y : B)\ \phi :\ \mathsf{prop}}{(\forall x : A)(\exists y : B)\phi \ \Rightarrow\ (\exists z : C)(\forall x : A)\phi[\mathsf{app}(z,x)/y]} \qquad\qquad \textbf{(AC)}$$

where C is $(\Pi x : A)B$

In order to state our result we assume that **T** is any standard pure theory that includes **ML**. For each raw judgement J of **T** + **IL** let J^{PaT} be the raw judgement of **T** that is just J except when J has either the form $(\Gamma)\ \phi :$ prop or $(\Gamma)\ \phi_1, \ldots, \phi_m \Rightarrow \phi$ and then J^{PaT} has the form $(\Gamma)\ Pr(\phi) :$ type or $(\Gamma)\ Pr(\phi_1), \ldots, Pr(\phi_m) \Rightarrow Pr(\phi)$ respectively. For the next result we need the rule $(0\bot)$.

$$\frac{a : 0}{\bot} \qquad\qquad (0\bot)$$

Note that, given our abbreviatory conventions, the rule $(0\bot)$ allows us to infer $(\Gamma)\ \phi_1, \ldots, \phi_n \Rightarrow \bot$ from $(\Gamma)\ a : 0$ and $(\Gamma)\ \phi_i :$ prop for $i = 1, \ldots, n$.

Theorem 2. *Let* **T** *be as above. Then*

1. **T** + **IL** + (**PaT**) $\vdash J$ *implies* **T** $\vdash J^{\text{PaT}}$.
2. *In* **T** + **IL**, *the rule* (**PaT**) *is equivalent to the combination of rules* (**AC**) *and* $(0\bot)$.

The main work in proving part 1 is first to show how each logical inference rule translates into a derived rule of **ML** following the well-known propositions-as-types idea and second to observe that the instances of (**PaT**) translate into instances of the following derived rule

$$\frac{A : \text{type}}{A \leftrightarrow 1 \times A}$$

where, for types A, B, $A \leftrightarrow B \stackrel{\text{def}}{=} (A \to B) \times (B \to A)$. Provided that **T** does not have any additional rules for forming the types of **ML** the converse implication to part 1 holds. For part 2, the rule (**PaT**) can be used to prove (**AC**) as in Martin-Löf's original proof of the type theoretic axiom of choice in his type theory and the rule $(0\bot)$ is derived using the instance of (**PaT**) when ϕ is $\bot$. For the other direction of part 2, (**PaT**) is proved by induction on the formation of the formula ϕ. The rule $(0\bot)$ is needed to deal with $\bot$ and (**AC**) is needed to deal with the implication and universal quantification cases.

Theorem 3. *The induction rules for the inductive types of* **ML** *or* **MLW** *can be derived in* **ML** + **IL** + (**PaT**) *or* **MLW** + **IL** + (**PaT**), *respectively.*

This result expresses the familiar observation that each instance of the induction rule for an inductive type comes from an instance of the elimination rule for that type when treating propositions as types.

2.2   A Proposition Universe and Its PaT Translation

When adding logic to a standard pure type theory **T** that includes **ML₁** it is natural to also add a proposition universe $\mathbb{P}$ to match the type universe $\mathbb{U}$. The formation rule for this type is

$$\mathbb{P} : \text{type} .$$

Elements of this type are to be thought of as representatives for propositions whose quantifiers range over small types. Introduction rules for $\mathbb{P}$ are straightforward. Each object $a : \mathbb{P}$ represents a proposition $\tau(a) :$ prop. The elimination rule for the type $\mathbb{P}$ is the following:

$$\frac{a : \mathbb{P}}{\tau(a) : \mathsf{prop}}.$$

For the type $\mathbb{P}$, it seems convenient to avoid the use of an equality form of judgement for propositions in order to express that $\mathbb{P}$ reflects logic. Instead we use logical equivalence. As examples of these rules, in Table 3 we give rules for disjunction and existential quantification.

Table 3. Disjunction and existential quantification in the proposition universe

$$\frac{p_1 : \mathbb{P} \qquad p_2 : \mathbb{P}}{p_1 \dot{\vee} p_2 : \mathbb{P}} \qquad\qquad \frac{a : \mathbb{U} \qquad (x : \tau(a))\, p : \mathbb{P}}{(\dot{\exists} x : a) p : \mathbb{P}}$$

$$\frac{p_1 : \mathbb{P} \qquad p_2 : \mathbb{P}}{\tau(p_1 \dot{\vee} p_2) \equiv \tau(p_1) \vee \tau(p_2)} \qquad\qquad \frac{a : \mathbb{U} \qquad (x : \tau(a))\, p : \mathbb{P}}{\tau(\dot{\exists} x : a) p \equiv (\exists x : \tau(a)) \tau(p)}$$

When the pure type theory $\mathbf{T}$ includes $\mathbf{ML_1}$ then we write $\mathbf{T} + \mathbf{IL_1}$ for the enrichment of $\mathbf{T}$ with intuitionistic predicate logic and also the rules for $\mathbb{P}$, as illustrated in Table 3. We now wish to give a translation of $\mathbf{T} + \mathbf{IL_1}$ into $\mathbf{T} + \mathbf{IL}$ by interpreting $\mathbb{P}$ as $\mathbb{U}$ following the propositions-as-types idea. Each new symbol of $\mathbf{T} + \mathbf{IL_1}$ that was added to $\mathbf{T} + \mathbf{IL}$ is reinterpreted as a primitive or defined symbol of $\mathbf{T} + \mathbf{IL}$ according to the following correspondence:

$$\frac{\mathbb{P} \quad \tau \quad \dot{\bot} \quad \dot{\top} \quad \dot{\wedge} \quad \dot{\vee} \quad \dot{\supset} \quad \dot{\forall} \quad \dot{\exists}}{\mathbb{U} \quad \tau^* \quad \dot{0} \quad \dot{1} \quad \dot{\times} \quad \dot{+} \quad \dot{\rightarrow} \quad \dot{\Pi} \quad \dot{\Sigma}}$$

where the symbols $\tau^*, \dot{\times}, \dot{+}, \dot{\rightarrow}$ are defined in $\mathbf{T} + \mathbf{IL}$ as follows:

$$(x : \mathbb{U}) \qquad \tau^*(x) \overset{\mathrm{def}}{=} \,!\mathbb{T}(x) : \mathsf{prop},$$
$$(x_1, x_2 : \mathbb{U}) \qquad x_1 \dot{\times} x_2 \overset{\mathrm{def}}{=} (\dot{\Sigma}_- : x_1) x_2 : \mathbb{U},$$
$$(x_1, x_2 : \mathbb{U}) \qquad x_1 \dot{+} x_2 \overset{\mathrm{def}}{=} (\dot{\Sigma} z : \dot{2}) R_2(x_1, x_2, z) : \mathbb{U},$$
$$(x_1, x_2 : \mathbb{U}) \qquad x_1 \dot{\rightarrow} x_2 \overset{\mathrm{def}}{=} (\dot{\Pi}_- : x_1) x_2 : \mathbb{U}.$$

For each expression M of $\mathbf{T} + \mathbf{IL_1}$ let us write M^* for the result of this reinterpretation of the symbols occuring in M. For each raw judgement J of $\mathbf{T} + \mathbf{IL_1}$ let J^* be the raw judgement of $\mathbf{T} + \mathbf{IL}$ obtained by this reinterpretation of the symbols in J.

The PaT translation of $\mathbf{T} + \mathbf{IL}$ into $\mathbf{T}$ extends to a translation of $\mathbf{T} + \mathbf{IL_1}$ into $\mathbf{T}$ if we define

$$Pr(\tau(a)) \overset{\text{def}}{=} \mathbb{T}(a^*)$$

for each raw term a of $\mathbf{T} + \mathbf{IL_1}$. To state our next result we need the rule $(\mathbb{P}^*)$.

$$\frac{a : \mathbb{P}}{\tau(a) \equiv \tau^*(a^*)\,.} \qquad (\mathbb{P}^*)$$

Note that the instances of this rule are given by the instances of $(\mathbf{PaT})$ where ϕ has the form $\tau(a)$ for $a : \mathbb{P}$.

Theorem 4. *Let $\mathbf{T}$ be any standard pure type theory that includes $\mathbf{ML_1}$ and let J be any raw judgement of $\mathbf{T} + \mathbf{IL_1}$. Then*

1. $\mathbf{T} + \mathbf{IL_1} + (\mathbb{P}^*) \vdash J$ *implies* $\mathbf{T} + \mathbf{IL} \vdash J^*$.
2. $\mathbf{T} + \mathbf{IL_1} + (\mathbf{PaT}) \vdash J$ *implies* $\mathbf{T} + \mathbf{IL} + (\mathbf{PaT}) \vdash J^*$.
3. *In* $\mathbf{T} + \mathbf{IL_1}$ *the rule* $(\mathbf{PaT})$ *is equivalent to the combination of rules* $(\mathbf{AC}) + (\mathbb{P}^*)$.

Combining the first part of theorem 2 with the second part of theorem 4 we get the following result.

Corollary 5. *Let $\mathbf{T}$ be a standard pure type theory that includes $\mathbf{ML_1}$. Then $\mathbf{T} + \mathbf{IL_1} + (\mathbf{PaT}) \vdash J$ implies $\mathbf{T} \vdash (J)^{(\mathrm{PaT_1})}$, where $J^{(\mathrm{PaT_1})}$ is defined to be the judgement $(J^*)^{(\mathrm{PaT})}$.*

3   Types-as-Classes

In setting up our standard type theories for the purpose of giving a translation into a set theory it will be convenient to have a raw syntax that categorises each expression into one of the three categories of

- individual expression (i.e. term),
- type expression,
- proposition expression (i.e. formula).

These raw expressions need not be well-formed expressions of the type theory. In fact it is exactly the three judgement forms

- (Γ) $a : A$,
- (Γ) $A : \mathsf{type}$,
- (Γ) $\phi : \mathsf{prop}$,

that we use to express that, in the context Γ,

- a is a well-formed term of type A,
- A is a well-formed type,
- ϕ is a well-formed formula.

It will be convenient to call the terms, type expressions and formulae the 0-expressions, 1-expressions and 2-expressions respectively. The raw expressions will be built up from an unlimited supply of individual variables and a signature of constant symbols according to the rules given below. We assume that each constant symbol of the signature has been assigned an arity $(n_1^{\epsilon_1} \cdots n_k^{\epsilon_k})^\epsilon$ where $k \geq 0$, $n_1, \ldots, n_k \geq 0$ and each of $\epsilon, \epsilon_1, \ldots, \epsilon_k$ is one of $0, 1, 2$. A symbol of such an arity is k-**place**. The rules for forming raw expressions of the three kinds are as follows.

1. Every variable is a 0-expression.
2. If κ is a constant symbol of arity $(n_1^{\epsilon_1} \cdots n_k^{\epsilon_k})^\epsilon$ and, for $i = 1, \ldots, k$, M_i is an ϵ_i-expression and $\boldsymbol{x}_i$ is a list of n_i distinct variables then

$$\kappa((\boldsymbol{x}_1)M_1, \ldots, (\boldsymbol{x}_k)M_k)$$

is an ϵ-expression.

Some Conventions. When $k = 0$ then we just write κ rather than $\kappa(\ )$. Also, if some $n_i = 0$ then we write just M_i rather than $(\ )M_i$.

Free and Bound Occurrences. These are defined in the standard way when the $(\boldsymbol{x}_i)$ are treated as variable binding operations, so that free occurrences in M_i of variables from the list $\boldsymbol{x}_i$ become bound in $(\boldsymbol{x}_i)M_i$ and so also bound in the whole expression $\kappa((\boldsymbol{x}_1)M_1, \ldots, (\boldsymbol{x}_k)M_k)$.

Substitution. The result $M[M_1, \ldots, M_k/y_1, \ldots, y_k]$ of simultaneously substituting M_i for free occurrences of y_i in M for $i = 1, \ldots, k$, where $y_1, \ldots, y_k$ are distinct variables, is defined in the standard way, relabelling bound variables as usual so as to avoid variable clashes. This is only uniquely specified up to α-convertibility; i.e. up to suitable relabelling of bound variables. In general expressions will be identified up to α-convertibility.

3.1   The Symbols for the Raw Syntax of ML(CZF)

We will eventually be interested in a standard type theory **ML(CZF)** which will be obtained from $\mathbf{ML_1^-} + \mathbf{W^-} + \mathbf{IL_1}$ by adding some additional rules including the induction rules for its inductive types, but without adding to its raw syntax.

We now present the symbols of $\mathbf{ML_1^-} + \mathbf{W^-} + \mathbf{IL_1}$, together with their arities. In an arity a missing superscript will be taken to be 0 by default.

0-place symbols
- $0_1, 1, 2, 0, \dot{0}, \dot{1}, \dot{2}, \mathbb{N}, \bot, \top$ of arity $()$,
- $\mathbb{0}, \mathbb{1}, \mathbb{2}, \mathbb{N}, \mathbb{U}, \mathbb{P}$ of arity $()^1$ and $\bot, \top$ of arity $()^2$.

1-place symbols
- succ, R_0 of arity (0), $\mathbb{T}$ of arity $(0)^1$ and τ of arity $(0)^2$.

2-place symbols
- $R_1, \mathsf{pair}, \mathsf{app}, \mathsf{sup}, \dot{\wedge}, \dot{\vee}, \dot{\supset}$ of arity (00) snd $\lambda, \dot{\Sigma}, \dot{\Pi}, \dot{\forall}, \dot{\exists}$ of arity (01),

- split of arity (20) and R_{W^-} of arity (30),
- Σ, Π^-, W^- of arities $(0^1 1^1)^1, (01^1)^1, (0^1 1)^1$ respectively and $\vee, \wedge, \supset$ of arity $(0^2 0^2)^2$,
- $\forall, \exists$ of arity $(0^1 1^2)^2$.

3-place symbols

- $R_2, R_{\mathbb{N}}, \mathbb{R}_{W^-}^+, \mathbb{R}_2$ of arities $(000), (030), (500), (0^1 0^1 0)^1$ respectively.

4-place symbols

- $\mathbb{R}_2^+$ of arity (1100).

Special Conventions. If $\star$ is one of the 2-place symbols $\vee, \wedge, \supset, \dot{\wedge}, \dot{\vee}, \dot{\supset}$ then we use infix notation and write $(M_1 \star M_2)$ rather than $\star(M_1, M_2)$ and if ∇ is any one of $\lambda, \Sigma, \Pi^-, W^-, \forall, \exists, \dot{\Sigma}, \dot{\Pi}, \dot{\forall}, \dot{\exists}$ then we use quantifier notation and write $(\nabla x : M)M'$ rather than $\nabla(M, (x)M')$

3.2   The Set Theoretical Interpretation of Raw Syntax

In the rest of sec. 3 we will work informally in **CZF**. By a **set theoretical sentence** we mean a sentence in the language of **CZF** that may have sets as parameters. By a **variable assignment**, ξ, we mean an assignment of a set $\xi(x)$ to each variable x. The following terminology will be useful. We define a 0-class to be a set, a 1-class to be a class and a 2-class to be a set theoretical sentence. Also, for $n \geq 0$ and $\epsilon = 0, 1, 2$, an n^ϵ-class is a definable operator F assigning an ϵ-class $F(a_1, \ldots, a_n)$ to each n-tuple $(a_1, \ldots, a_n)$ of sets.

Given a signature as above for determining a raw syntax for a type theory we will want to give a set theoretical denotation $[[M]]_\xi$ to each expression M and each variable assignment ξ, so that for each term a its denotation, $[[a]]_\xi$, should be a set, for each type expression A its denotation, $[[A]]_\xi$, should be a class and, for each proposition expression ϕ, its denotation $[[\phi]]_\xi$, should be a set theoretical sentence. We will use structural induction on the way expressions are built up. To do this we will need to have a set theoretical interpretation $\mathcal{F}_\kappa$ for each symbol κ of the signature. Each $\mathcal{F}_\kappa$ has to be a suitable operator so that the second clause of the following definition by structural induction makes sense.

1. $[[M]]_\xi = \xi(x)$ if M is a variable x.
2. $[[M]]_\xi = \mathcal{F}_\kappa(F_1, \ldots, F_k)$ if M is the expression $\kappa((x_1)M_1, \ldots, (x_k)M_k)$ where κ is a constant symbol of the signature of arity $(n_1^{\epsilon_1} \cdots n_k^{\epsilon_k})^\epsilon$ and, for $i = 1, \ldots, k$, F_i is the $n_i^{\epsilon_i}$-class such that $F_i(a_i) = [[M_i]]_{\xi(a_i/x_i)}$ for all n_i-tuples a_i of sets.

When κ has arity $(n_1^{\epsilon_1} \cdots n_k^{\epsilon_k})^\epsilon$ we will require that $\mathcal{F}_\kappa$ is a set operator of that arity. This means that whenever F_i is an $n_i^{\epsilon_i}$-class, for $i = 1, \ldots, k$ then $\mathcal{F}_\kappa(F_1, \ldots, F_k)$ should be an ϵ-class obtained 'uniformly' from $F_1, \ldots, F_k$.

3.3   Soundness

Given a set theoretical interpretation as above of the symbols of a signature that determines a set theoretical denotation $[[M]]_\xi$ to each expression M relative to a variable assignment ξ we can define the following semantic notions.

Definition 6. – *If Γ is $x_1 : A_1, \ldots, x_n : A_n$ then let*
  $\xi \models \Gamma$ *iff* $\xi(x_i) \in [[A_i]]_\xi$ *for* $i = 1, \ldots, n$.
– *Let* $\xi \models A :$ type *for any type expression A.*
– *Let* $\xi \models A = A' :$ type *iff* $[[A]]_\xi = [[A']]_\xi$.
– *Let* $\xi \models a : A :$ type *iff* $[[a]]_\xi \in [[A]]_\xi$.
– *Let* $\xi \models a = a' : A :$ type *iff* $[[a]]_\xi = [[a']]_\xi \in [[A]]_\xi$.
– *Let* $\xi \models \phi :$ prop *for any formula ϕ.*
– *Let* $\xi \models \phi_1 \ldots, \phi_m \Rightarrow \phi$ *if* $[[\phi_1]]_\xi \wedge \cdots \wedge [[\phi_m]]_\xi \supset [[\phi]]_\xi$ *is a set theoretical sentence that is true (in **CZF**).*

Definition 7. *The raw judgement $(\Gamma)\mathcal{B}$ is **valid** if $\xi \models \Gamma$ implies $\xi \models \mathcal{B}$ for every variable assignment ξ. A type theory rule is **sound** if whenever the premisses of an instance of the rule are valid then so is the conclusion.*

Along the lines of section 2.4 of [4] we can get the following result.

Theorem 8 (CZF). *There is an interpretation of the raw syntax of the type theory $\mathbf{ML_1^-} + \mathbf{W^-} + \mathbf{IL_1}$ in **CZF** so that each rule of inference of the type theory **T** is sound, where **T** is obtained from $\mathbf{ML_1^-} + \mathbf{W^-} + \mathbf{IL_1}$ by adding the induction rules for its inductive types.*

The interpretation given by this theorem can be rephrased as a syntactic translation into **CZF**.

Corollary 9. *There is a syntactic translation that assigns a sentence "J is valid" of **CZF** to each raw judgement J of the theory **T** of the theorem such that $\mathbf{T} \vdash J$ implies $\mathbf{CZF} \vdash$ "J is valid".*

4   Collection Principles

The original type theoretic interpretation of **CZF** in $\mathbf{ML_1 V}$ rests on two main components. The first component is the definition of a type V, called the type of constructive iterative sets, that is used to interpret the universe of sets of **CZF**. The second component is the propositions-as-types interpretation of logic. This interpretation of logic plays a role in proving the validity of the Restricted Separation, Strong Collection and Subset Collection axiom schemes of **CZF**. Validity of the Restricted Separation axiom scheme follows from the correspondence between restricted propositions and small types. Validity of the Strong Collection and Subset Collection axiom schemes follows instead from the type theoretic axiom of choice, that holds in the propositions-as-types interpretation of logic.

In the following we will present a type theoretic interpretation of **CZF** in a logic-enriched type theory that generalises the original type theoretic interpretation. The generalisation involves treating logic as primitive and not via the propositions-as-types interpretation. In order to do so, we will introduce a logic-enriched type theory called **ML(CZF)**. The type theory **ML(CZF)** extends the logic-enriched type theory **T**, of Theorem 8, with two collection rules, corresponding to the collection axiom schemes of **CZF**. Within the type theory **ML(CZF)** we define a type V, called the type of iterative small classes, that will be used to interpret the universe of sets of **CZF**. The definition of V allows us to prove the validity of the Restricted Separation axiom scheme without assuming the propositions-as-types interpretation of logic. The collection rules of **ML(CZF)** allow us to prove the validity of the Strong Collection and Subset Collection axiom schemes of **CZF** without assuming the type theoretic axiom of choice.

4.1 The Type of Subsets of a Type

Let us now introduce the type of subsets of a type and define some operations on this type that will be useful in the following. For A : **type** we define the **type of subsets** of A, $\mathsf{Sub}(A)$, as follows:

$$\mathsf{Sub}(A) \overset{\text{def}}{=} (\Sigma x : \mathsf{U})\left((x \to \mathbb{P}) \times (x \to A)\right).$$

For $a : \mathsf{Sub}(A)$ we define

$$\dot{\mathsf{el}}(a) \overset{\text{def}}{=} a.1 : \mathsf{U},$$
$$\mathsf{el}(a) \overset{\text{def}}{=} \mathsf{T}(\dot{\mathsf{el}}(a)) : \mathsf{type},$$

and for $x : \mathsf{el}(a)$ we define

$$\dot{\mathsf{dom}}(a, x) \overset{\text{def}}{=} \mathsf{app}(a.2.1, x) : \mathbb{P},$$
$$\mathsf{dom}(a, x) \overset{\text{def}}{=} \tau(\dot{\mathsf{dom}}(a, x)) : \mathsf{prop},$$
$$\mathsf{val}(a, x) \overset{\text{def}}{=} \mathsf{app}(a.2.2, x) : A.$$

Using these definitions, we can informally think of $a : \mathsf{Sub}(A)$ as the 'set' of all objects $\mathsf{val}(a, x) : A$ with $x : \mathsf{el}(a)$ such that $\mathsf{dom}(a, x)$. If $(x : A)\ p : \mathbb{P}$ we define

$$(\dot{\forall} x \in a)\, p \overset{\text{def}}{=} (\dot{\forall} x : \dot{\mathsf{el}}(a))\, \dot{\mathsf{dom}}(a, x) \dot{\supset} p[\mathsf{val}(a, x)/x] : \mathbb{P},$$
$$(\dot{\exists} x \in a)\, p \overset{\text{def}}{=} (\dot{\exists} x : \dot{\mathsf{el}}(a))\, \dot{\mathsf{dom}}(a, x) \dot{\wedge} p[\mathsf{val}(a, x)/x] : \mathbb{P}.$$

If $(x : A)\ \phi : \mathsf{prop}$ we define

$$(\forall x \in a)\, \phi \overset{\text{def}}{=} (\forall x : \mathsf{el}(a))\, \mathsf{dom}(a, x) \supset \phi[\mathsf{val}(a, x)/x] : \mathsf{prop},$$
$$(\exists x \in a)\, \phi \overset{\text{def}}{=} (\exists x : \mathsf{el}(a))\, \mathsf{dom}(a, x) \wedge \phi[\mathsf{val}(a, x)/x] : \mathsf{prop}.$$

4.2   The Collection Rules of ML(CZF)

Type theoretic rules corresponding to the collection axiom schemes of **CZF** will now be introduced. The Strong Collection rule corresponds to the Strong Collection axiom scheme and the Subset Collection rule corresponds to the Subset Collection axiom scheme. We will refer to these two rules as Collection rules.

In order to present these rules as simply as possible, let us introduce some definitions. For A, B : type, a : $\mathsf{Sub}(A), b$: $\mathsf{Sub}(B)$ and $(x : A, y : B)\,\phi$: prop we define:

$$\mathsf{coll}(a, b, (x, y)\phi) \stackrel{\mathrm{def}}{=} (\forall x \in a)\,(\exists y \in b)\,\phi \,\wedge\, (\forall y \in b)\,(\exists x \in a)\,\phi \quad : \mathsf{prop}\,.$$

Strong Collection Rule.

$$\frac{A, B : \mathsf{type} \qquad a : \mathsf{Sub}(A) \qquad (x : A, y : B)\,\phi : \mathsf{prop}}{(\forall x \in a)\,(\exists y : B)\,\phi \Rightarrow (\exists v : \mathsf{Sub}(B))\,\mathsf{coll}(a, v, (x, y)\phi)}$$

Subset Collection Rule.

$$\frac{A, B, C : \mathsf{type} \quad a : \mathsf{Sub}(A) \quad b : \mathsf{Sub}(B) \quad (x : A, y : B, z : C)\,\psi : \mathsf{prop}}{(\exists u : \mathsf{Sub}(\mathsf{Sub}(B)))(\forall z : C)\big((\forall x \in a)\,(\exists y \in b)\psi \supset (\exists v \in u)\mathsf{coll}(a, v, (x, y)\psi)\big)}$$

We define the type theory **ML(CZF)** as the extension of the type theory $\mathbf{ML_1^-} + \mathbf{W^-} + \mathbf{IL_1}$, obtained by adding the induction rules for its inductive types and the Strong Collection and the Subset Collection rules. Recall that **CZF⁻** is the subsystem of **CZF** obtained from **CZF** by leaving out the Subset Collection axiom scheme [3, 6]. We define **ML(CZF⁻)** as the type theory obtained from **ML(CZF)** by leaving out the Subset Collection rule.

4.3   The Types-as-Classes Interpretation of the Collection Rules

We now work informally in **CZF**. To interpret the operation Sub on types we introduce the corresponding operation Sub on classes, where for each class A we let

$$Sub(A) \stackrel{\mathrm{def}}{=} \Sigma_{I \in V}(Pow(1)^I \times A^I)\,.$$

If $b = (I, (f, g)) \in Sub(A)$ then let

$$set(b) \stackrel{\mathrm{def}}{=} \{g(i) \mid i \in I \,\wedge\, 0 \in f(i)\} \in Sub(A)\,.$$

Conversely, if $a \in Pow(A)$ then let

$$sub(a) \stackrel{\mathrm{def}}{=} (a, ((\lambda_- \in a)1, (\lambda x \in a)x)) \in Sub(A)\,.$$

Then $set(sub(a)) = a$ for all $a \in Pow(A)$. Recall that, for sets a, b and any set theoretical formula ψ in the two variables x, y we define

$$coll(a, b, (x, y)\psi) \stackrel{\mathrm{def}}{=} (\forall x \in a)(\exists y \in b)\psi \,\wedge\, (\forall y \in b)(\exists x \in a)\psi\,.$$

Theorem 10. *There is an interpretation of the raw syntax of* **ML(CZF)** *in* **CZF** *so that each rule of inference is sound.*

Corollary 11. *There is a syntactic translation that assigns a sentence "J is valid" of* **CZF** *to each raw judgement J of* **ML(CZF)** *such that* **ML(CZF)** $\vdash J$ *implies* **CZF** $\vdash$ *" J is valid".*

4.4   Propositions-as-Types Interpretation of the Collection Rules

The (**AC**) rule cannot be fully formulated when we only have the restricted Π types, as in **ML(CZF)**. So we need the weakening (**AC⁻**) in order to state the next result.

$$\frac{a : \mathbb{U} \quad (x : \mathbb{T}(a))\ B : \mathsf{type} \quad (x : \mathbb{T}(a), y : B)\ \phi : \mathsf{prop}}{(\forall x : \mathbb{T}(a))(\exists y : B)\phi \;\Rightarrow\; (\exists z : C)(\forall x : \mathbb{T}(a))\phi[\mathsf{app}(z,x)/y]} \quad (\mathbf{AC^-})$$

where C is $(\Pi^- x : a)B$.

Theorem 12. *Let* **T** *be any standard pure type theory that includes* **ML₁⁻**. *Then the Strong Collection and Subset Collection rules are derived rules of* **T** + **IL₁** + (**AC⁻**) + (**P***) *and so they have a* PaT_1 *translation into any standard pure type theory that includes both* **T** *and* **ML₁**, *where the* PaT_1 *translation was defined in Corollary 5.*

Corollary 13. *The type theory* **ML(CZF)** *has a* PaT_1 *translation into the type theory* **ML₁** + **W⁻**.

5   A Generalised Type Theoretic Interpretation

In this section we work informally within the type theory **ML(CZF)**. Our aim is to define an interpretation of **CZF**. In order to do so, we define the type **V** of **iterative small classes** as follows:

$$\mathsf{V} \overset{\mathrm{def}}{=} \left(W^- y : (\Sigma x : \mathbb{U})(x \to \mathbb{P}) \right)\ y.1 \,.$$

A canonical iterative small class consists of a small type, a small predicate on the type and a function from the small type to **V**. A canonical iterative small class $\mathsf{sup}(\mathsf{pair}(a,p), f)$ can be thought of as the 'set' of all $f(x) : \mathsf{V}$ with $x : \mathbb{T}(a)$ such that $p(x)$. By recursion on **V** and on $\mathsf{Sub}(\mathsf{V})$ we can define $(x : \mathsf{Sub}(\mathsf{V}))\ \mathsf{set}(x) : \mathsf{V}$ and $(y : \mathsf{V})\ \mathsf{sub}(y) : \mathsf{Sub}(\mathsf{V})$ such that, for $a : \mathbb{U}, b : a \to \mathbb{P}, c : a \to \mathsf{V}$

$$\mathsf{set}(\mathsf{pair}(a, \mathsf{pair}(b, c))) = \mathsf{sup}(\mathsf{pair}(a, b), c) : \mathsf{V}\,,$$
$$\mathsf{sub}(\mathsf{sup}(\mathsf{pair}(a, b), c) = \mathsf{pair}(a, \mathsf{pair}(b, c)) : \mathsf{Sub}(\mathsf{V})\,.$$

Given these definitions, the introduction rule for the type **V** can be derived from the following rule:

$$\frac{d : \mathsf{Sub}(\mathsf{V})}{\mathsf{set}(d) : \mathsf{V}}$$

For $(x : \mathsf{V})\ \phi : \mathsf{prop}$ and $(y : \mathsf{Sub}(\mathsf{V}))\ \psi : \mathsf{prop}$ the judgements

$$(\nabla x : \mathsf{V})\ \phi \equiv (\nabla y : \mathsf{Sub}(\mathsf{V}))\ \phi[\mathsf{set}(y)/x]\,,$$

where ∇ is either $\forall$ or $\exists$, are derivable. For $x : \mathsf{V}$, $(y : \mathsf{V})\ p : \mathbb{P}$, and $(y : \mathsf{V})\ \phi : \mathsf{prop}$ we define

$$(\nabla y \in x)\ p \overset{\mathrm{def}}{=} (\dot\forall y \in \mathsf{sub}(x))p : \mathbb{P}$$

$$(\nabla y \in x)\ \phi \overset{\mathrm{def}}{=} (\forall y \in \mathsf{sub}(x))\phi : \mathsf{prop}$$

where ∇ is $\forall$ or $\exists$. By double recursion on V it is possible to define $(x, y : \mathsf{V})\ x \mathbin{\dot\approx} y : \mathbb{P}$ such that if we let $x \approx y \overset{\mathrm{def}}{=} \tau(x \mathbin{\dot\approx} y)$ then the judgement

$$x \approx x' \equiv \forall y \in x\, \exists y' \in x'\,(y \approx y') \wedge \forall y' \in x'\, \exists y \in x\,(y \approx y')$$

is derivable. We now define the generalised type theoretic interpretation of **CZF**. We assume that **CZF** is formulated in a language with equality, with primitive restricted quantifiers but no membership relation. Membership can easily be defined using equality and existential quantification. In the following, we assume that the symbols for variables for sets of **CZF** coincide with the symbols for variables of type V. We now define two interpretations. A first interpretation, indicated with $[\![\cdot]\!]$, applies to arbitrary formulas, and another interpretation, indicated with $(\!|\cdot|\!)$, applies only to restricted formulas. Both interpretations are defined in table 4, where $\star$ is $\wedge$, $\vee$ or $\supset$, and ∇ is $\forall$ or $\exists$.

Table 4. Interpretation of the language of **CZF**

$$[\![x = y]\!] \overset{\mathrm{def}}{=} x \approx y \qquad\qquad (\!|x = y|\!) \overset{\mathrm{def}}{=} x \mathbin{\dot\approx} y\,,$$

$$[\![\phi_1 \star \phi_2]\!] \overset{\mathrm{def}}{=} [\![\phi_1]\!] \star [\![\phi_2]\!]\,, \qquad\qquad (\!|\phi_1 \star \phi_2|\!) \overset{\mathrm{def}}{=} (\!|\phi_1|\!) \mathbin{\dot\star} (\!|\phi_2|\!)\,,$$

$$[\![(\nabla x \in y)\ \phi_0]\!] \overset{\mathrm{def}}{=} (\nabla x \in y)\,[\![\phi_0]\!]\,, \qquad\qquad (\!|(\nabla x \in y)\ \phi_0|\!) \overset{\mathrm{def}}{=} (\dot\nabla x \in y)\,(\!|\phi_0|\!)\,,$$

$$[\![(\nabla x)\ \phi_0]\!] \overset{\mathrm{def}}{=} (\nabla x : \mathsf{V})\,[\![\phi_0]\!]\,,$$

Lemma 14. *If ϕ_1 and ϕ_2 are formulas of the language of set theory with free variables x, and ϕ_2 is restricted, then the judgements*

$$(x : \mathsf{V})\ [\![\,\phi_1\,]\!] : \mathsf{prop}\,,$$
$$(x : \mathsf{V})\ (\!|\,\phi_2\,|\!) : \mathbb{P}\,,$$
$$(x : \mathsf{V})\ \tau\big((\!|\,\phi_2\,|\!)\big) \equiv [\![\,\phi_2\,]\!]$$

are derivable.

A formula ϕ with free variables x will be said to be **valid** if the judgement

$$(x : \mathsf{V}) \quad \llbracket \phi \rrbracket$$

is derivable. We say that the generalised type theoretic interpretation of **CZF** is sound if each axiom and each instance of each axiom scheme of **CZF** is valid.

Theorem 15 (ML(CZF)). *The generalised type theoretic interpretation of the set theory* **CZF** *is sound.*

Corollary 16. CZF *and* **ML(CZF)** *are mutually interpretable.*

6   Reinterpreting Logic

We now describe how both the logic-enriched type theories **ML(CZF⁻)** and **ML(CZF)** can accommodate reinterpretations of the logic. We focus our attention on reinterpretations of the logic as determined by an operator j on the type $\mathbb{P}$ that satisfies a type theoretic version of the properties of a Lawvere-Tierney topology in an elementary topos [16] or of a nucleus on a frame [14]. We will call such an operator j a **topology**. The reinterpretation of logic determined by j will be called the j-interpretation.

In discussing j-interpretations, it seems appropriate to consider **ML(CZF⁻)** initially, and **ML(CZF)** at a later stage. There are two main reasons for doing so. A first reason is that the Strong Collection rule is sufficient to prove the basic properties of j-interpretations. A second reason is that the Strong Collection rule is preserved by the j-interpretation determined by any topology j, while the Subset Collection rule does not seem to be. In order to obtain the derivability of the j-interpretation of the Subset Collection rule, we will introduce a further assumption.

6.1   Topologies in ML(CZF⁻)

To introduce topologies, for $a, b : \mathbb{P}$ we define $a \leq b \stackrel{\text{def}}{=} \tau(a) \supset \tau(b) : \mathsf{prop}$.

Definition 17. *Let j be an explicitly defined operator on $\mathbb{P}$, i.e. there is an explicit definition of the form $jx \stackrel{\text{def}}{=} e : \mathbb{P}$, for $x : \mathbb{P}$, where $(x : \mathbb{P}) e : \mathbb{P}$. We say that j is a **topology** if the following hold for $a_1, a_2 : \mathbb{P}$:*

1. $a_1 \leq ja_1$,
2. $a_1 \leq a_2 \Rightarrow ja_1 \leq ja_2$,
3. $ja_1 \wedge ja_2 \leq j(a_1 \wedge a_2)$,
4. $j(ja_1) \leq ja_1$.

From now on we assume given an arbitrary topology j. For $\phi : \mathsf{prop}$, we define

$$J\phi \stackrel{\text{def}}{=} \exists x : \mathbb{P}\big(\tau(jx) \wedge \tau(x) \supset \phi\big) .$$

Proposition 18. *For* $a : \mathbb{P},\ J(\tau(a)) \equiv \tau(ja)$.

The properties of j can be lifted to J. It is worth pointing out that the Strong Collection rule is used to prove the fourth part of proposition 19.

Proposition 19. *For* $\phi_1, \phi_2 : \mathsf{prop}$, *the following hold:*

1. $\phi_1 \supset J\phi_1$,
2. $\phi_1 \supset \phi_2 \Rightarrow J\phi_1 \supset J\phi_2$,
3. $J\phi_1 \wedge J\phi_2 \supset J(\phi_1 \wedge \phi_2)$,
4. $J(J\phi_1) \supset J\phi_1$.

We now define the j-interpretation of of **ML(CZF$^-$)** into itself determined by the topology j. This interpretation acts solely on the logic, leaving types unchanged. We define the j-interpretation $\langle \cdot \rangle_j$ by structural induction on the raw syntax of the type theory. First of all type expressions are left unchanged. Table 5 contains the definition of the interpretation of formulae, where $\star$ is either $\wedge, \vee$ or $\supset$ and ∇ is either $\forall$ or $\exists$, and of judgement bodies.

Table 5. Definition of the j-interpretation of formulae and judgement bodies

$$\langle \top \rangle_j \stackrel{\text{def}}{=} \top,$$
$$\langle \bot \rangle_j \stackrel{\text{def}}{=} \bot,$$
$$\langle \phi_1 \star \phi_2 \rangle_j \stackrel{\text{def}}{=} J\langle \phi_1 \rangle_j \star J\langle \phi_2 \rangle_j,$$
$$\langle (\nabla x : A)\, \phi_0 \rangle_j \stackrel{\text{def}}{=} (\nabla x : A)\, J\langle \phi_0 \rangle_j,$$
$$\langle \tau(a) \rangle_j \stackrel{\text{def}}{=} \tau(a).$$

$$\langle A : \mathsf{type} \rangle_j \stackrel{\text{def}}{=} A : \mathsf{type},$$
$$\langle A = A' : \mathsf{type} \rangle_j \stackrel{\text{def}}{=} A = A' : \mathsf{type},$$
$$\langle a : A \rangle_j \stackrel{\text{def}}{=} a : A,$$
$$\langle a = a' : A \rangle_j \stackrel{\text{def}}{=} a = a' : A,$$
$$\langle \phi : \mathsf{prop} \rangle_j \stackrel{\text{def}}{=} \langle \phi \rangle_j : \mathsf{prop},$$
$$\langle \phi_1, \ldots, \phi_n \Rightarrow \phi \rangle_j \stackrel{\text{def}}{=} J\langle \phi_1 \rangle_j, \ldots, J\langle \phi_n \rangle_j \Rightarrow J\langle \phi \rangle_j.$$

Finally, we define the j-interpretation of judgements as follows.

$$\langle\, (\Gamma)\, \mathcal{B}\, \rangle_j \stackrel{\text{def}}{=} (\Gamma)\, \langle \mathcal{B} \rangle_j\,.$$

Definition 20. *The j-interpretation of a rule*

$$\frac{(\Gamma_1)\, \mathcal{B}_1 \quad \cdots \quad (\Gamma_n)\, \mathcal{B}_n}{(\Gamma)\, \mathcal{B}}$$

is said to be **sound** *if the judgement* $\langle (\Gamma)\, \mathcal{B} \rangle_j$ *is derivable from the judgements* $\langle (\Gamma_1)\, \mathcal{B}_1 \rangle_j,\ \ldots,\ \langle (\Gamma_n)\, \mathcal{B}_n \rangle_j$.

It is worth pointing out that the Strong Collection rule implies that its j-interpretation is sound. However, it does not seem possible to prove that the j-interpretation of the Subset Collection rule is sound for an arbitrary topology j. We therefore introduce the following definition.

Definition 21. *A topology j on $\mathbb{P}$ is said to be* **set presented** *by $R : \mathsf{Sub}(\mathbb{P})$ if the judgement*

$$(\forall p : \mathbb{P}) \left(\tau(jp) \equiv (\exists q \in R)\, q \leq p \right)$$

is derivable.

The notion of set presented topology is closely related to the notions of cover algebra with basic covers [11], set presented meet semilattice and frame [3, 10] and inductively generated formal topology [8, 20].

Theorem 22 (ML(CZF$^-$)). *Let j be a topology.*

1. *The j-interpretation of each rule of* **ML(CZF$^-$)** *is sound.*
2. *Assuming the Subset Collection rule of* **ML(CZF)**, *if j is set presented, then the j-interpretation of the Subset Collection rule is sound.*

6.2   Double Negation Interpretation

As an application of the results just described we present a type theoretic version of the double negation interpretation. We define the double negation topology as follows:

$$(x : \mathbb{P})\ jx \stackrel{\text{def}}{=} \dot{\neg}\dot{\neg} x : \mathbb{P},$$

where $\dot{\neg} x \stackrel{\text{def}}{=} x \mathbin{\dot{\supset}} \bot : \mathbb{P}$, for $x : \mathbb{P}$. It is easy to prove that j is a topology. Let us point out that the operator J determined by the double negation topology need not to be logically equivalent to double negation. In fact, for $\phi : \mathsf{prop}$ it holds

$$J\phi \equiv (\exists p : \mathbb{P})\big(\neg\neg\tau(p) \wedge \tau(p) \supset \phi\big),$$

where $\neg\phi \stackrel{\text{def}}{=} \phi \supset \bot$, for $\phi : \mathsf{prop}$. In general it will hold only that $J\phi$ implies $\neg\neg\phi$ but not vice versa. This fact seems to be one of the reasons for which it is possible to prove the soundness of the j-interpretation of the Strong Collection rule. These observations first arose in connection with the development of frame-valued semantics for **CZF** [10]. The double-negation nucleus on the frame of truth values corresponds closely to a double-negation interpretation [9, 12].

Since the j-interpretation acts as the double negation only on small propositions, it is natural to consider the following principle of **restricted excluded middle**.

$$(x : \mathbb{P})\ \tau(x) \vee \neg\tau(x) \qquad\qquad \textbf{(REM)}$$

Theorem 23. *The $\neg\neg$-interpretation of* **ML(CZF$^-$)** *$+$* **(REM)** *in* **ML(CZF$^-$)** *is sound.*

Let us now point out that the type theory **ML(CZF$^-$)** and the theory **CZF$^-$** are mutually interpretable. It is easy to see that **(REM)** allows to prove in type theory the validity of the principle of excluded middle for restricted formulas. Theorem 23 gives then as a corollary an interpretation of the set theory **CZF$^-$** + REM, obtained from **CZF$^-$** by adding the law of excluded middle for restricted formulae, into **CZF$^-$**, a result originally obtained in [7].

We can now consider a type theoretic principle asserting that the double negation topology is set presented:

$$\exists R : \mathrm{Sub}(\mathbb{P}) \; \forall p : \mathbb{P}\big(\neg\neg\tau(p) \equiv \exists q \in R\,(q \supset \tau(p))\big)\,. \qquad \text{(DNSP)}$$

Theorem 24. *The $\neg\neg$-interpretation of* $\mathbf{ML(CZF)} + \mathbf{(REM)}$ *in* $\mathbf{ML(CZF)}+$ **(DNSP)** *is sound.*

The type theory $\mathbf{ML(CZF)} + \mathbf{(REM)}$ and the set theory $\mathbf{CZF + REM}$ are mutually interpretable. Recall that the set theory $\mathbf{CZF + REM}$ has proof theoretic strength at least above that of Bounded Zermelo set theory, which is obtained from Zermelo set theory by limiting the separation axiom scheme to restricted formulas. This is because the power set axiom is derivable in $\mathbf{CZF} + \mathbf{REM}$ and Bounded Zermelo set theory has a double-negation interpretation into its intuitionistic counterpart. That set theory is in fact a subsystem of $\mathbf{CZF} + \mathbf{REM}$. The addition of the type theoretic principle **(DNSP)** to $\mathbf{ML(CZF)}$ pushes therefore the proof theoretic strength of the type theory above that of second-order arithmetic.

Acknowledgements

We wish to thank Helmut Schwichtenberg for helpful suggestions. The first author wishes to thank Christopher Nix for pointing out some inaccuracies in a preliminary version of the paper. The second author wishes to thank Steve Awodey, Andrej Bauer, Maria Emilia Maietti, Giovanni Sambin and Alex Simpson for useful discussions.

References

1. Peter Aczel, The Type Theoretic Interpretation of Constructive Set Theory, in: A. MacIntyre, L. Pacholski and J. Paris (eds.), *Logic Colloquium'77*, (North-Holland, Amsterdam, 1978).
2. Peter Aczel, The Type Theoretic Interpretation of Constructive Set Theory: Inductive Definitions, in: R.B. Barcan Marcus, G. J. W. Dorn and P. Weingartner (eds.) *Logic, Methodology and Philosophy of Science, VII*, (North-Holland, Amsterdam, 1986).
3. Peter Aczel, Notes on Constructive Set Theory, Draft manuscript. Available at http://www.cs.man.ac.uk/~petera, 1997.
4. Peter Aczel, On Relating Type Theories and Set Theories, in T. Altenkirch, W. Naraschewski and B. Reus (eds.) *Types for Proofs and Programs, Proceedings of Types'98*, LNCS 1657, (1999).
5. Peter Aczel, The Russell-Prawitz Modality, *Mathematical Structures in Computer Science*, vol. 11, (2001) 1 – 14.
6. Peter Aczel and Michael Rathjen, Notes on Constructive Set Theory, Preprint no. 40, Institut Mittag-Leffler, The Royal Swedish Academy of Sciences, 2001. Available at http://www.ml.kva.se.
7. Thierry Coquand and Erik Palmgren, Intuitionistic Choice and Classical Logic, *Archive for Mathematical Logic*, vol. 39 (2000) 53 – 74.

8. Thierry Coquand, Giovanni Sambin, Jan Smith and Silvio Valentini, Inductively Generated Formal Topologies, Submitted for publication, 2000.
9. Harvey M. Friedman, The Consistency of Classical Set Theory Relative to a Set Theory with Intuitionistic Logic, *Journal of Symbolic Logic*, vol. 38 (1973) 315 – 319.
10. Nicola Gambino and Peter Aczel, Frame-Valued Semantics for Constructive Set Theory, Preprint no. 39, Institut Mittag-Leffler, The Swedish Royal Academy of Sciences, 2001. Available at http://www.ml.kva.se.
11. Robin Grayson, Forcing in Intuitionistic Theories without Power Set, *Journal of Symbolic Logic*, vol. 48 (1983) 670 – 682.
12. Robin Grayson, Heyting-valued models for Intuitionistic Set Theory, in M. Fourman, C. Mulvey and D.S. Scott (eds.) *Applications of Sheaves*, vol. 743, SLNM (1979).
13. Edward Griffor and Michael Rathjen, The Strength of Some Martin-Löf's Type Theories, *Archiv for Mathematical Logic* vol. 33, (1994) 347 – 385.
14. Peter T. Johnstone, *Stone Spaces*, (Cambridge University Press, Cambridge, 1982).
15. Per Martin-Löf, *Intuitionistic Type Theories*, (Bibliopolis, Napoli, 1984).
16. Saunders MacLane and Ieke Moerdijk, *Sheaves in Geometry and Logic. A First Introduction to Topos Theory*, (Springer, Berlin, 1992).
17. Ieke Moerdijk and Erik Palmgren, Wellfounded Trees in Categories, *Annals of Pure and Applied Logic*, vol. 104, (2000) 189 – 218.
18. Ieke Moerdijk and Erik Palmgren, Type Theories, Toposes and Constructive Set Theory: Predicative Aspects of AST, *Annals of Pure and Applied Logic*, to appear.
19. Michael Rathjen, Edward Griffor and Erik Palmgren, Inaccessibility in Constructive Set Theory and Type Theory, *Annals of Pure and Applied Logic*, vol. 94, (1998) 181 – 200.
20. Giovanni Sambin, Intuitionistic Formal Spaces, in D. Skordev (ed.) *Mathematical Logic and its Applications* (Plenum, New York, 1987), 187 – 204.
21. Anne Troelstra and Dirk van Dalen, *Constructivism in Mathematics*, Vol. 1, Studies in Logic No. 121, (North Holland, 1988).

Executing Higher Order Logic

Stefan Berghofer* and Tobias Nipkow

Technische Universität München
Institut für Informatik, Arcisstraße 21, 80290 München, Germany
http://www.in.tum.de/~berghofe/
http://www.in.tum.de/~nipkow/

Abstract. We report on the design of a prototyping component for the theorem prover Isabelle/HOL. Specifications consisting of datatypes, recursive functions and inductive definitions are compiled into a functional program. Functions and inductively defined relations can be mixed. Inductive definitions must be such that they can be executed in Prolog style but requiring only matching rather than unification. This restriction is enforced by a mode analysis. Tail recursive partial functions can be defined and executed with the help of a `while` combinator.

1 Introduction

Executing formal specifications has been a popular research topic for some decades, covering every known specification formalism. Executability is essential for validating complex specifications by running test cases and for generating code automatically ("rapid prototyping"). In the theorem proving community executability is no less of an issue. Two prominent examples are the Boyer-Moore system and its successor ACL2 [11] or constructive type theory, both of which contain a functional programming language. In contrast, HOL specifications can be highly non-executable, and various approaches to their execution have been reported in the literature (see §5 for references). The aim of our paper is to give a precise definition of an executable subset of HOL and to describe its compilation into a functional programming language.

The basic idea is straightforward: datatypes and recursive functions compile directly into their programming language equivalents, and inductive definitions are executed like Prolog programs. Things become interesting when functions and relations are mixed.

We are the first to acknowledge that very few of the ideas in this paper are genuinely original. Instead we flatter ourselves by believing we have achieved a new blend of HOL and functional-logic programming that may serve as the basis for many future approaches to executing HOL. In particular we have precisely identified a subset of HOL definitions that allow efficient execution (§2), outlined a compilation schema for inductive definitions (§3), and devised a method for the

---

* Supported by DFG Graduiertenkolleg *Logic in Computer Science*, and IST project 29001 *TYPES*.

P. Callaghan et al. (Eds.): TYPES 2000, LNCS 2277, pp. 24–40, 2002.

definition and execution of tail recursive functions without tears (§4). Our aim
has not been to reach or extend the limits of functional-logic programming but
to design a lightweight and efficient execution mechanism for HOL specifications
that requires only a functional programming language and is sufficient for typical applications like execution of programming language semantics or abstract
machines.

2   An Executable Subset of Isabelle/HOL

As promised in the introduction, we now give a more precise definition of the
executable subset of the specification language *Isabelle/HOL*, which is based on
Church's simple theory of types. The main ingredients of HOL specifications are:

inductive datatypes can be defined by specifying their *constructors*, e.g.

> **datatype** nat = 0 | Suc nat

recursive functions can be defined by specifying several characteristic equations, e.g.

> **primrec**
> add 0 y = y
> add (Suc x) y = Suc (add x y)

All functions in HOL must be terminating. Supported recursion schemes are
primitive recursion (**primrec**) and *well-founded recursion* (**recdef**) [20].

inductive relations (or predicates) can be defined by specifying a set of *introduction rules*, e.g.

> **inductive**
> 0 ∈ even
> x ∈ even ⟹ Suc (Suc x) ∈ even

Introduction rules are essentially *Horn Clauses*, which are also used in logic
programming languages such as Prolog.

Recursive functions and inductive definitions may also be intermixed: For
example, an inductive predicate may refer to a recursive function and vice versa.

Executable Elements of HOL Specifications. We now inductively define
the elements an *executable* HOL specification may consist of:

- **Executable terms** contain only executable constants
- **Executable constants** can be one of the following
  - executable inductive relations
  - executable recursive functions
  - constructors, recursion and case combinators of executable datatypes
  - operators on executable primitive types such as bool, i.e. the usual propo-
    sitional operators ∧, ∨ and ¬, as well as if _ then _ else _.

- **Executable datatypes**, where each constructor argument type is again an executable datatype or an executable primitive type such as bool or $\rightarrow$.
- **Executable inductive relations**, whose introduction rules have the form

$$(u_1^1, \ldots, u_{n_1}^1) \in q_1 \implies \ldots \implies (u_1^m, \ldots, u_{n_m}^m) \in q_m \implies (t_1, \ldots, t_k) \in p$$

where u_j^i and t_i are executable terms and q_i is either p or some other executable inductive relation. In addition, also arbitrary executable terms not of the form $(\ldots) \in p_i$, so-called *side conditions*, which may not contain p, are allowed as premises of introduction rules.

- **Executable recursive functions**, i.e. sets of rewrite rules, whose left-hand side contains only constructor patterns with distinct variables, and the right-hand side is an executable term.

In the sequel, we write $\mathcal{C}$ to denote the set of datatype constructors. Note that in the above definition, we view $t \in p$, where p is an inductive relation, as synonymous with $p(t)$. Thus, the term $t \in p$ is executable, provided that t and p are, whereas a term of the form $t \in u$ is not executable in general.

The non-executable elements of HOL are, among others, arbitrary universal and existential quantification, equality of objects having higher-order types, Hilbert's selection operator ε, arbitrary type definitions (other than datatypes) or inductive definitions whose introduction rules contain quantifiers, like

$$(\forall y. \, (y, \, x) \in r \implies y \in \mathsf{acc} \; r) \implies x \in \mathsf{acc} \; r$$

Execution. What exactly do we mean by *execution* of specifications? Essentially, execution means finding solutions to queries. A *solution* σ is a mapping of variables to *closed solution terms*. A term t is called a *solution term* iff

- t is of function type, or
- $t = c \; t_1 \ldots t_n$, where the t_i are solution terms and $c \in \mathcal{C}$.

Let solve be a function that returns for each query a set of solutions. We distinguish two kinds of queries:

Functional queries have the form $t = X$, where t is a closed executable term and X is a variable. Queries of this kind should return at most one solution, e.g. solve(add 0 (Suc 0) = X) = $\{[X \mapsto \mathsf{Suc} \; 0]\}$

Relational queries have the form $(t_1, \ldots, t_n) \in r$, where r is an executable inductively defined relation and t_i is either a closed executable term or a variable. A query Q of this kind returns a set of solutions solve(Q). Note that the set returned by solve may also be empty, e.g. solve(Suc 0 $\in$ even) = $\{\}$, or infinite, e.g. solve($X \in$ even) = $\{[X \mapsto 0], [X \mapsto \mathsf{Suc} \; (\mathsf{Suc} \; 0)], \ldots\}$.

It is important to note that all relational queries have to be *well-moded* in order to be executable. We will make this notion more precise in §3.1.

The restriction to ground terms in queries can be relaxed at the expense of the complexity and efficiency of the solver. We consider this an optional extension not necessary for our primary application areas (see §6).

Correctness and Completeness. Function solve is called *sound* w.r.t. an executable HOL specification *Spec* iff

$$\sigma \in \mathsf{solve}(Q) \implies Spec \vdash \sigma(Q)$$

Here, $\vdash$ denotes derivability using introduction and elimination rules for $\forall$ and $\implies$ as well as the substitution rule. The former corresponds to the execution of logic programs, while the latter corresponds to functional execution.

Completeness is more subtle. We omit its definition as we will not be able to guarantee completeness anyway.

3   Compiling Functional Logic Specifications

Functional-logic programming languages such as Curry [9] should be ideal target languages for code generation from HOL specifications. But although such languages contain many of the required concepts and there is an impressive amount of research in this area, the implementations which are currently available are not always satisfactory. We therefore decided to choose ML, the implementation language of Isabelle, as a target language. Datatypes and recursive functions can be translated to ML in a rather straightforward way, with only minor syntactic modifications. Therefore, this section concentrates on the more interesting task of translating inductive relations to ML.[1] The translation is based on assigning *modes* to relations, a well-known standard technique for the analysis and optimization of logic programs [12].

3.1   Mode Analysis

In order to translate a predicate into a function, the direction of *dataflow* has to be analyzed, i.e. it has to be determined which arguments are *input* and which are *output*. Note that for a predicate there may be more than one possible direction of dataflow. For example, the predicate

(Nil, ys, ys) $\in$ append
(xs, ys, zs) $\in$ append $\implies$ (Cons x xs, ys, Cons x zs) $\in$ append

may be given two lists $xs = [1,\ 2]$ and $ys = [3,\ 4]$ as input, the output being the list $zs = [1,\ 2,\ 3,\ 4]$. We may as well give a list $zs = [1,\ 2,\ 3,\ 4]$ as an input, the output being a sequence of pairs of lists xs and ys, where zs is the result of appending xs and ys, namely $xs = [1,\ 2,\ 3,\ 4]$ and $ys = []$, or $xs = [1,\ 2,\ 3]$ and $ys = [4]$, or $xs = [1,\ 2]$ and $ys = [3,\ 4]$, etc.

Mode Assignment. A specific direction of dataflow is called a *mode*. We describe a mode of a predicate by a set of indices, which denote the positions of the input arguments. In the above example, the two modes described were

---

[1] Translation into Haskell looks simpler because of lazy lists and list comprehension, but has essentially the same intellectual and computational complexity.

$\{1, 2\}$ and $\{3\}$. Given a set of predicates P, a relation *modes* is called a *mode assignment* if

$$modes \subseteq \{(p, M) \mid p \in P \wedge M \subseteq \{1, \ldots, \text{arity } p\}\}$$

The set

$$modes\ p = \{M \mid (p, M) \in modes\} \qquad \subseteq \mathcal{P}(\{1, \ldots, \text{arity } p\})$$

is the set of modes assigned to predicate p.

Consistency of Modes. A mode M is called *consistent* with respect to a mode assignment *modes* and a clause

$$(u_1^1, \ldots, u_{n_1}^1) \in q_1 \implies \ldots \implies (u_1^m, \ldots, u_{n_m}^m) \in q_m \implies (t_1, \ldots, t_k) \in p$$

if there exists a permutation π and sets of variable names $v_0, \ldots, v_m$ such that

(1) $v_0 = \text{vars\_of (args\_of } M\ (t_1, \ldots, t_k))$

(2) $\forall 1 \leq i \leq m.\ \exists M' \in modes\ q_{\pi(i)}.\ M' \subseteq \text{known\_args } v_{i-1}\ (u_1^{\pi(i)}, \ldots, u_{n_{\pi(i)}}^{\pi(i)})$

(3) $\forall 1 \leq i \leq m.\ v_i = v_{i-1} \cup \text{vars\_of }(u_1^{\pi(i)}, \ldots, u_{n_{\pi(i)}}^{\pi(i)})$

(4) $\text{vars\_of }(t_1, \ldots, t_k) \subseteq v_m$

The permutation π denotes a suitable execution order for the predicates $q_1, \ldots, q_m$ in the body of p, where v_i is the set of variables whose value is known after the ith execution step. Condition (1) means that initially, when invoking mode M of predicate p, the values of all variables occurring in the input arguments of the clause head are known. Condition (2) means that in order to invoke a mode M' of a predicate $q_{\pi(i)}$, all of the predicate's input arguments which are specified by M' must be known. According to condition (3), the values of all arguments of $q_{\pi(i)}$ are known after its execution. Finally, condition (4) states that the values of all variables occurring in the clause head of p must be known. Here, function args_of M returns the tuple of input arguments specified by mode M, e.g.

$$\text{args\_of } \{1, 2\}\ (\text{Cons } x\ xs,\ ys,\ \text{Cons } x\ zs) = (\text{Cons } x\ xs,\ ys)$$

Function vars_of returns all variables occurring in a tuple, e.g.

$$\text{vars\_of }(\text{Cons } x\ xs,\ ys) = \{x,\ xs,\ ys\}$$

Given some set of variables and an argument tuple, known_args returns the indices of all arguments, whose value is fully known, provided the values of the variables given are known, e.g.

$$\text{known\_args } \{x,\ xs,\ ys\}\ (\text{Cons } x\ xs,\ ys,\ \text{Cons } x\ zs) = \{1, 2\}$$

Mode Inference. We write

$$\text{consistent }(p,\ M)\ modes$$

if mode M of predicate p is consistent with respect to all clauses of p, under the mode assignment *modes*. Let

$$\Gamma(modes) = \{(p,\ M) \mid (p,\ M) \in modes \wedge \text{consistent }(p,\ M)\ modes\}$$

Then the greatest set of allowable modes for a set of predicates P is the greatest fixpoint of Γ. According to Kleene's fixpoint theorem, since Γ is monotone and its domain is finite, this fixpoint can be obtained by finite iteration: we successively apply Γ, starting from the greatest mode assignment

$$\{(p, M) \mid p \in P \wedge M \subseteq \{1, \ldots, \text{arity } p\}\}$$

until a fixpoint is reached.

Example. For append, the allowed modes are inferred as follows:

$\{\}$ is illegal, because it is impossible to compute the value of ys in the first clause
$\{1\}$ is illegal for the same reason
$\{2\}$ is illegal, because it is impossible to compute the value of x in the second clause
$\{3\}$ is legal, because
  - in the first clause, we can compute the first and second argument (Nil, ys) from the third argument ys
  - in the second clause, we can compute x and zs from the third argument. By recursively calling append with mode $\{3\}$, we can compute the value of xs and ys. Thus, we also know the value of the first and second argument (Cons x xs, ys).
$\{1, 2\}$ is legal, because
  - in the first clause, we can compute the third argument ys from the first and second argument
  - in the second clause, we can compute x, xs and ys from the first and second argument. By recursively calling append with mode $\{1, 2\}$, we can compute the value of zs. Thus, we also have the value of the third argument Cons x zs
$\{1, 3\}$, $\{2, 3\}$, $\{1, 2, 3\}$ are legal as well (see e.g. $\{3\}$)

Well-Moded Queries. A query $(t_1, \ldots, t_n) \in p$ is called *well-moded* with respect to a mode assignment *modes* iff

$$\{i \mid t_i \text{ is not a variable}\} \in \text{modes } p$$

Mixing Predicates and Functions. The above conditions for the consistency of modes are sufficient, if the only functions occurring in the clauses are *constructor functions*. If we allow arbitrary functions to occur in the clauses, we have to impose some additional restrictions on the positions of their occurrence. Since non-constructor functions may not be inverted, they cannot appear in an *input position* in the clause head or in an *output position* in the clause body. Thus, we rephrase conditions (1) and (2) to

(1') $v_0 = \text{vars\_of (args\_of } \{i \in M \mid \text{funs\_of } t_i \subseteq C\} \, (t_1, \ldots, t_k)) \wedge$
       $\forall i \in M. \text{ funs\_of } t_i \not\subseteq C \longrightarrow \text{eqtype } t_i$

(2') $\forall 1 \leq i \leq m. \exists M' \in \text{modes } q_{\pi(i)}.$
       $M' \subseteq \text{known\_args } v_{i-1} \, (u_1^{\pi(i)}, \ldots, u_{n_{\pi(i)}}^{\pi(i)}) \wedge$
       $\text{funs\_of (args\_of } (\{1, \ldots, \text{arity } q_{\pi(i)}\} \backslash M') \, (u_1^{\pi(i)}, \ldots, u_{n_{\pi(i)}}^{\pi(i)})) \subseteq C$

where C is the set of constructor functions and funs_of returns the set of all functions occurring in a tuple. The intuition behind (1') is as follows: if some of the input parameters specified by M contain non-constructor functions, we try mode analysis with a subset of M that does not contain the problematic input parameters. After successful execution, we compare the computed values of t_j, where $j \in M \wedge$ funs_of $t_j \not\subseteq C$, with the values provided as input arguments to the predicate. For this to work properly, the terms t_j need to have an *equality type*, i.e. not be of a function type or a datatype involving function types. Note that any M_2 with $M_1 \subseteq M_2$ will be a valid mode, provided M_1 is a valid mode and $\forall j \in M_2 \backslash M_1$. funs_of $t_j \not\subseteq C \longrightarrow$ eqtype t_j. As condition (2') suggests, we can get around the restriction on the occurrence of non-constructor functions in the clause body by choosing modes M' which are sufficiently large, i.e. have sufficiently many input parameters.

3.2   Translation Scheme

In the following section, we will explain how to translate predicates given by a set of Horn Clauses into functional programs in the language ML. For each legal mode of a predicate, a separate function will be generated. Given a tuple of input arguments, a predicate may return a potentially infinite sequence of result tuples. Sequences are represented by the type 'a seq which supports the following operations:

```
Seq.empty  : 'a seq
Seq.single : 'a -> 'a seq
Seq.append : 'a seq * 'a seq -> 'a seq
Seq.map    : ('a -> 'b) -> 'a seq -> 'b seq
Seq.flat   : 'a seq seq -> 'a seq
```

In the sequel, we will write s1 ++ s2 instead of Seq.append (s1, s2). In addition, we define the operator

```
fun s :-> f = Seq.flat (Seq.map f s);
```

which will be used to compose subsequent calls of predicates. Using these operators, the modes {1, 2} and {3} of predicate append can be translated into the ML functions

```
append_1_2 : 'a list * 'a list -> 'a list seq
append_3   : 'a list -> ('a list * 'a list) seq
```

which are defined as follows:

```
fun append_1_2 inp =
  Seq.single inp :->
    (fn (Nil, ys) => Seq.single (ys) | _ => Seq.empty) ++
  Seq.single inp :->
    (fn (Cons (x, xs), ys) =>
       append_1_2 (xs, ys) :->
```

```
        (fn (zs) => Seq.single (Cons (x, zs)) | _ => Seq.empty)
      | _ => Seq.empty);

fun append_3 inp =
  Seq.single inp :->
    (fn (ys) => Seq.single (Nil, ys) | _ => Seq.empty) ++
  Seq.single inp :->
    (fn (Cons (x, zs)) =>
      append_3 (zs) :->
        (fn (xs, ys) => Seq.single (Cons (x, xs), ys)
         | _ => Seq.empty)
      | _ => Seq.empty);
```

In the above translation, every operand of ++ corresponds to one clause of the predicate. Initially, the input is converted into a one-element sequence using Seq.single, to which successively all predicates in the body of the clause are applied using :->. Therefore, the operator :-> can also be interpreted as a visualization of dataflow.

We will now describe the general translation scheme. Assume the predicate to be translated has the clause

$$(ipat_1, \; opat_1) \in q_1 \implies \ldots \implies (ipat_m, \; opat_m) \in q_m \implies (ipat_0, \; opat_0) \in p$$

To simplify notation, we assume without loss of generality that the predicates in the body of p are already sorted with respect to the permutation π calculated during mode analysis and that the arguments of the predicates are already partitioned into input arguments $ipat_i$ and output arguments $opat_i$. Then, p is translated into the function

```
fun p inp =
  Seq.single inp :->
    (fn ipat₀ => q₁ ipat₁ :->
        (fn opat₁ => q₂ ipat₂ :->
            ⋱
                (fn opatₘ => Seq.single opat₀
                 | _ => Seq.empty)
              ⋮
          | _ => Seq.empty)
        | _ => Seq.empty)
  ++
  ...;
```

where the ... after the operator ++ correspond to the translation of the remaining clauses of p. A characteristic feature of this translation is the usage of ML's built-in pattern matching mechanism instead of unification and logical variables. Before calling a predicate q_i in the body of the clause, the output pattern $opat_{i-1}$ of the preceeding predicate is checked. Before calling the first predicate q_1, the input pattern $ipat_0$ in the head of the clause is checked.

Example: Some λ-Calculus Theory Formalized. As an example of a program making use of both functional and logical features, we now consider a specification of β-reduction for λ-terms in de Bruijn notation, which is taken from [13]. First, the datatype term of λ-terms is defined, together with a function lift for incrementing indices in a term as well as a function subst for substituting a term for a variable with a given index:

datatype term = Var nat | App term term | Abs term

primrec
    lift (Var i) k = (if $i < k$ then Var i else Var $(i + 1)$)
    lift (App $s\ t$) k = App (lift $s\ k$) (lift $t\ k$)
    lift (Abs s) k = Abs (lift $s\ (k + 1)$)

primrec
    subst (Var i) $s\ k$ = (if $k < i$ then Var $(i - 1)$ else if $i = k$ then s else Var i)
    subst (App $t\ u$) $s\ k$ = App (subst $t\ s\ k$) (subst $u\ s\ k$)
    subst (Abs t) $s\ k$ = Abs (subst t (lift $s\ 0$) $(k + 1)$)

This is a purely functional specification, whose translation to ML is straightforward. It is therefore not shown here. Using subst, one can now define beta reduction $\rightarrow_\beta$ inductively:

inductive
    App (Abs s) $t \rightarrow_\beta$ subst $s\ t\ 0$
    $s \rightarrow_\beta t \Longrightarrow$ App $s\ u \rightarrow_\beta$ App $t\ u$
    $s \rightarrow_\beta t \Longrightarrow$ App $u\ s \rightarrow_\beta$ App $u\ t$
    $s \rightarrow_\beta t \Longrightarrow$ Abs $s \rightarrow_\beta$ Abs t

Note that $t \rightarrow_\beta u$ just abbreviates $(t,\ u) \in \rightarrow_\beta$. This specification of β-reduction is essentially a functional logic program. Using the translation scheme described above, the HOL specification of $\rightarrow_\beta$ can be translated to the following ML program for mode $\{1\}$:

```
fun beta_1 inp =
  Seq.single inp :->
    (fn (App (Abs s, t)) =>
      Seq.single (subst s t 0) | _ => Seq.empty) ++
  Seq.single inp :->
    (fn (App (s, u)) =>
      beta_1 (s) :->
        (fn (t) => Seq.single (App (t, u)) | _ => Seq.empty)
      | _ => Seq.empty) ++
  Seq.single inp :->
    (fn (App (u, s)) =>
      beta_1 (s) :->
        (fn (t) => Seq.single (App (u, t)) | _ => Seq.empty)
      | _ => Seq.empty) ++
  Seq.single inp :->
    (fn (Abs s) =>
      beta_1 (s) :->
        (fn (t) => Seq.single (Abs t) | _ => Seq.empty)
      | _ => Seq.empty);
```

Note that the recursive function `subst` can easily be called from within the logic program `beta_1`.

Running the Translated Program. We will now try out the compiled predicate on a small example: the sequence

```
val test = beta_1 (Abs (App (Abs (App (Var 0, Var 0)),
    App (Abs (App (Var 0, Var 0)), Var 0))));
```

contains the possible reducts of the term $\lambda x.\ (\lambda y.\ y\ y)\ ((\lambda z.\ z\ z)\ x)$. The first element of this sequence is

```
> Seq.hd test;
val it = Abs (App (App (Abs (App (Var 0, Var 0)), Var 0),
    App (Abs (App (Var 0, Var 0)), Var 0)))
```

There is yet another solution for our query, namely

```
> Seq.hd (Seq.tl test);
val it = Abs (App (Abs (App (Var 0, Var 0)), App (Var 0, Var 0)))
```

3.3   Extending the Mode System

The mode system introduced in §3.1 is not always sufficient: For example, it does not cover inductive relations such as

inductive
$(x,\ x) \in \mathsf{trancl}\ r$
$(x,\ y) \in r \implies (y,\ z) \in \mathsf{trancl}\ r \implies (x,\ z) \in \mathsf{trancl}\ r$

which take other inductive relations as arguments. This case can be covered by introducing so-called *higher-order modes*: a mode of a higher-order relation p taking n relations $r_1, \ldots, r_l$ as arguments and returning a relation as result is an $n+1$ tuple, where the first n components of the tuple correspond to the modes of the argument relations, and the last component corresponds to the mode of the resulting relation, i.e.

$$\mathit{modes}\ p \subseteq \mathcal{P}(\{1, \ldots, \mathsf{arity}\ r_1\}) \times \cdots \times \mathcal{P}(\{1, \ldots, \mathsf{arity}\ r_l\}) \times \mathcal{P}(\{1, \ldots, \mathsf{arity}\ p\})$$

For example, trancl has modes $\{(\{1\}, \{1\}), (\{2\}, \{2\}), (\{1, 2\}, \{1, 2\})\}$, i.e. if r has mode $\{1\}$ then trancl r has mode $\{1\}$ as well. A higher-order relation may have clauses of the form

$$(u_1^1, \ldots, u_{n_1}^1) \in Q_1 \implies \ldots \implies (u_1^m, \ldots, u_{n_m}^m) \in Q_m \implies (t_1, \ldots, t_k) \in p\ r_1 \ldots r_l$$

where $Q_{i'} = r_i \mid q_j\ Q'_{\varrho_1} \cdots Q'_{\varrho_{l'}}$

To describe the consistency of a higher order mode $(M_1,\ \ldots,\ M_l,\ M)$ with respect to a mode assignment *modes* and the above clause, we rephrase condition (2) of the definition of consistency given in §3.1 to

$(2')\ \forall 1 \leq i \leq m.\ \exists M' \in \mathit{modes}'\ Q_{\pi(i)}.\ M' \subseteq \mathsf{known\_args}\ v_{i-1}\ (u_1^{\pi(i)},\ \ldots,\ u_{n_{\pi(i)}}^{\pi(i)})$

where
$\mathit{modes}'\ r_i = \{M_i\}$
$\mathit{modes}'\ (q_j\ Q'_{\varrho_1}\ \cdots\ Q'_{\varrho_{l'}}) = \{M' \mid \exists M_1' \in \mathit{modes}'\ Q'_{\varrho_1} \ldots M_{l'}' \in \mathit{modes}'\ Q'_{\varrho_{l'}}.$
$(M_1',\ \ldots,\ M_{l'}',\ M') \in \mathit{modes}\ q_j\}$

Mode $\{1\}$ of trancl could be translated as follows

```
fun trancl_1 r inp =
  Seq.single inp ++ r inp :-> trancl_1 r;
```

Interestingly, the translation of mode {2} looks exactly the same.

3.4   Some Notes on Correctness and Completeness

We claim that our translation scheme described above is correct in the sense of §2, although a formal proof is beyond the scope of this paper. As far as completeness is concerned, there are two problems that have to be taken into consideration. One problem is due to ML's *eager evaluation* strategy. For example,

defs
$g \equiv \lambda x\, y.\, x$
$f_1 \equiv \lambda x.\, g\, x\, (\text{hd } [])$

recdef
$f_2\, 0\, =\, 0$
$f_2\, (\text{Suc } x)\, =\, g\, (f_2\, x)\, (f_2\, (\text{Suc } x))$

is an admissible HOL specification, but, if compiled naively, f_1 raises an exception, because the argument [] cannot be handled by hd, and f_2 loops. To avoid this, the definition of g could be expanded, or the critical arguments could be wrapped into dummy functions, to delay evaluation. Paulin-Mohring and Werner [16] discuss this problem in detail and also propose alternative target languages with *lazy evaluation*.

Another source of possible nontermination is the Prolog-style *depth first* execution strategy of the translated inductive relations. Moreover, some inferred modes or permutations of predicates in the body of a clause may turn out to be non-terminating. Termination of logic programs is an interesting problem in its own right, which we do not attempt to solve here.

4   Partial Functions

So far we have (implicitly) assumed that HOL is a suitable logic for defining recursive functions, i.e. that it is possible to define functions by recursion equations as in functional programming languages. This is indeed the case for primitive recursion and still works well for well-founded recursion. However, all HOL functions must be total. Hence we cannot hope to define truly partial functions. The best we can do are functions that are *underdefined*: for certain arguments we only know that a result exists, but we don't know what it is. When defining functions that are normally considered partial, underdefinedness turns out to be a very reasonable alternative. We will now discuss two issues: how to define such underdefined functions in the first place, and how to obtain recursion equations that allow efficient execution.

4.1   Guarded Recursion

Given a partial function f that should satisfy the recursion equation $f(x) = t$ over its domain $dom(f)$, we turn this into the guarded recursion

```
f x = (if x ∈ dom f then t else arbitrary)
```

where `arbitrary` is a constant of type `'a` which has no definition, i.e. its value is completely underspecified. As a simple example we define division on `nat`:

```
consts divi :: "nat × nat → nat"
recdef divi "measure(λ(m,n). m)"
  "divi(m,n) = (if n = 0 then arbitrary else
                if m < n then 0 else divi(m-n,n)+1)"
```

The keyword **consts** declares constants and **recdef** defines a function by well-founded recursion — the term `measure (λ(m, n). m)` is the well-founded relation [20]. For the sake of the example, we equate `divi (m, 0)` with `arbitrary` rather than some specific number, for example 0.

As a more substantial example we consider the problem of searching a graph. For simplicity our graph is given by a function (`f`) of type `'a → 'a` which maps each node to its successor, and the task is to find the end of a chain, i.e. a node pointing to itself. Here is a first attempt:

```
find (f, x) = (if f x = x then x else find (f, f x))
```

This may be viewed as a fixed point finder or as one half of the well known *Union-Find* algorithm. The snag is that it may not terminate if `f` has non-trivial cycles. Phrased differently, the relation

```
constdefs step1 :: "('a → 'a) → ('a × 'a)set"
  "step1 f ≡ {(y,x). y = f x ∧ y ≠ x}"
```

must be well-founded (in Isabelle/HOL: `wf`). Thus we define

```
consts find :: "('a → 'a) × 'a → 'a"
recdef find
  "find(f,x) = (if wf(step1 f)
                then if f x = x then x else find(f, f x)
                else arbitrary)"
```

The recursion equation should be clear enough: it is our aborted first attempt augmented with a check that there are no non-trivial cycles. We omit to show the accompanying termination relation, which is not germane to the subject of our paper.

Although the above definition of `find` is quite satisfactory from a theorem proving point of view, it is a disaster w.r.t. executability: the test `wf (step1 f)` is undecidable in general. This is the key problem with guarded recursion: unless the domain of the function is (efficiently) decidable, this scheme does not yield (efficiently) executable functions.

4.2   The while Combinator

Fortunately, tail recursive functions admit natural HOL definitions which can be executed efficiently. This insight was communicated to us by Wolfgang Goerigk [8]. To understand why, consider the following two "definitions":

$f(x) = f(x + 1)$ is perfectly harmless, as it merely asserts that all values of f are the same, but leaving the precise value open.

$f(x) = f(x) + 1$ must not be admitted because there is no total function that satisfies this equation and it implies $0 = 1$.

The key property of tail recursive function definitions is that they have total models. Instead of dealing with arbitrary tail recursive function definitions we introduce a while combinator. This is merely a notational variant (and simplification) of tail recursion. The definition of while follows the guarded recursion schema, but we omit to show the termination relation:

```
consts while :: "('a → bool) × ('a → 'a) × 'a → 'a"
recdef while
  "while(b,c,s) = (if ∃f. f 0 = s ∧ (∀i. b(f i) ∧ c(f i) = f(i+1))
                   then arbitrary
                   else if b s then while(b,c,c s) else s)"
```

The guard checks if the loop terminates or not. If it does, while(b,c,s) mimicks the imperative program

```
    x := s; while b(x) do x := c(x); return x
```

It appears we have not made much progress because the definition of while is certainly not executable. However, it is possible to derive the following unguarded recursion equation:

```
theorem while_rec: "while(b,c,s) = (if b s then while(b,c,c s) else s)"
```

The proof is easy. If the loop terminates, the left-hand side reduces by definition to the right-hand side. If it does not terminate, the left-hand side is arbitrary, as is the right-hand side: nontermination implies that b s must hold, in which case the right-hand side reduces to while (b, c, c s), which diverges as well, and thus to arbitrary.

What has happened here is well-known in the program transformation literature: starting from a total function f it is easy to derive a recursion equation for f which, if interpreted as a function definition, yields a partial version of the original f. As an extreme example, one can always prove $f(x) = f(x)$, which is as partial as can be.

This phenomenon is usually considered a problem, but for us it is the solution: in HOL we are forced to define a total version of while, but can recover the desired partial one by proof. And instead of generating code for while from its definition, we generate the code from theorem while_rec, which any decent compiler will then translate into a loop. In fact, the code generator is always driven by theorems. Thus the generated function definitions are necessarily partially correct, provided the code generator is correct. Basing everything upon theorems is possible because Isabelle/HOL follows the definitional approach: recursive functions are not axiomatized but the recursion equations are *proved* from a suitable non-recursive definition — the latter process is hidden from the user [20]. Thus the code generator takes an arbitrary list of theorems, checks they constitute a well-formed function definition (merely to avoid static errors later on) and translates them into ML. It makes no difference whether these theorems correspond to the initial definition of the function or are derived from it, as is the case for `while_rec`.

As an application of `while` we define the above function `find` without tears:

```
constdefs find2 :: "('a → 'a) → 'a → 'a"
  "find2 f x ≡
    fst(while (λ(x,x'). x' ≠ x, λ(x,x'). (x',f x'), (x,f x)))"
```

The loop operates on two "local variables" x and x' containing the "current" and the "next" value of function f. They are initalized with the global x and f x. At the end `fst` selects the local x.

Although the definition of `find2` was easy, there is no free lunch: when proving properties of functions defined by `while`, termination rears its ugly head again. Such proofs are best conducted by means of the derived `while_rule`, the well known proof rule for total correctness of loops expressed with `while`:

$$
\begin{aligned}
&[P \ s; \ \bigwedge s. \ [P \ s; \ b \ s] \Longrightarrow P \ (c \ s); \\
&\quad \bigwedge s. \ [P \ s; \ \neg \ b \ s] \Longrightarrow Q \ s; \ wf \ r; \\
&\quad \bigwedge s. \ [P \ s; \ b \ s] \Longrightarrow (c \ s, \ s) \in r] \\
&\Longrightarrow Q \ (while \ (b, \ c, \ s))
\end{aligned}
$$

P needs to be true of the initial state s and invariant under c (premises 1 and 2). The post-condition Q must become true when leaving the loop (premise 3). And each loop iteration must descend along a well-founded relation r (premises 4 and 5). In order to show that `find2` does indeed find a fixed point

theorem `"wf(step1 f) ⟹ f(find2 f x) = find2 f x"`

we prove the following lemma with the help of `while_rule`:

lemma `lem:` `"[ wf(step1 f); x' = f x ] ⟹`
    `∃y. while (λ(x,x'). x' ≠ x, λ(x,x'). (x',f x'), (x,x')) = (y,y) ∧`
        `f y = y"`

The proof is almost automatic after supplying the invariant x' = f x and the termination relation.

As we have seen, the `while` combinator has the advantage of enabling us to define functions without having to worry about termination. However, this merely delays the evil hour, which comes as soon as one wants to prove properties of a function thus defined. On top of that, tail recursive functions tend to be more complicated to reason about. Therefore `while` should only be used if executability is an issue or the function in question is naturally tail recursive.

5 Related Work

Previous Work on Executing HOL Specifications. There has already been some work on generating executable programs from specifications written in HOL. One of the first papers on this topic is by Rajan [19] who translates HOL datatypes and recursive functions to ML. However, inductive definitions are not covered in this paper. Andrews [2] has chosen λProlog as a target language. His translator for a higher order specification language called S can also handle specifications of transition rules of programming languages such as CCS, although these are given in a somewhat different way than the inductive definitions of Isabelle/HOL. In contrast to our approach, all functions have to be translated into predicates in order to be executable by λProlog. On the other hand it is possible to execute a wider range of specifications and queries, as λProlog allows embedded universal quantifiers and implications and supports higher-order unification, with all the performance penalties this entails. Similar comments apply to Elf [18], a type theoretic higher-order logic programming language.

Other Theorem Provers. Aagaard et al [1] introduce a functional language called fl, together with a suitable theorem prover. Thanks to a "lifting" mechanism, their system supports both *execution* of fl functions as well as *reasoning* about fl functions in a seamless way.

Coq [4] is a type-theoretic proof assistant based on the *Calculus of Inductive Constructions*. Type theory allows for the uniform treatment of both proofs and programs within the same framework. In contrast to HOL, where computable and non-computable objects can be arbitrarily mixed, Coq strictly distinguishes between types that have a *computational content* and types that don't. This is done by introducing two *universes* called Set and Prop, the first of which contains types that have a *computational content*. Coq can directly generate code from definitions of recursive datatypes and functions such as nat and add from §2. To obtain a program from an inductive predicate such as append, an approach substantially different from ours described in §3 is used: one builds a constructive proof of

$$\forall (xs :: \alpha \text{ list}) \ (ys :: \alpha \text{ list}). \ \exists (zs :: \alpha \text{ list}). \ \text{append } xs \ ys \ zs$$

from which a functional program of type α list $\rightarrow$ α list $\rightarrow$ α list can be *extracted* by erasing the parts of the proof which are in Prop. Note that this method relies on the inductive predicate to be decidable. Paulin-Mohring and Werner [16] describe how to obtain efficient functional programs from Coq specifications.

The latest version of the theorem prover *PVS* [15] includes a procedure for evaluating ground terms. The PVS ground evaluator essentially consists of a translator from an executable subset of PVS into Common Lisp. The unexecutable fragments are uninterpreted functions, non-bounded quantification and higher-order equalities.

The *Centaur* system [10] is an environment for specifying programming languages. One of its components is a Prolog-style language called *Typol* [6], in which transition rules of natural semantics can be specified. Attali et al [3] have used Typol to specify a formal, executable semantics of a large subset of the programming language *Java*. Originally, Typol specifications were compiled to Prolog in order to execute them. Recently, Dubois, and Gayraud [7] have proposed a translation of Typol specifications to ML. The consistency conditions for modes described in §3.1 are inspired by this paper.

6 Conclusion

We conclude the paper with a survey of the intended applications. Our primary aim has been to validate the rather complex specifications arising in our Java modeling efforts [14]. Such language and machine specifications are an important application area for theorem provers. They have the pleasant property that their execution requires no logical variables: results are not synthesized by unification but computed by evaluation. Hence these specifications meet the requirements imposed by our mode system. No major changes were required to make our Java semantics, an inductive definition, executable. Note that this was possible only because our Java model is based on a first-order abstract syntax. Higher-order abstract syntax requires a richer language of inductive definitions than what we can currently compile.

Of course prototypes can be used to debug all kinds of specifications. One further promising application area that we intend to study is that of cryptographic protocols. Paulson [17] has shown how to specify protocols as inductive definitions and how to verify them. Basin [5] has shown that functional languages are well-suited for prototyping such protocols and for finding bugs. Thus it is highly desirable to have one integrated environment for specification, debugging, and verification, which our prototyping facility could turn Isabelle into.

Finally we intend to pursue *reflection*, i.e. re-importing the result of an external execution into the theorem prover. A simple example would be the efficient evaluation of arithmetic or other expressions.

References

[1] M. D. Aagaard, R. B. Jones, and C.-J. H. Seger. Lifted-FL: A Pragmatic Implementation of Combined Model Checking and Theorem Proving. In Y. Bertot, G. Dowek, A. Hirschowitz, C. Paulin, and L. Théry, editors, *Theorem Proving in Higher Order Logics, 12th International Conference (TPHOLs'99)*, volume 1690 of *Lect. Notes in Comp. Sci.*, pages 323–340. Springer-Verlag, 1999.

[2] J. H. Andrews. Executing formal specifications by translation to higher order logic programming. In E. L. Gunter and A. Felty, editors, *10th International Conference on Theorem Proving in Higher Order Logics*, volume 1275 of *Lect. Notes in Comp. Sci.*, pages 17–32. Springer-Verlag, 1997.

[3] I. Attali, D. Caromel, and M. Russo. A formal and executable semantics for Java. In *Proceedings of Formal Underpinnings of Java, an OOPSLA'98 Workshop, Vancouver, Canada*, 1998. Technical report, Princeton University.

[4] B. Barras, S. Boutin, C. Cornes, J. Courant, Y. Coscoy, D. Delahaye, D. de Rauglaudre, J.-C. Filliâtre, E. Giménez, H. Herbelin, G. Huet, H. Laulhère, C. Muñoz, C. Murthy, C. Parent-Vigouroux, P. Loiseleur, C. Paulin-Mohring, A. Saïbi, and B. Werner. The Coq proof assistant reference manual – version 6.3.1. Technical report, INRIA, 1999.

[5] D. Basin. Lazy infinite-state analysis of security protocols. In *Secure Networking — CQRE [Secure] '99*, volume 1740 of *Lect. Notes in Comp. Sci.*, pages 30–42. Springer-Verlag, 1999.

[6] T. Despeyroux. Typol: a formalism to implement natural semantics. Technical Report 94, INRIA, 1988.

[7] C. Dubois and R. Gayraud. Compilation de la sémantique naturelle vers ML. In *Proceedings of journées francophones des langages applicatifs (JFLA99)*, 1999. Available via http://pauillac.inria.fr/~weis/jfla99/ps/dubois.ps.

[8] W. Goerigk, July 2000. Personal communication.

[9] M. Hanus, H. Kuchen, and J. Moreno-Navarro. Curry: A truly functional logic language. In *Proc. ILPS'95 Workshop on Visions for the Future of Logic Programming*, pages 95–107, 1995.

[10] I. Jacobs and L. Rideau-Gallot. A Centaur tutorial. Technical Report 140, INRIA Sophia-Antipolis, July 1992.

[11] M. Kaufmann, P. Manolios, and J. S. Moore. *Computer-Aided Reasoning: An Approach.* Kluwer Academic Publishers, June 2000.

[12] C. S. Mellish. The automatic generation of mode declarations for Prolog programs. Technical Report 163, Department of Artificial Intelligence, University of Edinburgh, August 1981.

[13] T. Nipkow. More Church-Rosser proofs (in Isabelle/HOL). *Journal of Automated Reasoning*, 26, 2001.

[14] D. v. Oheimb and T. Nipkow. Machine-checking the Java specification: Proving type-safety. In J. Alves-Foss, editor, *Formal Syntax and Semantics of Java*, volume 1523 of *Lect. Notes in Comp. Sci.*, pages 119–156. Springer-Verlag, 1999.

[15] S. Owre, N. Shankar, J. M. Rushby, and D. W. J. Stringer-Calvert. PVS System Guide version 2.3. Technical report, SRI International Computer Science Laboratory, Menlo Park CA, September 1999.

[16] C. Paulin-Mohring and B. Werner. Synthesis of ML programs in the system Coq. *Journal of Symbolic Computation*, 15:607–640, 1993.

[17] L. C. Paulson. The inductive approach to verifying cryptographic protocols. *J. Computer Security*, 6:85–128, 1998.

[18] F. Pfenning. Logic programming in the LF Logical Framework. In G. Huet and G. Plotkin, editors, *Logical Frameworks*, pages 66–78. Cambridge University Press, 1991.

[19] P. S. Rajan. Executing HOL specifications: Towards an evaluation semantics for classical higher order logic. In L. J. M. Claesen and M. J. C. Gordon, editors, *Higher order Logic Theorem Proving and its Applications*, Leuven, Belgium, September 1992. Elsevier.

[20] K. Slind. *Reasoning about Terminating Functional Programs.* PhD thesis, Institut für Informatik, TU München, 1999.

A Tour with Constructive Real Numbers[*]

Alberto Ciaffaglione and Pietro Di Gianantonio

Dipartimento di Matematica e Informatica, Università di Udine
via delle Scienze, 206 - 33100 Udine (Italy)
{ciaffagl,pietro}@dimi.uniud.it

Abstract. The aim of this work is to characterize constructive real numbers through a minimal axiomatization. We introduce, discuss and justify 16 constructive axioms. Then we address their expressivity considering the alternative axiomatizations.

1 Overview of the Work

This work tries to understand (again) constructive real numbers. Our main contribution is a new system of axioms, synthesized with the aim of being minimal, i.e. of assuming the least number of primitive notions and properties. Such a system is consistent with respect to reference models we have in mind — (equivalence classes of) Cauchy sequences [TvD88] and co-inductive streams of digits [CDG00] — and will be compared to other proposals of the literature [Bri99,GN01]. In particular we will prove that our axiomatization has a sufficient deductive power.

We have formalized and used our axioms inside the Logical Framework Coq [BB^{+}01]. However, the axioms can be stated and worked with in a general constructive logical setting, because we do not need all the richness of the Calculus of Constructions [CH88], the logic beneath Coq. In particular we do not require the use of dependent inductive types and universes. On the contrary, we should have available a logical system that accommodates second-order quantification (in order to axiomatize the existence of limit) and the Axiom of Choice (for defining the "reciprocal" function on reals different from zero).

We define constructive real numbers through sixteen axioms organized in four groups: arithmetic operations, ordering, Archimedes' postulate and completeness. Our axiomatization uses only three basic concepts: addition $(+)$, multiplication $(\times)$ and strict order $(<)$.

In most of the constructive approaches to analysis [Bis67,Bee85,Wei00], real numbers are defined as a quotient of a set of representations (e.g. equivalence classes of Cauchy sequences, digit expansions, etc.). Hence, also in an axiomatic approach, it is necessary to see the reals as a set provided with an equivalence relation. In our proposal, this equivalence relation $(\sim)$ is not a primitive notion, but it is derived together with its fundamental properties through the strict

---

[*] Research partially supported by EEC Working Group "TYPES" and Italian MIUR COFIN "TOSCA".

order relation. We define the equivalence by $(x \sim y) \triangleq \neg[(x < y) \vee (y < x)]$. Similarly, it is not necessary to assume as basic the apartness relation $(\#)$ — which is a semi-decidable version of the inequality $(\neq)$ — as it is definable in terms of the order relation.

The paper has the following structure. Section 2 introduces the constructive axioms, which are then explained and motivated. In Section 3 we start deducing some elementary consequences of the axioms. The following Section is devoted to a digression concerning possible models for the constructive real numbers. We conclude by articulating a detailed comparison between our axiomatization and other similar works in the literature.

2   Constructive Axioms

We introduce constructive real numbers as the mathematical entities satisfying four groups of axioms. The basic notions are the following:

- a representation set R, with two elements 0_R (zero) and 1_R (one);
- a binary relation $<$ (strict order) over R;
- two binary operations $+$ (addition) and $\times$ (multiplication) over R.

We do not assume the negation $(-)$ and reciprocal $(^{-1})$ as primitive functions. The main reason for this choice is that the reciprocal function cannot be defined in Coq: in fact, in Coq, each function has to be totally defined; and, in a constructive setting, each function has to be continuous (w.r.t. the Euclidean topology). On the contrary, it is impossible to make continuous by extension the reciprocal function.

In order to state the axioms, it is convenient to define two relations and two functions:

- a binary relation $\sim$ (equivalence) over R tells that two names represent the same number. It expresses the redundancy of the representation;
- two recursively defined functions $inj, exp : \mathbb{N} \to R$ $(inj(n) = n, exp(n) = 2^n)$ are used in the Archimedean and completeness axioms;
- a ternary relation $near \subseteq R \times R \times \mathbb{N}$ $(near(x, y, n) \Leftrightarrow |x-y| \leq 2^{-n})$ expresses the Euclidean metric.

Our axiomatization is parametric with respect to a set $\mathbb{N}$ of the natural numbers, that we suppose to be given. In our formalization in Coq, $\mathbb{N}$ is taken as the set of the inductive natural numbers. In a different context, $\mathbb{N}$ could be defined as a set satisfying the Peano's arithmetic axioms. Finally, we claim that constructive real numbers are captured by the following axioms.

Definition 1. *(Axioms for Constructive Real Numbers)*

$Consts: R, \ \{0_R, 1_R\} \in R \quad < \ \subseteq R \times R \quad + : R \times R \to R \quad \times : R \times R \to R$

$Defs:$
$\quad \sim \ \subseteq R \times R \qquad (x \sim y) \triangleq \neg ((x < y) \vee (y < x))$
$\quad inj : \mathbb{N} \to R \qquad inj(0) \triangleq 0_R, \ inj(n+1) \triangleq inj(n) + 1_R$
$\quad exp : \mathbb{N} \to R \qquad exp(0) \triangleq 1_R, \ exp(n+1) \triangleq exp(n) \times (1_R + 1_R)$
$\quad near \subseteq R \times R \times \mathbb{N} \ \ near(x, y, n) \triangleq \forall \epsilon \in R. \ (1_R < \epsilon \times exp(n)) \to$
$\qquad\qquad\qquad\qquad\qquad\qquad\qquad\qquad\qquad (x < y + \epsilon) \wedge (y < x + \epsilon)$

$Add:$

$+$-associativity	$\forall x, y, z \in R. \ (x + (y + z)) \sim ((x + y) + z)$	
$+$-unit	$\forall x \in R. \ (x + 0_R) \sim x$	
negation	$\forall x \in R. \ \exists y \in R. \ (x + y) \sim 0_R$	
$+$-commutativity	$\forall x, y \in R. \ (x + y) \sim (y + x)$	

$Mult:$

$\times$-associativity	$\forall x, y, z \in R. \ (x \times (y \times z)) \sim ((x \times y) \times z)$
$\times$-unit	$\forall x \in R. \ (x \times 1_R) \sim x$
reciprocal	$\forall x \in R. \ (0_R < x) \to \exists y \in R. \ (x \times y) \sim 1_R$
$\times$-commutativity	$\forall x, y \in R. \ (x \times y) \sim (y \times x)$
distribuitivity	$\forall x, y, z \in R. \ (x \times (y + z)) \sim (x \times y) + (x \times z)$

$Order:$

non triviality	$0_R < 1_R$
$<$-asymmetry	$\forall x, y \in R. \ (x < y) \to \neg(y < x)$
$<$-co-transitivity	$\forall x, y, z \in R. \ (x < y) \to (x < z) \vee (z < y)$
$+$-reflects-$<$	$\forall x, y, z \in R. \ (x + z < y + z) \to (x < y)$
$\times$-reflects-$<$	$\forall x, y \in R. \ (x \times z < y \times z) \to$
	$\qquad\qquad (x < y) \vee ((y < x) \wedge (z < 0))$

Archimedean	$\forall x \in R. \ \exists n \in \mathbb{N}. \ x < inj(n)$
completeness	$\forall f : \mathbb{N} \to R. \ \exists x \in R.$
	$(\forall n \in \mathbb{N}. \ near(f(n), f(n+1), n+1)) \to$
	$(\forall m \in \mathbb{N}. \ near(f(m), x, m))$

Arithmetic Operations. As the reader can see, the properties required for the arithmetic operations are just the same characterizing a classical Abelian field; in [Bri99] this same set of properties is named "Heyting field". Note that it is sufficient to assume the existence of the reciprocal only for positive reals.

As we have already remarked, we do not assume the "negation" and the "reciprocal" functions: instead we assume the existence, for each real x, of its negation and, if $0 < x$, of its reciprocal elements. In this way we have to postulate the Axiom of Choice for extracting effectively from a number x its negation and its reciprocal.

The necessity of the Axiom of Choice can be seen as a weakness of our axiomatization. However, there is no simple way to avoid it: in fact, without Choice, the reciprocal function cannot be defined inside Coq (whereas the negation function and the limit functional are definable).

An alternative axiomatization that does not require the Axiom of Choice could be obtained as follows. One postulates the existence of the negation and limit functions and, instead of a single inversion function, the existence of a series of approximations of the inversion function, $inv : (\mathbb{N} \times R) \rightarrow R$, satisfying the axiom:

$$\forall n \in \mathbb{N}. \ \forall x \in R. \ (1_R < x \times exp(n)) \rightarrow (x \times inv(n, x) \sim 1_R)$$

that is, the function $\lambda x. \ inv(n, x)$ behaves as the reciprocal function for all the real numbers bigger than 2^{-n}. Given a suitable representation for the reals, the function inv is Coq-definable, and allows the evaluation of the reciprocal of any real number x for which it is possible to find a natural number n such that $2^n < x$. We did not pursue this alternative axiomatization for simplicity reasons.

Order Relation. Concerning the ordering, we make the following remarks.

First, the classical trichotomy of total order $(x < y) \vee (x = y) \vee (y < x)$ is not a constructive property: its substitute in the constructive setting is the property $(x < y) \rightarrow (x < z) \vee (z < y)$, which we name "co-transitivity".

Secondly, we have thought that it is cleaner to define only the relation of order: in fact, in constructive mathematics [TvD88,Bri99], the order is universally considered the most fundamental relation for the real numbers. The alternative would have been to start from the apartness relation — the constructive non equality — then to assume axioms for it, and further to introduce the order itself with its proper axioms. But this increases the length of the presentation of the constructive reals, and moreover introduces some redundancy, thus not permitting to carry out our declared purpose of being minimal.

The equivalence and the apartness relations are defined using the basic strict order, therefore their properties are derived from the axioms. We define the equivalence ($\sim$) and the apartness ($\#$) in the following way:

$$(x \sim y) \triangleq \neg[(x < y) \vee (y < x)] \qquad (x \ \# \ y) \triangleq (x < y) \vee (y < x)$$

There is still more, because we have been careful in the design of the relationship between the order and the operations. We are able to deduce all the basic properties relating the equivalence and the operations from the two reflection axioms:

$$(x + z < y + z) \rightarrow (x < y)$$
$$(x \times z < y \times z) \rightarrow (x < y) \vee [(y < x) \wedge (z < 0)]$$

The fact that the equivalence is preserved by the basic notions (addition, multiplication and order) is an immediate consequence of these two axioms and the $<$-*co-transitivity* one. Notice that, on the contrary, the preservation of the equivalence does not follow from the more usual preservation axioms [Bri99,GN01]:

$$(x < y) \rightarrow (x + z < y + z)$$
$$(0 < x) \wedge (0 < y) \rightarrow (0 < x \times y)$$

This particular phenomenon relies on the fact that the reflection of the order is more powerful than its preservation, as will be argued in Section 3.

Archimedean Property. The Archimedean axiom links real numbers to natural numbers, stating that the reals are standard with respect to the naturals. This axiom does not exclude the existence of non-standard reals, but in this case also the naturals must be non-standard. That is, it is possible to conceive non-standard models for our axioms: these models would contain infinitary and infinitesimal real numbers as well as infinitary naturals.

Completeness. Finally, the completeness property for the field of the real numbers is postulated asking the existence of the limit of any Cauchy sequence $\langle s_n \rangle_{n \in \mathbb{N}}$ with an exponential convergency rate:

$$\forall n \in \mathbb{N}. \ |s_n - s_{n+1}| \leq 2^{-(n+1)}$$

Many others choices for capturing the completeness are possible, and our axiom could appear weak at a first glance. Anyway, in order to evaluate constructively the limit of a Cauchy sequence S, it is necessary to know its convergency rate: from this convergency rate it is possible to extract (constructively) a subsequence of S having an exponential convergency rate. It follows that starting from our axiom we are able to derive the completeness properties that are found in the literature [Bri99,GN01]. Our choice has been motivated by simplicity reasons.

The minimality of our axiomatization could be useful both for theoretical reasons — the mathematical curiosity about an essential characterization of the constructive reals is addressed — and practical ones — a simple test for possible models is provided. However, rather than pursuing a minimal set of axioms at all costs, we have chosen to axiomatize the different notions (order, addition, multiplication, etc.) separately, for the sake of the clarity of the axiomatization.

3   Axioms at Work

In this Section we point out the elementary mathematical theory arising from the axioms. Such results can be used for deducing more complex properties of the constructive real numbers.

We begin with some consequences concerning the order; successively we will address the addition, the multiplication and other defined notions. Notice that all the logic steps used in the following proofs are constructive. These proofs have been also carried out formally in the proof assistant Coq: this and further work is documented in [CDG01].

Proposition 1. *(Order)*
The following properties of the order and derived notions follow from the axioms:

1. *irreflexivity:* $\forall x.\ \neg(x < x)$
2. *transitivity:* $\forall x, y, z.\ (x < y) \wedge (y < z) \rightarrow (x < z)$
3. *preservation by equivalence:*
$$\forall x, y, z, w.\ (x < y) \wedge (x \sim z) \wedge (y \sim w) \rightarrow (z < w)$$
4. *co-transitivity of apartness:* $\forall x, y, z.\ (x \mathbin{\#} y) \rightarrow (x \mathbin{\#} z) \vee (z \mathbin{\#} y)$
5. *apartness preservation:* $\forall x, y, z, w.\ (x \mathbin{\#} y) \wedge (x \sim z) \wedge (y \sim w) \rightarrow (z \mathbin{\#} w)$

Proof. The proofs are quite easy, so we just sketch some of them. Point (2) is proved applying the $<$-*co-transitivity* to $(x < y)$; point (3) using twice the $<$-*co-transitivity*; point (5) by a double application of (4). □

We state now the main fact concerning the addition, i.e. it preserves the equivalence. Moreover we deduce also the preservation of order, which is assumed as an axiom in the alternative axiomatizations [Bri99,GN01].

Proposition 2. *(Addition)*
The following properties of the addition follow from the axioms:

1. *strong extensionality* : $\forall x, y, z, w.\ (x + y) \mathbin{\#} (z + w) \rightarrow (x \mathbin{\#} z) \vee (y \mathbin{\#} w)$
2. *equivalence preservation* : $\forall x, y, z, w.\ (x \sim y) \wedge (z \sim w) \rightarrow (x + z) \sim (y + w)$
3. *order preservation* : $\forall x, y, z.\ (x < y) \rightarrow (x + z < y + z)$
4. *equivalence reflection* : $\forall x, y, z.\ (x + z) \sim (y + z) \rightarrow (x \sim y)$

Proof. (1) The goal follows immediately from the two judgments $(x + y) \mathbin{\#} (z + y) \rightarrow (x \mathbin{\#} z)$ (left extensionality) and $(x + y) \mathbin{\#} (x + z) \rightarrow (y \mathbin{\#} z)$ (right extensionality). Left extensionality can be written $(x + y < z + y) \vee (z + y < x + y) \rightarrow (x < z) \vee (z < x)$ and proved by cases using the $+$-*reflects*-$<$ axiom. Right extensionality can be reduced to left extensionality by *co-transitivity of apartness*, *apartness preservation* and the $+$-*commutativity* axiom.

(3) Using *negation*, point (2), $+$-*unit* and $+$*associativity* we can derive $x \sim ((x + z) + (-z))$. By Proposition 1.3 it is then possible to deduce $(x + z) + (-z) < (y + z) + (-z)$, and from this the thesis via $+$-*reflects*-$<$.

Point (2) follows from (1); point (4) from (3). □

We consider similar results for the multiplication: we prove it preserves the equivalence and deduce additional consequences.

Proposition 3. *(Multiplication)*
The following properties of the multiplication follow from the axioms:

1. *strong extensionality* : $\forall x, y, z, w.\ (x \times y) \mathbin{\#} (z \times w) \rightarrow (x \mathbin{\#} z) \vee (y \mathbin{\#} w)$
2. *equivalence preservation* : $\forall x, y, z, w.\ (x \sim y) \wedge (z \sim w) \rightarrow (x \times z) \sim (y \times w)$
3. *positivity reflection* : $\forall x, y, z.\ ((x \times z) < (y \times z)) \wedge (0 < z) \rightarrow (x < y)$
4. *zero annuls multiplication:* $\forall x.\ (x \times 0 \sim 0)$
5. *reciprocal preserves positivity:* $\forall x.\ (0 < x) \rightarrow (0 < x^{-1})$
6. *positivity preservation* : $\forall x, y.\ (0 < x) \wedge (0 < y) \rightarrow (0 < x \times y)$

Proof. The proofs of points (1) and (2) use the same arguments of the corresponding proofs for the addition. Point (3) follows through axiom $\times$-*reflects*-$<$.

(4) From $(0 + 0) \sim 0$ we derive $((x \times 0) + (x \times 0)) \sim (0 + (x \times 0))$, and then the thesis through the Proposition 2.4.

(5) From the hypothesis $(0 < x)$ and axiom $(0 < 1)$ we obtain $0 < (x^{-1} \times x)$; then we have $(0 \times x) < (x^{-1} \times x)$, from which we conclude by point (3).

(6) Using the hypotheses we derive $x \sim ((x \times y) \times y^{-1})$, from which we deduce $(0 \times y^{-1}) < ((x \times y) \times y^{-1})$; the thesis follows by points (5) and (3). $\square$

It is easy to prove that the $+$-*reflects*-$<$ axiom is equivalent to the *equivalence preservation* plus *order preservation* of Proposition 2. In [Bri99] and [GN01] these two properties are taken as axioms: we prefer our choice for minimality reasons. A similar consideration applies to multiplication.

We also remark that two possible candidates for the $+$-*reflects*-$<$ axiom, namely $(x \times z < y \times z) \wedge (0 < z) \rightarrow (x < y)$ and $(x \times z < y \times z) \rightarrow (x < y) \vee (z < 0)$, are too weak.

We list now other typical and useful properties of the constructive real numbers. They involve also the auxiliary notion of non-strict order ($\leq$), which is formalized as follows:

$$(x \leq y) \triangleq \neg(y < x)$$

Proposition 4. *(Other Properties)*
The following judgments can be derived from the axioms and their corollaries:

$$
\begin{aligned}
Order: \quad & (x \leq y) \wedge (y \leq x) \rightarrow (x \sim y) \\
& (x \leq y) \wedge (y \leq z) \rightarrow (x \leq z) \\
& ((z < x) \rightarrow (z < y)) \rightarrow (x \leq y)
\end{aligned}
$$

$$
\begin{aligned}
Addition: \quad & (0 < x) \rightarrow (-x < 0) \\
& (0 < x + y) \rightarrow ((0 < x) \vee (0 < y)) \\
& (x \leq y) \leftrightarrow (x + z \leq y + z)
\end{aligned}
$$

$$
\begin{aligned}
Multiplication: \quad & (x \times (-y)) \sim -(x \times y) \\
& (x < y) \wedge (0 < z) \rightarrow (x \times z < y \times z) \\
& (x < y) \wedge (z < 0) \rightarrow (y \times z < x \times z) \\
& (0 < x \times y) \rightarrow (x \# 0) \wedge (y \# 0)
\end{aligned}
$$

4   Consistency and Completeness

The usual way for defining real numbers is the use of (equivalence classes of) Cauchy sequences of rational numbers [Bee85,BB85,TvD88], but there exist other constructions that can be easily proved equivalent to this one. An example is the approach that introduces the reals as infinite sequences of digits [PEE97,Wei00,CDG00]: in this case the equivalence follows from the fact that it is possible to transform effectively a representation of a number through a

Cauchy sequence in a representation of the same number through an infinite stream of digits, and vice-versa.

In the distillation of our set of axioms we have used as reference the definition of the real numbers via Cauchy sequences of rationals. This construction has been used as a model for testing the consistency of the axioms: in order to accept a judgment as an axiom we have first informally verified that it is satisfied by the Cauchy model. We are now developing a formal proof in Coq [CDG01] that our axioms are satisfied by a construction of the real numbers through "streams of digits".

A similar result is presented in [GN01]: the constructive reals are built as Cauchy sequences of rational numbers, and a different axiomatization is introduced. That work gives also a proof that the axiomatization is categorical — i.e. any two models are isomorphic. Since in the next Section we will show that our axiomatization is equivalent to that one [GN01], we can deduce that our axiomatization is complete; and we can claim that it is also categorical.

5   Comparison to the Related Literature

In this Section we compare our axioms to the other approaches of the literature [TvD88,Bri99,GPWZ00,GN01]. In particular, we will prove that our axiomatization has the same deductive power of the alternative ones.

The work of Troelstra and van Dalen [TvD88] is a contribution which gives a constructive treatment of the theory of the real numbers in the context of a constructive approach to mathematics. Although the authors do not address the quest for an axiomatization of the constructive reals, their approach focuses on aspects strictly related to the present work. They build the constructive reals as equivalence classes of fundamental (Cauchy) sequences of rationals; then they introduce the primitive strict order relation ($<$) and the arithmetic functions ($+, \cdot$). The basic properties of these notions are the same we have proved in Section 3, or follow simply from those results.

We will mainly refer our work to other two contributions.

5.1   FTA

The FTA approach [GPWZ00] is similar to ours in regard to the tool used for the formalization, i.e. the proof assistant Coq. There, constructive real numbers are just seen as a parameter contextually to a more extensive project concerning the mechanical certification of a theorem of constructive mathematics. In fact, the main aim is just to dispose of a collection of properties for describing the reals which is sufficiently powerful for proving the Fundamental Theorem of Algebra. The axiomatization is separately presented and focused in [GN01]: this work first introduces the algebraic structure of constructive setoids via an apartness relation; successively, step by step, it gains the notion of constructive real number. We are proving the equivalence between our axiomatization and this one.

The FTA approach uses 28 axioms. The essential differences with respect to ours are the introduction of the apartness relation as primitive and the assumption of strongly extensionality for the arithmetic functions of addition and multiplication.

The Structure. Constructive reals are introduced by the tuple:

$$\langle \mathbb{R}, 0, 1, +, *, -, ^{-1}, =, <, \# \rangle$$

Field. The following axioms are assumed in order to have the structure of constructive setoids:

$$\text{ap\_irr} : \forall x.\ \neg(x \mathbin{\#} x)$$
$$\text{ap\_sym} : \forall x, y.\ (x \mathbin{\#} y) \to (y \mathbin{\#} x)$$
$$\text{ap\_cot} : \forall x, y.\ (x \mathbin{\#} y) \to \forall z.(x \mathbin{\#} z) \vee (z \mathbin{\#} y)$$
$$\text{ap\_tight} : \forall x, y.\ \neg(x \mathbin{\#} y) \leftrightarrow (x = y)$$

The above properties can easily be derived from our axioms: ap_irr (irreflexivity) follows from the irreflexivity of the order $\neg(x < x)$; ap_cot (cotransitivity) is a consequence of the *<-co-transitivity* axiom. Instead ap_sym (symmetry) and ap_tight (tightness) are deduced just from the definitions. Some of the properties which lead to a constructive field coincide with ours:

$$\text{add\_assoc} : \forall x, y, z.\ (x + (y + z)) = ((x + y) + z)$$
$$\text{add\_unit} : \forall x.\ (x + 0) = x$$
$$\text{add\_commut} : \forall x, y.\ (x + y) = (y + x)$$
$$\text{minus\_proof} : \forall x.\ x + (-x) = 0$$
$$\text{mult\_assoc} : \forall x, y, z.\ (x * (y * z)) = ((x * y) * z)$$
$$\text{mult\_unit} : \forall x.\ (x * 1) = x$$
$$\text{mult\_commut} : \forall x, y.\ (x * y) = (y * x)$$
$$\text{dist} : \forall x, y, z.\ (x * (y + z)) = ((x * y) + (x * z))$$
$$\text{rcpcl\_proof} : \forall x.\ (x \mathbin{\#} 0) \to (x * (x^{-1})) = 1$$

Moreover, FTA uses the extra axioms:

$$\text{add\_strext} : \forall x, y, z, u.\ (x + y) \mathbin{\#} (z + u) \to (x \mathbin{\#} z) \vee (y \mathbin{\#} u)$$
$$\text{minus\_strext} : \forall x, y.\ ((-x) \mathbin{\#} (-y)) \to (x \mathbin{\#} y)$$
$$\text{mult\_strext} : \forall x, y, z, u.\ (x * y) \mathbin{\#} (z * u) \to (x \mathbin{\#} z) \vee (y \mathbin{\#} u)$$
$$\text{non\_triv} : 1 \mathbin{\#} 0$$
$$\text{rcpcl\_ap\_zero} : \forall x.\ (x \mathbin{\#} 0) \to (x^{-1} \mathbin{\#} 0)$$
$$\text{rcpcl\_strext} : \forall x, y.\ (x^{-1}) \mathbin{\#} (y^{-1}) \to (x \mathbin{\#} y)$$

We have proved **add_strext** and **mult_strext** (strong extensionality of addition and multiplication) in Propositions 2.1 and 3.1. We establish here rcpcl_strext (strong extensionality of the "reciprocal" function); simpler arguments suffice for proving the property minus_strext.

First we show `rcpcl_ap_zero`, i.e. the reciprocal respects the apartness with respect to zero. Starting from $x \,\#\, 0$ and the non triviality $1 \,\#\, 0$ — which follows from our axiom $0 < 1$ — we have both $(x * (x^{-1})) = 1$ by the $\times$-*unit* axiom and $(x * 0) = 0$ by Proposition 3.4. Next we have $(x * x^{-1}) \,\#\, (x * 0)$ by the Proposition 1.5, and then we conclude through the strong extensionality of the multiplication.

The main proof of `rcpcl_strext` works as follows. Since $x^{-1} = x^{-1} * (y * y^{-1})$ and $y^{-1} = y^{-1} * (x * x^{-1})$, by Proposition 1.5 and $\times$-*associativity* we deduce $(x^{-1} * y) * y^{-1} \,\#\, (x^{-1} * x) * y^{-1}$. Using the strong extensionality of the multiplication, we have first $(x^{-1} * y) \,\#\, (x^{-1} * x)$ and further $y \,\#\, x$, thus concluding via `ap_sym`.

Ordered Field. The only axiom shared with ours is the asymmetry:

$$\texttt{less\_asym} : \forall x, y. \ (x < y) \to \neg(y < x)$$

The extra axioms are the following:

> `less_strext` : $\forall x, y, z, u. \ (x < y) \to (z < u) \lor (x \,\#\, z) \lor (y \,\#\, u)$
> `less_trans` : $\forall x, y, z. \ (x < y) \land (y < z) \to (x < z)$
> `less_irr` : $\forall x. \ \neg(x < x)$
> `add_resp_less` : $\forall x, y, z. \ (x < y) \to (x + z < y + z)$
> `times_resp_pos` : $\forall x, y. \ (0 < x) \land (0 < y) \to (0 < x * y)$
> `less_conf_ap` : $\forall x, y, z. \ (x \,\#\, y) \leftrightarrow (x < y) \lor (y < x)$.

The properties `less_irr` (irreflexivity) and `less_trans` (transitivity) have been derived in Proposition 1.2; `less_strext` (strong extensionality of order) follows from $<$-*co-transitivity*; `less_conf_ap` just coincides with our definition of apartness. Finally the `add_resp_less` and `times_resp_pos` axioms have been derived in the Propositions 2.3 and 3.6.

Archimedes. This axiom coincide with ours:

$$\texttt{arch\_proof} : \forall x. \ \exists n \in \mathbb{N}. \ (x < (\texttt{nreal } n))$$

Limit. The axiom of completeness asks that every Cauchy sequence has a limit:

$$\texttt{lim\_proof} : \forall s : \mathbb{N} \to \mathbb{R}, \text{ cauchy}. \ \forall \epsilon > 0. \ \exists n \in \mathbb{N}.$$
$$(\forall m \in \mathbb{N}. \ (n \le m) \to |\, (s \ m) - \lim s \,| < \epsilon)$$

It is not difficult to prove that this axiom is at least as strong as the one we have used in our axiomatization. We can deduce the FTA axiom from ours as follows: in order to calculate the limit of a generic Cauchy sequence S we need to be able to extract a subsequence of S converging with an exponential convergency rate. This subsequence can be obtained easily once we know the convergency rate of S, which in turn can be extracted, using the Axiom of Choice, from the proposition stating that S satisfies the Cauchy condition.

Theorem 1. *(Equivalence between Axiomatizations)*
We conclude that our axiomatization and the FTA one are equivalent.

5.2   Bridges

Bridges [Bri99] uses the framework of Bishop's constructive mathematics for presenting a constructive axiomatization of the real line. His main motivation coincides partially with the ours: the curiosity about the properties that suffice to characterize the real numbers and to develop the real analysis. The way chosen is to capture the idea that a real could be approximated by arbitrarily close rational numbers.

The constructive axiomatization given by Bridges collects 20 axioms. The axioms are quite similar to ours apart the completeness one: we discuss briefly the main differences.

The equivalence relation $=$ is defined by $(x = y) \triangleq (x \geq y) \wedge (y \geq x)$, where $(x \geq y) \triangleq (\forall z.\ (y > z) \to (x > z))$. By Propositions 4.4 and 4.5, we have that $\forall z.\ (y > z) \to (x > z)$ if and only if $\neg (x < y)$. The same could be proved using the axioms of Bridges. And so, Bridges' approach is equivalent to ours up to the use of the more involved definition of the equivalence relation "$=$".

The axioms concerning the field structure coincide with ours. Bridges assumes as additional implicit axioms the "extensionality" of relations and operations, i.e. they preserve the equivalence. As already remarked, in our approach these are just derived properties.

As far as the axioms for the order are concerned, the main difference is that Bridges requires that the operations "preserve" the order, whereas we require that the operations "reflect" the order.

The completeness axiom chosen by Bridges is quite different from ours. In [Bri99] the completeness of the real line is postulated through a "least-upper-bound principle", which requires that every "strongly bounded" set of reals has a least upper bound; from such a principle, it is then derived that every Cauchy sequence has limit. In [BR99] and [GPWZ00] it is proved that the existence of the l.u.b. of "strongly bounded" sets can be deduced from the existence of the limit for Cauchy sequences. Hence Bridges' approach to completeness is equivalent to ours. We have preferred to state the completeness in terms of Cauchy sequences, because it is simpler.

6   Conclusion

In this work we have focused on the only two existing axiomatizations of the constructive real numbers and we have proposed a third one.

All the three axiomatizations have the same deductive power. As far as we know, Bridges [Bri99] and the FTA group [GPWZ00] have proposed their axioms independently; we too have stated our axioms without being aware of the work by Bridges.

We claim that our axiomatization has the advantage of being simpler and of using a minimal set of notions. In particular, we give a more direct treatment of the equivalence "$\sim$" and the non-strict order "$\leq$" relations than Bridges. Also our completeness axiom is simpler than the corresponding one in any of the other proposals, and in general the whole axiomatization is more compact.

A possible direction for future work is to consider an axiomatization which does not require the Axiom of Choice. In this perspective, it would be interesting to consider also an axiomatization of the constructive reals obtained by Dedekind cuts: in fact, Cauchy sequences and Dedekind cuts are equivalent constructions for the reals only if the Axiom of Choice is available [TvD88]. Results in this sense would help to characterize the fundamental differences between the two constructions.

Acknowledgments

The authors are grateful to Herman Geuvers, Furio Honsell, Marino Miculan, Ivan Scagnetto and the anonymous referees for their interesting remarks.

References

[BB85]   E. Bishop and D. Bridges. *"Constructive Analysis"*. Springer-Verlag, 1985.

[BB+01]   B. Barras, S. Boutin, et al. *"The Coq proof assistant Reference manual"*. INRIA, Project Logical, 2001.

[Bee85]   M. J. Beeson. *"Foundations of Constructive Mathematics"*. Springer-Verlag, 1985.

[Bis67]   E. Bishop. *"Foundations of constructive analysis"*. McGraw-Hill, New York, 1967.

[BR99]   D. Bridges and S. Reeves. *"Constructive mathematics in theroy and programming practice"*. Philosophia Mathematica 7, 1999.

[Bri99]   D. Bridges. *"Constructive mathematics: a foundation for computable analysis"*. TCS 219, 1999.

[CDG00]   A. Ciaffaglione and P. Di Gianantonio. *"A co-inductive approach to real numbers"*. In Proceedings of Types'99, LNCS 1956, 2000.

[CDG01]   A. Ciaffaglione and P. Di Gianantonio. *"Constructive Real Numbers in Coq"*. Dipartimento di Matematica e Informatica, Udine (Italy), http://www.dimi.uniud.it/~ciaffagl/C_Reals/index.html, 2001.

[CH88]   T. Coquand and G. Huet. *"The Calculus of Constructions"*. Information and Control 76, 1988.

[GN01]   H. Geuvers and M. Niqui. *"Constructive reals in Coq: axioms and categoricity"*. In Proceedings of Types 2000, this Volume, 2001.

[GPWZ00]   H. Geuvers, R. Pollack, F. Wiedijk, and J. Zwanenburg. *"The Fundamental Theorem of Algebra" Project*. Computing Science Institute, Nijmegen (The Netherlands), http://www.cs.kun.nl/~freek/fta/index.html, 2000.

[PEE97]   P.J. Potts, A. Edalat, and M.H. Escardo. *"Semantics of exact real arithmetic"*. In IEEE Symposium on Logic in Computer Science, 1997.

[TvD88]   A.S. Troelstra and D. van Dalen. *"Constructivism in Mathematics"*. North-Holland, 1988.

[Wei00]   K. Weihrauch. *"Computable Analysis, An Introduction"*. Springer-Verlag, 2000.

An Implementation of Type:Type

Thierry Coquand and Makoto Takeyama

Department of Computing Science
Chalmers University of Technology and Göteborg University

Abstract. We present a denotational semantics of a type system with dependent types, where types are interpreted as finitary projections. We prove then the correctness of a type-checking algorithm w.r.t. this semantics. In this way, we can justify some simple optimisation in this algorithm. We then sketch how to extend this semantics to allow a simple record mechanism with manifest fields.

1 Introduction

A quite elegant interpretation of dependent types in domain theory, where types are interpreted as finitary projections, has been known for a long time [10]. However we do not know of any proof of correctness of a type-checking algorithm with respect to such a semantics. This semantics is used in [4] to show that correctness of a typing system, but no type-checking algorithm is described in [4]. Similarly, the work [2], which describes a denotational semantics of dependent types, does not address the issue of correctness of a type-checking algorithm. However, some usual computation rules that are used in type-checking algorithms are a priori problematic if types are interpreted as finitary projections: in this semantics, if a type a is interpreted as a finitary projection p_a, the interpretation of typed-abstraction gives

$$[\![(\lambda x : a)e]\!]\, u = [\![e]\!]_{x = p_{[\![a]\!]}(u)}$$

while the computation rule that is used in type-checking is

$$(\lambda x : a)e\, e_1 = e(e_1)$$

that is, in the β-reduction rule, we forget the type of the abstracted variable. The intuitive justification of this rule is that we shall use it only for *well-typed* terms, and that if $e_1 : a$ we have

$$p_{[\![a]\!]}([\![e_1]\!]) = [\![e_1]\!]$$

so that we can forget about the explicit typing. But this intuitive justification may look somewhat circular (the typing itself uses computation rules), and we think it is important to justify rigorously this β-reduction rule. Another, related, motivation for our work is that we would like in the implementation of the conversion algorithm to use the following optimisation: when comparing

$$(\lambda x_1 : a_1)e_1 = (\lambda x_2 : a_2)e_2$$

P. Callaghan et al. (Eds.): TYPES 2000, LNCS 2277, pp. 53–62, 2002.

we need only to test if $e_1 = e_2$. The motivation here is that an invariant is that we compare only two terms of the same type, and that if we know that the two terms are of the same types, then we shall not need to check $a_1 = a_2$.

The goal of this note is to present a complete justification of a simple type-checking algorithm w.r.t. the semantics of dependent types in domain theory. We shall justify the untyped β-conversion rule, as well as the optimisation described above when comparing two typed abstraction. Our semantics will be such that we can justify as well the following form of η-conversion

$$(\lambda x_1 : a_1)e_1 = e$$

whenever $e_1(x) = e\ x$ for a fresh variable x.

In order to keep things as simple as possible we limit ourselves here to consider one of the simplest type system with dependent types, with $Type : Type$. We do not expect any further problems to adapt our treatment to the case of a system with for instance a cumulative hierarchy of universes $Type(i) : Type(i + 1)$, for which we expect to have also a normalisation property[1]. With respect to the previous work [6], there are two main differences. The first one is that we have *typed* abstraction. This seems to be crucial in order to have a semantics in domain theory, with types interpreted as finitary projections. The second one is that we present a calculus which allows type inference. To the usual construction ev which computes the value of an expression e in a semantical environment ν, we add a new construction $e \mid \nu$ which computes the *type* of an expression e in a semantical environment ν.

2   Denotational Semantics

We start by describing the semantics for the simplest type theory containing dependent product types and the type V of all types. For this we solve the domain equations

$$D = \qquad [D \to D] \oplus T$$
$$T = \text{Prod } T\ [D \to T] \oplus V.$$

Here $D_1 \oplus D_2$ denotes the "smash sum" of D_1 and D_2, where the $\perp$ of D_1 and D_2 are identified. V is the two-element domain whose only non-bottom element is written again by V. Prod T $[D \to T]$ is the lift of Cartesian product of T and $[D \to T]$, with typical non-bottom elements written Prod $u\ f$ for $u \in$ T and $f \in [D \to T]$.

Given any solution D of the equation, we take that a 'type' is a priori a subdomain of D. Semantics of a closed expression will be either a function $g \in [D \to D]$ or a code $u \in$ T for a type.

'Decoding' is defined in terms of *finitary projections*. Recall that, for any domain D, subdomains form the domain $(\{E \mid E \triangleleft D\}, \subseteq, \{\perp\})$, and that this

---

[1] This proof should for instance follow [5].

is isomorphic to the domain of finitary projections $(\mathsf{Fp}(D), \sqsubseteq, \perp_{[D \to D]})$, where $\mathsf{Fp}(D) = \{p : D \longrightarrow D \mid p = p \circ p,\ p \sqsubseteq \mathrm{id},\ \mathrm{range}(p)\ \text{is a domain}\}$.

We now define the map $u \longmapsto \mathsf{p}_u$ as the solution $\mathsf{p}_{(-)} \in [\mathsf{T} \to \mathsf{Fp}(D)]$ of the following recursive equations[2]:

$$
\begin{aligned}
\mathsf{p}_\mathsf{V}\, u &= u & (u \in \mathsf{T}\quad) \\
\mathsf{p}_\mathsf{V}\, g &= \perp & (g \in [D \to D]) \\
\mathsf{p}_{\mathsf{Prod}\, u\, f}\, u' &= \perp & (u' \in \mathsf{T}\quad) \\
\mathsf{p}_{\mathsf{Prod}\, u\, f}\, g &= \lambda d.\, \mathsf{p}_{(f d_1)}(g\, d_1)\ \text{where } d_1 = \mathsf{p}_u d & (g \in [D \to D])
\end{aligned}
$$

So any $u \in \mathsf{T}$ defines a subdomain $D(u) = \mathrm{range}(\mathsf{p}_u) \lhd D$; e.g. $D(\mathsf{V}) = \mathsf{T}$.

Note that the map $u \longmapsto \mathsf{p}_u$ can be defined for *any* solution of the domain equations and not necessarily the initial one.

We extend the notation p_d for all of $d \in D$ by defining $\mathsf{p}_d(x) = \perp$ if $d \in [D \to D]$. For $d, u \in D$, we write $d : u$ to mean that $d \in D(u)$, which is equivalent to $\mathsf{p}_u(d) = d$.

Remark that the relation $d : u$ is not decidable; also, we cannot expect uniqueness of types in general, i.e. we may have $d : u_1$ and $d : u_2$ and $u_1 \neq u_2$.

Lemma 1.

- $g : \mathsf{Prod}\, u\, f$ if and only if $g \in [D \to D]$, $g = g \circ \mathsf{p}_u$, and $g\, d : f(\mathsf{p}_u d)$ for all $d \in D$. If $f = f \circ \mathsf{p}_u$ then the last may be replaced by that $g\, d : f\, d$ for all $d : u$.
- For $g, g' : \mathsf{Prod}\, u\, f$, $g = g'$ if and only if $g\, d = g'\, d$ for all $d : u$.

We write $(-) \cdot (-)$ for the application in D: $d_1 \cdot d_2 = (\mathsf{p}_{[D \to D]} d_1)\, d_2$, where $\mathsf{p}_{[D \to D]}$ is the projection $[D \to D] \lhd D$. Further, we use 'type application' given by

$$
\begin{aligned}
(-) * (-) &: \mathsf{T} \times D \longrightarrow \mathsf{T} \\
\mathsf{Prod}\, u\, f * d &= f(\mathsf{p}_u d) \\
\mathsf{V} \quad * d &= \perp .
\end{aligned}
$$

Lemma 2. For $d_1, d_2 \in D$ and $u \in \mathsf{T}$, $d_1 : u$ implies that $d_1 \cdot d_2 : u * d_2$.

A similar application is used in [8], with a reference to previous works by D. van Daalen on Automath [11].

3   Expressions

We define (pseudo)*expressions* e by

$$ e = x \mid \mathsf{V} \mid [x : e]e \mid (x : e)e \mid e\, e $$

where x is taken from a set *Ide* of identifiers. Besides e, we let the letters a and b range over expressions (that are intended to be types). The forms $[x : a]e$ and $(x : a)b$ are for typed lambda abstraction and product types, respectively. We do not assume the usual variable convention, but consider an inner binding of the same identifier to shadow the outer ones.

---

[2] Maps on smash sum specified by cases is taken to be always strict.

4 Semantics

To give expressions a semantics in D, we consider the domain $\mathsf{Env} = (D \times T)^{Ide}$ of typed semantical environments. For $\nu \in \mathsf{Env}$, we write $\nu(x) = d : u$ to mean $\nu(x) = (d, u)$. Given $u \in T$ and $d \in D$, we define updating of environment by

$$(\nu,\ x : u = d)(x) = \mathsf{p}_u d : u$$
$$(\nu,\ x : u = d)(y) = \nu(y) \qquad (y \neq x).$$

The semantics of expression e is given by two continuous functions $\nu \longmapsto e\nu : \mathsf{Env} \longrightarrow D$ and $\nu \longmapsto (e\,|\,\nu) : \mathsf{Env} \longrightarrow T$. The former is the denotation of e under ν, the latter that of type of e. Both are defined by induction on e.

$$
\begin{array}{llll}
x & \nu & = & \mathsf{p}_u d \\
V & \nu & = & V \\
[x : a]e\ \nu & = & \lambda d.\ e\nu_d \\
(x : a)b\ \nu & = & \mathsf{Prod}\ \mathsf{p}_V(a\nu)\ \lambda d.\ \mathsf{p}_V(b\nu_d) \\
e_1 e_2 & \nu & = & e_1\nu \cdot e_2\nu
\end{array}
\qquad
\begin{array}{llll}
x & |\,\nu & = & u \qquad \text{where } \nu(x) = d : u \\
V & |\,\nu & = & V \\
[x : a]e\ |\,\nu & = & \mathsf{Prod}\ \mathsf{p}_V(a\nu)\ \lambda d.(e\,|\,\nu_d) \\
(x : a)b\ |\,\nu & = & V \\
e_1 e_2 & |\,\nu & = & (e_1\,|\,\nu) * e_2\nu
\end{array}
$$

where $\nu_d = (\nu,\ x : \mathsf{p}_V(a\nu) = d)$ (we abbreviate this to $(\nu,\ [x : a] = d)$ in the following.) Note that, by definition of environment update, we have that $g = g \circ \mathsf{p}_u$ and $f = f \circ \mathsf{p}_u$ for $u = \mathsf{p}_V(a\nu)$, $g = [x : a]e\,\nu$, and $f = \lambda d.(e\,|\,\nu, [x : a] = d)$.

Theorem 3. For any ν and e, $e\nu\ :\ e\,|\,\nu$.

Proof. By induction on e, using Lemmas 1 and 2.

As explained in Introduction, the untyped reduction applied to ill-typed expressions does not preserve this semantics. The first version of our 'type checking' separates out well-typed expressions for which the reduction becomes correct. This version mixes syntax and semantics, but it is merely a way to state induction hypothesis for the correctness proof of final type checking algorithm, which shall be purely syntactic.

$$
\frac{\nu \vdash e \quad e\,|\,\nu = u}{\nu \vdash e : u} \qquad \nu \vdash x
$$

$$
\frac{\nu \vdash a : V \quad \forall d.\ \nu, [x : a] = d \vdash e}{\nu \vdash [x : a]e} \qquad
\frac{\nu \vdash a : V \quad \forall d.\ \nu, [x : a] = d \vdash b : V}{\nu \vdash (x : a)b}
$$

$$
\frac{\nu \vdash e_1 : \mathsf{Prod}\ (e_2\,|\,\nu)\ f \quad \nu \vdash e_2}{\nu \vdash e_1 e_2}
$$

5   Values and Conversion

The goal of conversion checking algorithm is to syntactically determine whether (type-) denotation of two expressions agree under all semantical environments. For this, we introduce the syntactic categories of *values* v and syntactic environment ρ by the following grammar:

$$v = v_k \mid V \mid e\rho \mid (e \mid \rho) \mid vv \mid v * v$$
$$\rho = () \mid (\rho, x : v = v)$$

Generic values v_k $(k = 0, 1, \cdots)$ represents indeterminate constants, and will be used for conversion and type checking under binders. *Type closure* $e \mid \rho$ and type application $v_1 * v_2$ are syntactic counterparts of the evident semantic notions introduced earlier. Besides v, we use letter t for values that are intended to be types.

The semantics of values depends on a given valuation $\delta \in (D \times T)^N$ of generic values; we employ the same notation for updating δ as for $\nu \in$ Env. The denotation $v\delta \in D$ of v and that of its type $v \mid \delta \in T$ are defined as expected:

$$
\begin{array}{llll}
v_k & \delta = \mathsf{p}_u d & v_k & \mid \delta = u \quad (\delta(k) = d : u) \\
V & \delta = V & V & \mid \delta = V \\
(e\rho) & \delta = e(\rho\delta) & (e\rho) & \mid \delta = e \mid \rho\delta \\
(e \mid \rho) & \delta = e \mid \rho\delta & (e \mid \rho) & \mid \delta = V \\
(v_1 v_2) & \delta = (v_1\delta) \cdot (v_2\delta) & (v_1 v_2) & \mid \delta = (v_1 \mid \delta) * (v_2\delta) \\
(v_1 * v_2) \delta = (v_1\delta) * (v_2\delta) & & (v_1 * v_2) \mid \delta = V \\
() & \delta = \bot_{\mathsf{Env}} & (\rho, x : t = v)\delta = (\rho\delta, \, x : \mathsf{p}_V(t\delta) = v\delta)
\end{array}
$$

Using previous lemmas and Theorem 3, it is direct to show:

Lemma 4. For any δ and v, $v\delta : v \mid \delta$.

Our conversion checking algorithm shall be guided by the structure of values in weak head normal form, as in [6]. We present the reduction to weak head normal form in the following (deterministic) small step manner.

$$
\begin{array}{ll}
x(\rho, x : t = v) \longrightarrow v & x \mid \rho, x : t = v \longrightarrow t \\[4pt]
x(\rho, y : t = v) \longrightarrow x\rho & x \mid \rho, y : t = v \longrightarrow x \mid \rho \quad (x \neq y) \\[4pt]
V\rho \longrightarrow V & V \mid \rho \longrightarrow V \\[4pt]
([x : a]e\,\rho)\,v \longrightarrow e(\rho, [x : a] = v) & ([x : a]e \mid \rho) * v \longrightarrow e \mid \rho, [x : a] = v \\[4pt]
((x : a)b\,\rho) * v \longrightarrow b(\rho, [x : a] = v) & (x : a)b \mid \rho \longrightarrow V \\[4pt]
e_1 e_2 \, \rho \longrightarrow (e_1\rho)(e_2\rho) & e_1 e_2 \mid \rho \longrightarrow (e_1 \mid \rho) * (e_2\rho) \\[4pt]
\dfrac{v_1 \longrightarrow v_1'}{v_1 v_2 \longrightarrow v_1' v_2} & \dfrac{v_1 \longrightarrow v_1'}{v_1 * v_2 \longrightarrow v_1' * v_2}
\end{array}
$$

where $(\rho, [x : a] = v)$ abbreviates $(\rho, x : a\rho = v)$.

We say value v is well-typed under δ if judgement $\vdash_\delta v$ is derivable with the following inference rules.

$$\frac{\vdash_\delta v \quad v \mid \delta = u}{\vdash_\delta v : u} \qquad \vdash_\delta \mathsf{v}_k \qquad \vdash_\delta \mathsf{V} \qquad \frac{\vdash_\delta \rho \quad \rho\delta \vdash e}{\vdash_\delta e\rho} \qquad \frac{\vdash_\delta e\rho}{\vdash_\delta e \mid \rho}$$

$$\frac{\vdash_\delta v_1 : \mathsf{Prod}\ (v_2 \mid \delta)\ f \quad \vdash_\delta v_2}{\vdash_\delta v_1 v_2} \qquad \frac{\vdash_\delta v_1 \quad v_1\delta = \mathsf{Prod}\ (v_2 \mid \delta)\ f \quad \vdash_\delta v_2}{\vdash_\delta v_1 * v_2}$$

$$\vdash () \qquad \frac{\vdash_\delta \rho \quad \vdash_\delta t : \mathsf{V} \quad \vdash_\delta v : t\delta}{\vdash_\delta (\rho,\ x : t = v)}$$

Lemma 5. If $\vdash_\delta v$ and $v \longrightarrow v'$, then $\vdash_\delta v'$, $v\delta = v'\delta$, and $v \mid \delta = v' \mid \delta$.

Proof. By induction on v. We use the following trivial consequences of Theorem 3 and Lemma 4: $\nu \vdash e : u$ implies $e\nu : u$, and $\vdash_\delta v : u$ implies $v\delta : u$.

- Cases $v = x(\rho,\ y{:}t{=}\ v_0)$, $(x \mid \rho,\ y{:}t{=}\ v_0)$, $\mathsf{V}\rho$, $\mathsf{V} \mid \rho$, $e_1 e_2\ \rho$, $e_1 e_2 \mid \rho$, or $(x{:}a)b \mid \rho$. That $v\delta = v'\delta$ and $v \mid \delta = v' \mid \delta$ are by definition. That $\vdash_\delta v'$ directly follows inversion of $\vdash_\delta v$.
- Case $v = x(\rho,\ x : t = v')$. Inverting $\vdash_\delta v$, we see that $\vdash_\delta t : \mathsf{V}$ and $\vdash_\delta v' : t\delta$. Hence $v \mid \delta = \mathsf{p}_\mathsf{V}(t\delta) = t\delta = v' \mid \delta$ and $v\delta = \mathsf{p}_{\mathsf{p}_\mathsf{V}t\delta}\ v'\delta = v'\delta$.
- Case $v = (x \mid \rho,\ x : t = v_0)$. As in the previous case.
- Case $v = v_1 v_2$
  - Subcase $v_1 = [x : a]e\,\rho$. v' must be $e(\rho, [x : a]{=}\ v_2)$. That $v\delta = v'\delta$ and $v \mid \delta = v' \mid \delta$ are by definition, using the remark before Theorem 3 for the latter. Among the conditions necessary for $\vdash_\delta v'$, only $a\rho\delta = v_2 \mid \delta$ is missing from those obtained by inverting $\vdash_\delta v$. For this, we note the latter include $\rho\delta \vdash a : \mathsf{V}$ and $[x : a]e\rho \mid \delta = \mathsf{Prod}\ (v_2 \mid \delta)\ f$. Since Prod is injective[3], $a\rho\delta = \mathsf{p}_\mathsf{V}(a\rho\delta) = v_2 \mid \delta$ as desired.
  - Subcase $v_1 \neq [x : a]e\,\rho$. v' must be $v_1' v_2$ where $v_1 \longrightarrow v_1'$. The desired conclusions follow inversion of $\vdash_\delta v$ and induction hypothesis on v_1.
- Case $v = v_1 * v_2$: Similarly to the above.
- Other forms of v cannot stand in relation $v \longrightarrow v'$.

Type-correct values in weak head normal form are contained in the following grammar[4]:

$$h = \mathsf{V} \mid [x : a]e\,\rho \mid ([x : a]e \mid \rho) \mid (x : a)b\,\rho \mid w$$
$$w = \mathsf{v}_k \mid wv \mid w * v$$

---

[3] i.e., $\mathsf{Prod}\ u_1\ f_1 = \mathsf{Prod}\ u_2\ f_2\ \Rightarrow\ u_1 = u_2 \wedge f_1 = f_2$.

[4] We ignore $x()$ and $x \mid ()$ for alternatives of w, assuming a prior syntactic check for unbound variables. Their presence or absence does not affect the argument.

The following inductive definition presents our conversion checking algorithm.

$$\frac{v_1 \longrightarrow^* h_1 \quad v_2 \longrightarrow^* h_2 \quad h_1 \text{ conv } h_2}{v_1 \text{ conv } v_2} \qquad (v_1 \text{ or } v_2 \text{ not in whnf})$$

$$\text{V conv V}$$

$$\frac{e_1\rho_1' \text{ conv } e_2\rho_2'}{[x_1 : a_1]e_1\rho_1 \text{ conv } [x_2 : a_2]e_2\rho_2} \qquad \frac{e_1\rho_1' \text{ conv } wv_k}{[x_1 : a_1]e_1\rho_1 \text{ conv } w} \qquad \frac{wv_k \text{ conv } e_2\rho_2'}{w \text{ conv } [x_2 : a_2]e_2\rho_2}$$

$$\frac{a_1\rho_1 \text{ conv } a_2\rho_2 \quad b_1\rho_1' \text{ conv } b_2\rho_2'}{(x_1 : a_1)b_1\rho_1 \text{ conv } (x_2 : a_2)b_2\rho_2} \qquad \frac{a_1\rho_1 \text{ conv } a_2\rho_2 \quad e_1 \mid \rho_1' \text{ conv } e_2 \mid \rho_2'}{[x_1 : a_1]e_1 \mid \rho_1 \text{ conv } [x_2 : a_2]e_2 \mid \rho_2}$$

$$\frac{a_1\rho_1 \text{ conv } a_2\rho_2 \quad e_1 \mid \rho_1' \text{ conv } b_2\rho_2'}{[x_1 : a_1]e_1 \mid \rho_1 \text{ conv } (x_2 : a_2)b_2\rho_2} \qquad \frac{a_1\rho_1 \text{ conv } a_2\rho_2 \quad b_1\rho_1' \text{ conv } e_2 \mid \rho_2'}{(x_1 : a_1)b_1\rho_1 \text{ conv } [x_2 : a_2]e_2 \mid \rho_2}$$

$$\text{v}_k \text{ conv v}_k \qquad \frac{w_1 \text{ conv } w_2 \quad v_1 \text{ conv } v_2}{w_1 v_1 \text{ conv } w_2 v_2} \qquad \frac{w_1 \text{ conv } w_2 \quad v_1 \text{ conv } v_2}{w_1 * v_1 \text{ conv } w_2 * v_2}$$

where $\rho_i' = (\rho_i, [x_i:a_i]= \text{v}_k)$ with v_k not occurring in respective rule's conclusion.

Lemma 6. If $\vdash_\delta v_1$, $\vdash_\delta v_2$, $v_1 \mid \delta = v_2 \mid \delta$, and $v_1 \text{ conv } v_2$, then $v_1\delta = v_2\delta$.

Proof. We prove it together with the following strengthening in case v_i 's are of the form w_i:

- If $\vdash_\delta w_1$, $\vdash_\delta w_2$, $w_1 \text{ conv } w_2$, then $w_1 \mid \delta = w_2 \mid \delta$ and $w_1\delta = w_2\delta$.

The proof proceeds by induction on the derivation of $v_1 \text{ conv } v_2$. If it is concluded by the first rule, the statement follows from induction hypothesis and Lemma 5. Identifying the rest of cases by its conclusion, we have:

- Cases V conv V and $\text{v}_k \text{ conv v}_k$. Immediate.
- Case $[x_1:a_1]e_1\rho_1 \text{ conv } [x_2:a_2]e_2\rho_2$. Inversion of $\vdash_\delta [x_i:a_i]e_i\rho_i$ gives (1) $\vdash_\delta \rho_i$, (2) $\rho_i\delta \vdash a_i : \text{V}$, and (3) $\forall d. \ \rho_i\delta, [x_i : a_i]= d \vdash e_i$. Since Prod is injective, $[x_1:a_1]e_1\rho_1 \mid \delta = [x_2:a_2]e_2\rho_2 \mid \delta$ implies (4) $a_1\rho_1\delta = \text{pv}(a_1\rho_1\delta) = \text{pv}(a_2\rho_2\delta) = a_2\rho_2\delta$, for which we write u, and (5) $\forall d. \ (e_1 \mid \rho_1\delta, x_1:u= d) = (e_2 \mid \rho_2\delta, x_2:u= d)$. We need to show $[x_1 : a_1]e_1\rho_1\delta = [x_2 : a_2]e_2\rho_2\delta$, i.e., $e_1(\rho_1\delta, x_1 : u = d_0) = e_2(\rho_2\delta, x_2 : u = d_0)$ for each d_0.
  Consider $\delta' = (\delta, \text{v}_k : u= d_0)$. Freshness of v_k implies $\vdash_\delta \rho_i \Leftrightarrow \vdash_{\delta'} \rho_i$ and $\rho_i\delta = \rho_i\delta'$. Hence $\vdash_{\delta'} e_i\rho_i'$ and $e_1\rho_1' \mid \delta' = e_2\rho_2' \mid \delta'$ by (1) – (5). Now induction hypothesis on $e_1\rho_1' \text{ conv } e_2\rho_2'$ gives $e_1(\rho_1\delta, x_1 : u= d_0) = e_1\rho_1'\delta' = e_2\rho_2'\delta' = e_2(\rho_2\delta, x_2 : u= d_0)$ as desired.
- Cases $[x_1:a_1]e_1\rho_1 \text{ conv } w$, $(x_1:a_1)b_1\rho_1 \text{ conv } (x_2:a_2)b_2\rho_2$, $[x_1:a_1]e_1 \mid \rho_1 \text{ conv } [x_2:a_2]e_2 \mid \rho_2$, and $[x_1:a_1]e_1 \mid \rho_1 \text{ conv } (x_2:a_2)b_2\rho_2$. Similarly to the above.

- Case $w_1 v_1$ conv $w_2 v_2$. Inversion of $\vdash_\delta w_i v_i$ gives $\vdash_\delta w_i :$ Prod $(v_i \mid \delta)$ f_i and $\vdash_\delta v_i$. By induction hypothesis on w_1 conv w_2, we have $w_1 \delta = w_2 \delta$ and Prod $(v_1 \mid \delta)$ $f_1 = (w_1 \mid \delta) = (w_2 \mid \delta) =$ Prod $(v_2 \mid \delta)$ f_2. Since Prod is injective, $v_1 \mid \delta = v_2 \mid \delta$, so induction hypothesis applies to v_1 conv v_2. Hence $v_1 \delta = v_2 \delta$. It follows immediately that $(w_1 v_1)\delta = (w_2 v_2)\delta$ and $w_1 v_1 \mid \delta = w_2 v_2 \mid \delta$.
- Case $w_1 * v_1$ conv $w_2 * v_2$. Similarly to the above.

6   Type-Checking

It is now straightforward to give our type-checking algorithm as syntax directed typing rules. The forms of judgements are $\rho \vdash e$, read 'expression e is well-typed under given syntactic environment ρ', and $\rho \vdash e : t$, which merely abbreviates '$\rho \vdash e$ and $e \mid \rho$ conv t' with a syntactic value t.

$$\frac{\rho \vdash e \quad e \mid \rho \text{ conv } t}{\rho \vdash e : t} \qquad \rho \vdash V \qquad \frac{x \in \text{dom } \rho}{\rho \vdash x}$$

$$\frac{\rho \vdash a : V \quad \rho, [x : a] = v_k \vdash e}{\rho \vdash [x : a]e} \qquad \frac{\rho \vdash a : V \quad \rho, [x : a] = v_k \vdash b : V}{\rho \vdash (x : a)b}$$
(with v_k not occurring in the conclusions)

$$\frac{\rho \vdash e_1 \quad e_1 \mid \rho \longrightarrow^* [x : a]e \mid \rho', \text{ or } \longrightarrow^* (x : a)e\rho' \quad \rho \vdash e_2 : a\rho'}{\rho \vdash e_1 e_2}$$

where the domain $\text{dom } \rho$ of an environment ρ is defined as usual: $\text{dom}() = \{\}$, $\text{dom}(\rho, x : t = v) = \text{dom } \rho \cup \{x\}$.

Lemma 7.
- If $\vdash_\delta \rho$ and $\rho \vdash e$, then $\vdash_\delta e\rho$.
- If $\vdash_\delta \rho$, $\vdash_\delta t : V$, and $\rho \vdash e : t$, then $\vdash_\delta e\rho : t\delta$.

Proof. By induction on derivations.

- Case $\rho \vdash e : t$. Immediate from induction hypothesis and Lemma 6.
- Cases $\rho \vdash V$ and $\rho \vdash x$. Immediate.
- Case $\rho \vdash [x : a]e$. We need to show (1) $\rho\delta \vdash a : V$ and (2) $\rho\delta, [x : a] = d \vdash e$ for each d. Induction hypothesis on $\rho \vdash a : V$ gives $\vdash_\delta a\rho : V$, hence (1) by inversion. For (2), consider $\delta' = (\delta, v_k : a\rho\delta = d)$, as in the proof of Lemma 6. Freshness of v_k, assumption $\vdash_\delta \rho$, and (1) gives $\vdash_{\delta'} (\rho, [x : a] = v_k)$. Hence induction on $\rho, [x : a] = v_k \vdash e$ gives $\vdash_{\delta'} e(\rho, [x : a] = v_k)$. Inverting this last and noting $\rho\delta = \rho\delta'$, we obtain (2).
- Case $\rho \vdash (x : a)b$. Similarly to the above.
- Case $\rho \vdash e_1 e_2$. We consider the case where the second premise is $\longrightarrow^*$ $[x : a]e \mid \rho$; the other case is similar. We need to show that (1) $\vdash_\delta e_1\rho$, (2) $e_1\rho \mid \delta = $ Prod $(e_2\rho \mid \delta)$ f for some f, and (3) $\vdash_\delta e_2\rho$. (1) holds by induction on $\rho \vdash e_1$. So Lemma 5 applies to the second premise, giving

$\vdash_\delta [x : a]e \mid \rho'$ and (2') $e_1 \mid \rho\delta = \mathsf{Prod}\ \mathsf{p_V}(a\rho'\delta)\ (\cdots)$. Inverting the first, we see that $\vdash_\delta a\rho' : \mathsf{V}$. Now induction on the third premise $\rho \vdash e_2 : a\rho'$ gives (3) and $e_2 \mid \rho\delta = a\rho'\delta = \mathsf{p_V}(a\rho'\delta)$, where (2') combined with this last is (2).

Theorem 8. For all a, e, and ν, $() \vdash a : \mathsf{V}$ and $() \vdash e : a()$ implies $e\nu\ :\ a\nu$.

Proof. Immediate from Lemma 7 and 4, observing the following : (1) if $\rho \vdash e_0$ then free variables of e_0 is in dom ρ, hence e and a above are closed; (2) $e_0\nu = e_0\nu'$ if ν and ν' agree on free variables of e_0, so in particular $e\nu$ and $a\nu$ are constant in ν and equal to $e \perp_{\mathsf{Env}}$ and $a \perp_{\mathsf{Env}}$, respectively.

7   Extensions and Applications

7.1   Partial Correctness

A first application of this work is to ensure that well-typed program "cannot go wrong" in the style of [9]. For this we have for instance to extend our domain with some primitive types, like **int** and **bool**. The correctness theorem shows then that if a program of type **bool** terminates, it will indeed compute a Boolean. Such a theorem is not so easy to establish syntactically, given that we use some form of η-conversion in the type-checking algorithm.

7.2   Addition of Definitions

It would be direct to extend the language with let expressions $[x = e_1]e$. The new computation rules would then be

$$([x{=}e_1]e)\rho = e\,(\rho,\,x{=}e) \qquad ([x{=}e_1]e) \mid \rho = e \mid (\rho,\,x{=}e)$$

with $(\rho,\,x{=}e) = (\rho,\,x{:}(e \mid \rho)) = e\rho)$.

7.3   Very Dependent Types

Another possible application, suggested to us by Thorsten Altenkirch, may be in the semantics of "very dependent types" [7]. We want to make sense of a recursive definition $x = (e : a)$ where x may occur not only in e like an usual recursive definition, but also in its type a. Semantically, the function $x \longmapsto (e, a)$ is interpreted by a function $u \longmapsto (f\ u, g\ u)$ of type $\mathsf{D} \to \mathsf{D} \times \mathsf{T}$. We can then solve in the domain D the recursive equation $u_0 = \mathsf{p}_{g\ u_0}(f\ u_0)$. We have always $g\ u_0 : \mathsf{V}$ since $g\ u \in \mathsf{T}$. If we have furthermore $v : g\ u \to f\ v : g\ u$ we get that $u_0 = f\ u_0$, and we make sense of the recursive definition $x = (e : a)$.

7.4   Record Types with Manifest Fields

The type-checker we present has for input a *segment*, which is a list of typed variable declarations $x : a$ and definitions $x = e$. This terminology comes from Automath [3]. Thus, a segment is an object of the form

$$x_1 : a_1, x_2 = e_2, x_3 : a_3, x_4 : a_4, x_5 = e_5$$

It is quite natural to *internalise* this notion, and we get in this way a simple extension of our calculus with a notion of records with manifest fields, like in Cayenne [1].

References

1. Augustsson, L. Cayenne, A language with dependent types. in Proceedings of ICFP'98, , ACM Press, 1998, 239-250.
2. Barendregt, H. and Rezus, A. Semantics for classical AUTOMATH and related systems. Inform. and Control 59 (1983), no. 1-3, 127–147.
3. Selected papers on Automath. Edited by R. P. Nederpelt, J. H. Geuvers and R. C. de Vrijer with the assistance of L. S. van Benthem Jutting and D. T. van Daalen. Studies in Logic and the Foundations of Mathematics, 133. North-Holland Publishing Co., Amsterdam, 1994
4. Cardelli, L. A polymorphic lambda-calculus with Type:Type. SRC Research Report 10, Digital Equipment Corporation Systems Research Center, May 1, 1986.
5. Coquand, C. A realizability interpretation of Martin-Löf's type theory. Twenty-five years of constructive type theory (Venice, 1995), 73–82, Oxford Logic Guides, 36, Oxford Univ. Press, New York, 1998.
6. Coquand, Th. An algorithm for type-checking dependent types. Mathematics of program construction (Kloster Irsee, 1995). Sci. Comput. Programming 26 (1996), no. 1-3, 167–177.
7. Hickey J. Formal Objects in Type Theory Using Very Dependent Types. in Informal proceedings of Third Workshop on Foundations of Object-Oriented Languages (FOOL 3), 1996.
8. Jutting, L.S. van Benthem, McKinna J. and Pollack R. Checking Algorithms for Pure Type Systems. In Types for Proofs and Programs, LNCS 806, H. Barendregt and T. Nipkow (Eds.) 1993, 19–61.
9. Milner, R. A theory of type polymorphism in programming. J. Comput. System Sci. 17 (1978), no. 3, 348–375.
10. Scott, D. Lectures on a Mathematical Theory of Computation. in Theoretical foundations of programming methodology. Papers presented at the NATO Summer School, Munich, 1981. Edited by Manfred Broy and Gunther Schmidt. NATO Advanced Study Institute Series C: Mathematical and Physical Sciences, 91. D. Reidel Publishing Co., Dordrecht-Boston, Mass., 1982, 145–292.
11. van Daalen, D. *The language theory of Automath*, PhD thesis, Eindhoven, 1980.

On the Logical Content of Computational Type Theory: A Solution to Curry's Problem*

Matt Fairtlough and Michael Mendler

Department of Computer Science, Sheffield University
211 Portobello Street, Sheffield S1 4DP, UK

Abstract. In this paper we relate the lax modality $\bigcirc$ to Intuitionistic Propositional Logic (IPL) and give a complete characterisation of inhabitation in Computational Type Theory (CTT) as a logic of constraint contexts. This solves a problem open since the 1940's, when Curry was the first to suggest a formal syntactic interpretation of $\bigcirc$ in terms of contexts.

1   Introduction

Recently, modal extensions to type theory have received a lot of attention as a natural enrichment suggested by a variety of typing problems that occur in programming and encodings of logic in type-theoretical frameworks.

Modalities may be used for enriching intuitionistic type theories, which traditionally focus on pure function and data, so as to include also various forms of reactive program features. In type theories for functional programming, for instance, modalities are added to accommodate non-functional ("impure") semantic features such as non-termination, side-effects or different operational modes [21, 20]. In applications to strictness analysis [3] and program optimisation [12, 6] modalities have been used to integrate static and dynamic types, while in other applications they provide a rigorous interface between inductive and non-inductive data-types, for instance in type systems to internalise higher-order abstract syntax [8, 7]. In strong functional programming [19] modalities occur (implicitly) as least or greatest fixpoint operators constructing data and codata types. Modalities have also proved useful in studying the relationship between second order encodings of logic in type theory [1].

In this paper we take a fresh look at one of the oldest modal extensions of type theory which arose with Moggi's influential work on computational monads [18]. This system, which we call *Computational Type Theory* (CTT), is an extension of simple type theory that features a single modality $\bigcirc$ satisfying the axioms

$$
\begin{aligned}
&\bigcirc I \;\; : \varphi \supset \bigcirc\varphi \\
&\bigcirc M : \bigcirc\bigcirc\varphi \supset \bigcirc\varphi \\
&\bigcirc S \;\; : (\bigcirc\varphi \wedge \bigcirc\psi) \supset \bigcirc(\varphi \wedge \psi)
\end{aligned}
$$

* This work is supported by EPSRC grants GR/L86180 and GR/M99637, by the EU Types Working Group IST-EU-29001 and by the British Council.

P. Callaghan et al. (Eds.): TYPES 2000, LNCS 2277, pp. 63–78, 2002.

together with the rule of Modus Ponens as well as the Extensionality Rule *Ext*:

$$\vdash \varphi \supset \psi \;\Rightarrow\; \vdash \bigcirc\varphi \supset \bigcirc\psi.$$

CTT can be given an alternative formulation as an axiomatic extension of IPL by $\bigcirc I + \bigcirc L : (\varphi \supset \bigcirc\psi) \supset (\bigcirc\varphi \supset \bigcirc\psi)$, which is extensionally equivalent to CTT in the sense that $\text{CTT} \vdash \varphi$ iff $\text{IPC} + \bigcirc I + \bigcirc L \vdash \varphi$. We prefer our presentation of CTT because it fits better with other algebraic and categorical treatments of the modality $\bigcirc$. This point is taken up in the concluding remarks to this paper. The logic of inhabitation of CTT is also known as *Propositional Lax Logic* (PLL) [9, 10] or *Computational Logic* [2]. Our analysis illuminates the characteristic rôle and the duality between two particular computational monads: the state readers monad $\bigcirc\varphi \equiv K \supset \varphi$ and the exceptions monad $\bigcirc\varphi \equiv \varphi \vee L$. Specifically, we define a notion of *standard context* as a context $C[\cdot]$ where $C[\varphi]$ has the form $\bigwedge_{i \in I} K_i \supset (\varphi \vee L_i)$ with K_i and L_i arbitrary formulas of IPL. We then show that the set $\mathbb{S}$ of standard contexts forms a Boolean algebra. Each such context provides a sound interpretation of CTT under the correspondence $\varphi \mapsto \varphi^C$, where φ^C is the formula φ with every subformula $\bigcirc\psi$ replaced[1] by $C[\psi]$. The contexts $K \supset \cdot$ and $\cdot \vee L$ have a simple computational interpretation in type theory: $K \supset \varphi$ represents the *state reader lifting* of type φ. Terms of $K \supset \varphi$ may be thought of as elements of type φ which depend on an additional state variable of type K which can be read from but not written to. The weakening $\varphi \vee L$ corresponds to an *exception lifting* of type φ. Terms of type $\varphi \vee L$ either denote proper elements of type φ or raise an exception of type L.

Our main theorem is that standard contexts are sound and complete for CTT, that is, $\text{CTT} \vdash \varphi$ iff for any standard context C, $\text{IPL} \vdash \varphi^C$. We go on to show that no finite set of standard contexts is complete for CTT. These results answer a question raised by Curry, as we shall shortly show.

Our paper concentrates on the logic of inhabitation of CTT, namely PLL. Of course, type theory is not simply concerned with whether or not types are inhabited. For any concrete interpretation of $\bigcirc$ in CTT, there must be λ-terms corresponding to the axioms and rules. Before proceeding further, we therefore present a set of such terms for every standard context C. For the "state readers plus exceptions" interpretation $C[\varphi] = K \supset (\varphi \vee L)$ we have $(\varphi \supset \bigcirc\varphi)^C = A \supset (K \supset (A \vee L))$ with $A = \varphi^C$, and this type is inhabited by $C_I =_{df} \lambda a.\, \lambda k.\, \iota_1(a)$. Next we have $(\bigcirc\bigcirc\varphi \supset \bigcirc\varphi)^C = (K \supset ((K \supset (A \vee L)) \vee L)) \supset (K \supset (A \vee L))$, and this type is inhabited *inter alia* by

$$C_M =_{df} \lambda c.\, \lambda k.\, \text{case } c\, k \text{ of } [\iota_1(e) \to \begin{pmatrix} \text{case } e\, k \text{ of} \\ [\iota_1(a) \to \iota_1(a), \\ \iota_2(l_1) \to \iota_2(l_1)] \end{pmatrix},\ \iota_2(l_2) \to \iota_2(l_2)]\,.$$

Now let B be ψ^C. For the axiom $\bigcirc S$, $((\bigcirc\varphi \wedge \bigcirc\psi) \supset \bigcirc(\varphi \wedge \psi))^C$ is the type $((K \supset (A \vee L)) \wedge (K \supset (A \vee L))) \supset K \supset ((A \wedge B) \vee L)$ which is inhabited *inter*

---

[1] Note that *all* occurrences of $\bigcirc$ must be replaced by the *same* constraint context. It would not be sound to permit independent expansion of different $\bigcirc$ occurrences by different contexts. Consider the trivial theorem $\bigcirc\varphi \supset \bigcirc\varphi$. Replacing the first $\bigcirc\varphi$ by $K \supset \varphi$ and the second by $\varphi \vee L$ does not give a theorem of IPL.

alia by

$$C_S =_{df} \lambda(c_1, c_2). \lambda k. \text{case } c_1 \ k \text{ of } [\iota_1(a) \to \begin{pmatrix} \text{case } c_2 \ k \text{ of} \\ [\iota_1(b) \to \iota_1(a, b), \\ \iota_2(l_1) \to \iota_2(l_1)] \end{pmatrix}, \ \iota_2(l_2) \to \iota_2(l_2)].$$

Finally, if $f : A \supset B$ then

$$C_{Ext}(f) =_{df} \lambda c. \lambda k. \text{case } c \ k \text{ of } [\iota_1(a) \to \iota_1(f \ a), \ \iota_2(l) \to \iota_2(l)]$$

inhabits $(\bigcirc A \supset \bigcirc B)^C = (K \supset (A \vee L)) \supset (K \supset (B \vee L))$. In fact, as PLL has the deduction property, the term $\lambda f. C_{Ext}(f)$ inhabits $((A \supset B) \supset (\bigcirc A \supset \bigcirc B))^C$. Suppose we have a suitable set of λ-terms for C_1 and C_2. We now wish to find λ-terms for the conjunction of constraints $C_1 \sqcap C_2$, defined by $(C_1 \sqcap C_2)[\varphi] = C_1[\varphi] \wedge C_2[\varphi]$. Suppose therefore that for $j = 1, 2$ and $f : A \supset B$ we have terms

$$C_{j\,I} : A \supset C_j[A],$$
$$C_{j\,M} : C_j[C_j[A]] \supset C_j[A],$$
$$C_{j\,S} : C_j[A] \wedge C_j[B] \supset C_j[A \wedge B],$$
$$C_{j\,Ext}(f) : C_j[A] \supset C_j[B].$$

Then we may define the following λ-terms

$$(C_1 \sqcap C_2)_I =_{df} \lambda a. (C_{1I}(a), C_{2I}(a)),$$
$$(C_1 \sqcap C_2)_M =_{df} \lambda(c_1, c_2).(C_{1M}(C_{1F}(\pi_1) \ c_1), C_{2M}(C_{2F}(\pi_2) \ c_2)),$$
$$(C_1 \sqcap C_2)_S =_{df} \lambda((c_{11}, c_{21}), (c_{12}, c_{22})).(C_{1S}(c_{11}, c_{12}), (C_{2S}(c_{21}, c_{22}))),$$
$$(C_1 \sqcap C_2)_{Ext}(f) =_{df} \lambda(c_1, c_2).(C_{1Ext}(f) \ c_1, C_{2Ext}(f) \ c_2),$$

and assign types to them as follows:

$$(C_1 \sqcap C_2)_I : A \supset (C_1 \sqcap C_2)[A] = A \supset C_1[A] \wedge C_2[A],$$
$$(C_1 \sqcap C_2)_M : (C_1 \sqcap C_2)^2[A] \supset (C_1 \sqcap C_2)[A],$$
$$(C_1 \sqcap C_2)_S : (C_1 \sqcap C_2)[A] \sqcap (C_1 \sqcap C_2)[B] \supset (C_1 \sqcap C_2)[A \wedge B],$$
$$(C_1 \sqcap C_2)_{Ext}(f) : (C_1 \sqcap C_2)[A] \supset (C_1 \sqcap C_2)[B].$$

In the above we assume the following syntax for typed λ-terms:

$$\frac{\Gamma \vdash p : A \quad \Gamma \vdash q : B}{\Gamma \vdash (p, q) : A \wedge B} \qquad \frac{\Gamma \vdash r : A \wedge B}{\Gamma \vdash \pi_1(r) : A} \qquad \frac{\Gamma \vdash r : A \wedge B}{\Gamma \vdash \pi_2(r) : B}$$

$$\frac{\Gamma \vdash r : A \vee B \quad \Gamma, y : A \vdash p : C \quad \Gamma, z : B \vdash q : C}{\Gamma \vdash \text{case } r \text{ of } [\iota_1(y) \to p, \iota_2(z) \to q] : C}$$

$$\frac{\Gamma \vdash p : A}{\Gamma \vdash \iota_1(p) : A \vee B} \qquad \frac{\Gamma \vdash q : B}{\Gamma \vdash \iota_2(q) : A \vee B}$$

$$\frac{\Gamma, z : A \vdash p : B}{\Gamma \vdash \lambda z. p : A \supset B} \qquad \frac{\Gamma \vdash p : A \supset B \quad \Gamma \vdash q : A}{\Gamma \vdash p q : B}$$

The interpretation of the results of this paper with respect to these λ-terms is left as future work. '

2   Curry's Problem

In [16, 17, 9] we proposed to read $\bigcirc\varphi$ as a weakened notion of validity, *viz.* "φ up to constraints." This constraint interpretation is related to an idea originally suggested by Curry in his 1948 Notre Dame Lectures on a *Theory of Formal Deducibility* reprinted as [5]. Indeed, Curry was probably the first to study the constructive modality $\bigcirc$ (see also [4]) and to suggest a formal interpretation inside IPL in terms of constraints and hidden assumptions.

Curry's proposal was to take $\bigcirc\varphi$ as the statement "*in some outer (stronger) theory, φ holds.*" As examples of such nested systems of reasoning (with two levels) he suggested Mathematics as the inner and Physics as the outer system, or Physics as the inner system and Biology as the outer. In both examples the outer system is more encompassing than the inner system where reasoning follows a more rigid notion of truth and deduction. The modality $\bigcirc$, which Curry conceived of as a modality of possibility, is a way of reflecting the relaxed, outer, notion of truth within the inner system.

Assuming that the outer theory can be axiomatised by some, possibly very complex, formula K inside the inner system the formal semantics of $\bigcirc$ would come down to

$$\bigcirc_K\varphi = \text{``}\varphi \text{ under the assumption } K\text{''} = K \supset \varphi.$$

It is evident that this interpretation within IPL provably satisfies the axioms of $\bigcirc$, hence provides a sound semantics of PLL. In fact, $\bigcirc_K\varphi$ is a particular constraint interpretation in the sense of [17], with fixed implicational constraint K. However, Curry's guess [4, § 5,p.261] that the interpretation $\bigcirc_K\varphi$ generates the theory PLL is unjustified: $\bigcirc_K\varphi$ validates $(\bigcirc_K\varphi \supset \bigcirc_K\psi) \supset \bigcirc_K(\varphi \supset \psi)$ which is not a theorem of PLL (see [9]).

So, is Curry's idea ill-conceived? Let us follow the constraint paradigm a bit further. Surely, implicational contexts are not the only way to weaken a proposition by constraints. Another, dual, way for doing this are the *disjunctive constraint contexts* $\bigcirc^L\varphi = \varphi \vee L$. Again, it is not difficult to show that this interpretation provably satisfies the axioms of $\bigcirc$. The contexts $K \supset \cdot$ and $\cdot \vee L$ are dual in the following semantical sense: Let $M(\psi)$, for proposition ψ and Kripke model M, denote the set of worlds of M where ψ is true. Then $M \models K \supset \varphi$ iff $M(K) \subseteq M(\varphi)$, while $M \models \varphi \vee L$ iff $\overline{M(L)} \subseteq M(\varphi)$, where $\overline{M(L)}$ is the complement of $M(L)$. Intuitively speaking, the difference is that the weakening of φ by $K \supset \varphi$ is obtained by "switching φ on" only in those worlds where the constraint K is *true*, while in $\varphi \vee L$ the weakening is obtained by switching φ on where L is *false*.

Let us call any syntactic context $C[\cdot]$ for which the axioms and rule for $\bigcirc$ are provable in IPL a *constraint context* or simply a *constraint*. Let the set of constraint contexts be $\mathbb{C}$. Now if $\bigcirc$ is to be the modality of truth under constraints then there should be a natural collection of constraints that completely characterise PLL. A suitable refinement of the original problem posed by Curry, then, is this:

Does there exist a class $\mathbb{S}$ of constraint contexts
such that $\mathsf{PLL} \vdash \varphi$ iff $\forall C \in \mathbb{S}.\,\mathsf{IPL} \vdash \varphi^C$?

To see that this question is non-trivial we first observe that neither one of the simple constraints $K \supset \cdot$ or $\cdot \vee L$ is sufficient as a semantics for PLL. We already saw that Curry's implicational contexts $K \supset \cdot$ validate the scheme $(\bigcirc\varphi \supset \bigcirc\psi) \supset \bigcirc(\varphi \supset \psi)$, which is not a theorem of PLL. The disjunctive contexts $\cdot \vee L$ on the other hand give rise to the theorem $\bigcirc(\varphi \vee \psi) \supset \bigcirc\varphi \vee \bigcirc\psi$, which is not part of PLL either (see [9]). Similarly, it is not enough to take for $\mathbb{S}$ the collection of implicational and disjunctive constraints, since then the disjunction

$$\bigcirc(\varphi \vee \psi) \supset \bigcirc\varphi \vee \bigcirc\psi \vee (\bigcirc\varphi \supset \bigcirc\psi) \supset \bigcirc(\varphi \supset \psi)$$

would be validated. For whatever context we choose, $\bigcirc[\cdot] = K \supset \cdot$ or $\bigcirc[\cdot] = \cdot \vee L$, one of the two disjuncts would be provable, and thus the disjunction itself. Yet the disjunction is not a theorem of PLL since neither disjunct is and PLL satisfies · the disjunction property (see [9]).

3   An Algebra of Constraint Contexts

The solution lies in considering combinations of implicational and disjunctive contexts. For instance, we can combine both in the contexts

$$[K, L)\,\varphi =_{df} K \supset (\varphi \vee L),$$

called *basic constraints*, which also provide a sound interpretation of $\bigcirc$. The interval notation $[K, L)$ is suggested by the fact that the constraint weakening $K \supset (\varphi \vee L)$ switches φ on in the "interval" of worlds between K (inclusive) and L (exclusive) as illustrated in Fig. 1. The shaded area, indicating this interval, measures the extent to which φ needs to be true in the model to validate $[K, L)\,\varphi$.

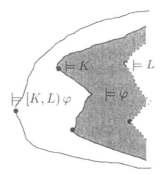

Fig. 1. Interval constraint $[K, L)$

It is not difficult to see that basic constraints by themselves are still not quite enough to characterise PLL. They are a model of $\bigcirc(\bigcirc(\varphi \vee \psi) \supset (\bigcirc\varphi \vee \bigcirc\psi))$, which is not a theorem of PLL. However, so it turns out, finite conjunctions of basic constraints do the job. Specifically, if C_1 and C_2 are constraints, then

$$(C_1 \sqcap C_2)[x] =_{df} C_1[x] \wedge C_2[x]$$

is a constraint, too. A finite conjunction of basic constraints $\sqcap_{i=0}^{n-1}[K_i, L_i)$ is called a *standard constraint*. We refer to n as the *depth* of the constraint. It will be expedient to include the degenerate case $n = 0$, and put $\bigcap_{i=0}^{0-1} C_i =_{df}$ $[false, false)$. It will turn out that the collection of standard constraints of arbitrary depth, henceforth called $\mathbb{S}$, is a suitable class of contexts to characterise PLL. But before we go into the details of the proof it is worthwhile to study some of the algebraic properties of $\mathbb{S}$.

We take constraints up to equivalence, $i.e.$ consider two constraints identical if they have the same action on all propositions. Formally, $C_1 = C_2$ iff $\forall x.\ C_1[x] \equiv$ $C_2[x]$. We shall now show that $\mathbb{S}$ forms an infinite Boolean algebra. To stress the algebraic standpoint let us write $x \leq y$ to indicate that $x \supset y$ is a theorem of IPL, while for constraints we take $C \leq D$ to abbreviate the statement that $C[x] \leq D[x]$ for all propositions x. Since $\mathbb{S}$ is a Boolean algebra one could equally well adopt a dual view swapping $\leq$ with $\geq$ and $\sqcap$ with $\sqcup$.

We begin by mentioning a few obvious facts about the algebra $\mathbb{S}$. The bottom element is the constraint $\bot =_{df} [true, false)$, $i.e.$ the identity modality acting as $\bot[x] = true \supset (x \vee false) \equiv x$. The top element is $\top =_{df} [false, false)$, $i.e.$ the modality $\top[x] = false \supset (x \vee false) \equiv true$ that forces everything true. As defined above $\top$ is the (unique) standard constraint of depth 0. Note that $\top = [K, K)$ for arbitrary propositions K. Top and bottom elements satisfy $C \sqcap \top = C = \top \sqcap C$ and $C \sqcap \bot = \bot = \bot \sqcap C$. Since $\sqcap$ is essentially conjunction $\wedge$ it is commutative, associative, idempotent, and satisfies $C \leq D$ iff $C \sqcap D = C$. Thus it is the real meet of the algebra.

Generally, a formula L may be identified with the disjunctive constraint $L =_{df} [true, L)$ so that $L[x] = true \supset (x \vee L) \equiv x \vee L$. Its complement is the dual implicational constraint $\overline{L} =_{df} [L, false)$, so that $\overline{L}[x] = L \supset (x \vee false) \equiv L \supset x$. We call these (positive and negative) $atomic$ constraints. With this lifting of formulas to constraints being understood, we have $\bot = false$ and $\top = true$. One also easily verifies that $\bot = \overline{true}$, $\top = \overline{false}$, and $L \sqcap \overline{L} = \bot$. Indeed as we will see, constraints L and their complements $\overline{L}$ are the generators of $\mathbb{S}$.

The join $C_1 \sqcup C_2$ in $\mathbb{S}$ is given by the following definition.

Definition 1. *Let* $C_1 = \bigcap_{i=0}^{m-1} [K_{1i}, L_{1i})$ *and* $C_2 = \bigcap_{j=0}^{n-1} [K_{2j}, L_{2j})$ *be standard constraints of depths* m *and* n, *respectively. Then,* $C_1 \sqcup C_2$ *is the standard constraint* $\bigcap_{i<m, j<n} [K_{1i} \wedge K_{2j}, L_{1i} \vee L_{2j})$.

Definition 1 gives an explicit representation of $C_1 \sqcup C_2$. For basic constraints we get $[K_1, L_1) \sqcup [K_2, L_2) = [K_1 \wedge K_2, L_1 \vee L_2)$, in particular $\overline{K} \sqcup L = [K, false) \sqcup$ $[true, L) = [K, L)$. In other words, basic constraints are conjunctions of atomic constraints. If atomic constraints are the literals of our algebra then basic constraints are the minterms. Generally, we have

$$\overline{K_1} \sqcup \cdots \sqcup \overline{K_m} \sqcup L_1 \sqcup \cdots \sqcup L_n = [K_1 \wedge \cdots \wedge K_m, L_1 \vee \cdots \vee L_n).$$

Next, let us consider the algebraic properties of $\sqcup$. From the definition of $\sqcup$ we easily get $C \sqcup \bot = C = \bot \sqcup C$ and $C \sqcup \top = \top = \top \sqcup C$ as well as the associativity and commutativity of $\sqcup$. Idempotency $C \sqcup C = C$ and the inequations $C_1 \leq$ $C_1 \sqcup C_2$ and $C_2 \leq C_1 \sqcup C_2$ require a little more work but are straightforward. We show $C_1 \leq C_1 \sqcup C_2$. Suppose as before that $C_1 = \bigcap_{i=0}^{m-1} [K_{1i}, L_{1i})$ and

$C_2 = \prod_{j=0}^{n-1}[K_{2j}, L_{2j})$. Note that if $K' \supset K$ and $L \supset L'$ then $[K, L) \leq [K', L')$. It follows that for a given $i < m$, $[K_{1i}, L_{1i}) \leq [K_{1i} \wedge K_{2j}, L_{1i} \vee L_{2j})$ for every $j < n$ and thus $[K_{1i}, L_{1i}) \leq \prod_{j=0}^{n-1}[K_{1i} \wedge K_{2j}, L_{1i} \vee L_{2j})$. Hence

$$C_1 = \prod_{i=0}^{m-1}[K_{1i}, L_{1i}) \leq \prod_{i=0}^{m-1}\left(\prod_{j=0}^{n-1}[K_{1i} \wedge K_{2j}, L_{1i} \vee L_{2j})\right) = C_1 \sqcup C_2$$

Also Definition 1 immediately implies the dual distributivity law $C \sqcup (D \sqcap E) = (C \sqcup D) \sqcap (C \sqcup E)$. From this the distributivity law $C \sqcap (D \sqcup E) = (C \sqcap D) \sqcup (C \sqcap E)$ and the characterisation of inequality $C \leq D$ iff $C \sqcup D = D$ follow. All this shows that $\mathbb{S}$ is a distributive lattice.

What remains is to define complements. For atoms L we obtained complementation as $\overline{L} = [L, false)$. In view of Definition 1 this suffices to define complements for arbitrary standard constraints "by duality." More precisely, given a standard constraint $C = \prod_{i \in I}[K_i, L_i)$ we define its complement as

$$\overline{C} =_{df} \prod_{A \subseteq I}[\bigwedge_{a \in A} L_a, \bigvee_{b \in I \setminus A} K_b).$$

This definition is simply an application of DeMorgan's rule: $\overline{C} = \overline{\prod_{i \in I}[K_i, L_i)} = \bigsqcup_i \overline{[K_i, L_i)} = \bigsqcup_i \overline{K_i \sqcup L_i} = \bigsqcup_i (K_i \sqcap \overline{L_i}) = \prod_{A \subseteq I}[\bigwedge_{a \in A} L_a, \bigvee_{b \in I \setminus A} K_b)$, where the last equation is by virtue of the dual distributivity law and Definition 1. We use the convention that empty disjunctions are false and empty conjunctions true. Using this convention one verifies $\overline{[K, false)} = [true, K)$ and $\overline{[true, L)} = [L, false)$, which confirms that the constraints $K \supset \cdot$ and $K \vee \cdot$ are indeed Boolean complements. It is a routine matter to check that C and $\overline{C}$ are complements for arbitrary C.

We can finally state the main theorem of this section, whose proof is obvious from the discussions above.

Theorem 2. *The collection $\mathbb{S}$ of standard constraints is a Boolean algebra generated by formulas as atomic constraints.*

In the remainder of this paper we show that $\mathbb{S}$ provides an adequate interpretation of PLL, while at the same time no finite subset of $\mathbb{S}$ is sufficient. The proof depends on a model-theoretic characterisation of PLL to be discussed next.

4 Kripke Model-Theory for PLL

Our model-theory is built on Kripke *constraint models* introduced in [9]. The main definitions and the completeness results are as follows:

Definition 3 (Kripke Constraint Models). *A (Kripke) constraint model for PLL is an intuitionistic modal model $(W, \sqsubseteq_i, \sqsubseteq_m, V, F)$, in which $\sqsubseteq_i$ and $\sqsubseteq_m$ are partial orderings, $\sqsubseteq_m$ is a subrelation of $\sqsubseteq_i$, V is a valuation, i.e. a mapping* *assigning a set of propositional variables to each $w \in W$, and $F \subseteq W$. V and F are hereditary in the sense that V is monotone in $\sqsubseteq_i$ and F is upper closed under $\sqsubseteq_i$, in other words, $w \in F$ and $w \sqsubseteq_i v$ imply $v \in F$.*

The interpretation of $w \sqsubseteq_m v$ put forward in [9] is that v is a *constraining* of w, or v is reachable from w up to a *constraint*. Elements of F are *fallible worlds* and if $w \sqsubseteq_m v$ and $v \in F$, then intuitively the constraint leading to v is inconsistent with world w. Using the "creative mathematician" interpretation of $w \sqsubseteq_i v$ as a construction step, the difference between $\sqsubseteq_i$ and $\sqsubseteq_m$ should be thought of as relating to some intensional feature of the world, such as the resources (time spent, energy dissipated, waste produced, *etc.*) used up in the constructions made. A modal step $w \sqsubseteq_m v$ then amounts to the stronger statement that v may be constructed from w *up to bounded resources*, whereas in a step $w \sqsubseteq_i v$ no such bound can be guaranteed.

Definition 4 (Validity). *Let $M = (W, \sqsubseteq_m, \sqsubseteq_i, V, F)$ be a constraint model. Given a proposition φ and $w \in W$, φ is valid at w in M, written $M, w \models \varphi$ iff*

- *φ is a propositional constant α and $\alpha \in V(w)$;*
- *φ is $\varphi_1 \wedge \varphi_2$ and both $M, w \models \varphi_1$ and $M, w \models \varphi_2$;*
- *φ is $\varphi_1 \vee \varphi_2$ and $M, w \models \varphi_1$ or $M, w \models \varphi_2$;*
- *φ is true; or φ is false and $w \in F$;*
- *φ is $\varphi_1 \supset \varphi_2$ and for all $v \in W$ such that $w \sqsubseteq_i v$, $M, v \models \varphi_1$ implies $M, v \models \varphi_2$;*
- *φ is of form $\bigcirc\psi$ and for all $v \in W$ such that $w \sqsubseteq_i v$, there exists $u \in W$ such that $v \sqsubseteq_m u$ and $M, u \models \psi$.*

A proposition φ is valid in M, written $M \models \varphi$, if for all $w \in W$, φ is valid at w in M; φ is valid, written $\models \varphi$, if φ is valid in any constraint model M.

Constraint models are not the only possible adequate Kripke semantics for PLL. A quite different kind of semantics was given by Goldblatt [11], in which only the intuitionistic part $\supset$ is represented by a frame, while the modality is realized by some extra topological information on the intuitionistic frame. Another kind of topological model has been introduced in [13]. To obtain the results of this paper, however, essential use of the structure of Kripke constraint models will be made, and of the following completeness theorem.

Theorem 5 ([9]). *Let a constraint model $M = (W, \sqsubseteq_i, \sqsubseteq_m, V, F)$ be called finite if W is finite, and $V(w)$ is finite for all $w \in W$. Then, PLL $\vdash \varphi$ iff for all finite constraint models M, $M \models \varphi$.*

5   Adequacy of Standard Constraints

We can now state and prove our version of Curry's conjecture. We show that a proposition is provable in PLL iff all its instantiations by standard constraints are provable in IPL. This gives a precise sense in which, proof-theoretically, a lax proposition is properly stronger than its $\bigcirc$-stripped version.

Theorem 6. *Let φ be a lax proposition. Then,* PLL $\vdash \varphi$ *iff for all standard constraints $C \in \mathbb{S}$,* IPL $\vdash \varphi^C$.

The soundness direction of the theorem is straightforward. For every standard constraint context $C[x]$ the axioms and rules of PLL are derivable in IPL, as shown by the λ-terms given in the introduction. Hence, PLL $\vdash \varphi$ implies IPL $\vdash \varphi^C$. The challenge lies in the completeness direction. In view of Theorem 5 it suffices to find for every finite model M a standard constraint C such that for all φ, $M \models \varphi$ iff $M \models \varphi^C$. In other words, we are done if we can expand the meaning of $\bigcirc$ relative to a fixed finite model in terms of a single standard constraint. This is indeed possible, as stated in Lemma 7 below. The proof makes essential use of the particular structure of constraint models.

To state the lemma we need three constructions for finite constraint models $M = (W, \sqsubseteq_i, \sqsubseteq_m, V, F)$: First, we call a world $w \in W$ *stable* if it has no proper modal successors, *i.e.* for all $v \in W$, $w \sqsubseteq_m v$ implies $v = w$. We denote by $Stab \subseteq W$ the set of stable worlds of M. As one shows without difficulty, the semantics of φ and $\bigcirc\varphi$ coincides on stable worlds $w \in Stab$, i.e. $M, w \models \varphi$ iff $M, w \models \bigcirc\varphi$. Second, for any $w \in W$ let $iSucc(w) \subseteq W$ be the (finite) set of immediate successors of w, in other words, $iSucc(w) =_{df} \{ v \in W \mid (w \sqsubset_i v)$ & $(\forall u. w \sqsubseteq_i u \sqsubseteq_i v \Rightarrow u = v) \}$. Third, for any finite set of propositional variables U, let $M_U^* = (W, \sqsubseteq_i, \sqsubseteq_m, V^*, F)$ be a *semantic completion of M avoiding U*, where V^* is determined by a choice of new propositional variables $\{ \alpha_w \mid w \in W \}$ such that $M^*, v \models \alpha_w$ iff $w \sqsubseteq_i v$. We construct M_U^* to ensure that each new variable α_w does not occur in the range of V or in U and we drop the subscript U from M_U^* when it is clear from the context. The model M^* generates the same theory as M with respect to propositions whose variables are disjoint from $\{\alpha_w \mid w \in W\}$ except that it also has every one of its worlds w explicitly represented by a propositional variable α_w.

Lemma 7. *Let M be a finite constraint model and the sets $Stab$ and $iSucc$ defined as above. Then, if φ is any formula whatsoever, we have $M^* \models \bigcirc\varphi \equiv \bigwedge_{w \in Stab} [K_w, L_w]\varphi$, where $K_w =_{df} \alpha_w$ and $L_w =_{df} \bigvee_{w' \in iSucc(w)} \alpha_{w'}$. If $iSucc(w) = \emptyset$ then $\bigvee_{w' \in iSucc(w)} \alpha_{w'} = false$.*

Proof. Let M be a finite constraint model and $w \in W$ an arbitrary world. We show that

$$M^*, w \models \bigcirc\varphi \Leftrightarrow \forall u \in Stab. \ M^*, w \models \alpha_u \supset (\varphi \lor \bigvee_{u' \in iSucc(u)} \alpha_{u'}).$$

We prove direction ($\Rightarrow$) first, and assume $M^*, w \models \bigcirc\varphi$. Let $u \in Stab$ and $v \in W$ such that $w \sqsubseteq_i v$ and $M^*, v \models \alpha_u$ be given. The latter, $M^*, v \models \alpha_u$, means $u \sqsubseteq_i v$. There are two possibilities for this: either (i) $u = v$, or (ii) there exists an immediate successor $u' \in iSucc(u)$ of u such that $u' \sqsubseteq_i v$. In case (ii) we have $M^*, v \models \alpha_{u'}$, which implies $M^*, v \models \varphi \lor \bigvee_{u' \in iSucc(u)} \alpha_{u'}$ as desired. In the first case, (i), we get $v = u \in Stab$ and also, by hereditariness of truth, $M^*, v \models \bigcirc\varphi$. But since $v \in Stab$, this implies $M^*, v \models \varphi$, which in turn entails $M^*, v \models \varphi \lor \bigvee_{u' \in iSucc(u)} \alpha_{u'}$. This completes direction ($\Rightarrow$).

Next, we show the other direction ($\Leftarrow$), where we assume that for all $u \in Stab$,

$$M^*, w \models \alpha_u \supset (\varphi \vee \bigvee_{u' \in iSucc(u)} \alpha_{u'}).$$

We must demonstrate $M^*, w \models \bigcirc\varphi$, i.e. that for all u, $w \sqsubseteq_i u$, there exists some v, $u \sqsubseteq_m v$, such that $M^*, v \models \varphi$. To this end, let such u be given. Because of the finiteness of the model there must exists a world v, $u \sqsubseteq_m v$ and $v \in Stab$. Otherwise, by definition of $Stab$ and the properties of $\sqsubseteq_m$ we could construct an infinite sequence $u = u_0, u_1, u_2, \ldots$ of worlds with $u_i \neq u_{i+1}$ and $u_i \sqsubseteq_m u_{i+1}$. Since $w \sqsubseteq_i v$, $M^*, v \models \alpha_v$, and $v \in Stab$, our assumption gives us $M^*, v \models \varphi \vee \bigvee_{v' \in iSucc(v)} \alpha_{v'}$. However, since $v' \not\sqsubseteq_i v$ for every proper (immediate) successor v' of v, $M^*, v \not\models \bigvee_{v' \in iSucc(v)} \alpha_{v'}$ which implies $M^*, v \models \varphi$ as desired. This, finally, proves $M^*, w \models \bigcirc\varphi$. $\square$

We are now ready to prove the completeness direction of Theorem 6.

Proof. (Theorem 6) Let φ be given such that PLL $\not\vdash \varphi$. Then, by Theorem 5 there exists a finite constraint model M such that $M \not\models \varphi$. But then we must have $M^* \not\models \varphi$ for any semantic completion M^* of M that avoids the variables of φ, since M^* will coincide with M on all propositional variables contained in φ. From Lemma 7 we get $M^* \models \bigcirc\psi \equiv \bigwedge_{w \in Stab}[K_w, L_w]\psi$ for arbitrary ψ, where $K_w = \alpha_w$ and $L_w = \bigvee_{v \in iSucc(w)} \alpha_v$. Thus, because of the extensionality of PLL,—namely that whenever $M' \models \psi \equiv \varphi$ then $M' \models C[\psi] \equiv C[\varphi]$ for arbitrary contexts $C[\cdot]$—we have $M^* \not\models \varphi'$ where φ' is obtained from φ by replacing each occurrence of a subproposition $\bigcirc\psi$ of φ by $\bigwedge_{w \in Stab}[K_w, L_w]\psi$. Thus, we have found a single standard context $C =_{df} \bigsqcap_{w \in Stab}[K_w, L_w]$ such that $\not\models \varphi^C$. $\square$

To illustrate the constructions involved in the proof let us look at an example. Consider the propositional scheme

$$\theta =_{df} \bigcirc((\bigcirc\beta \supset \beta) \vee (\bigcirc\beta \supset (\beta \vee \bigcirc false))).$$

If θ is stripped of all $\bigcirc$ it turns into a trivial theorem of IPL. However, because of the way the modalities are placed it is not a theorem of PLL. To explain this in the light of Theorem 6 we expect to find a context $C[\cdot]$ such that the constraint expansion of θ,

$$\theta^C = C[((C[\beta] \supset \beta) \vee (C[\beta] \supset (\beta \vee C[false])))]$$

is not provable in IPL. Observe that none of the simple contexts $C[x] = K \supset x$, $C[x] = x \vee L$, or $C[x] = K \supset x \vee L$ will work, since for all of them θ^C in fact *is* a theorem of IPL. This means we need a proper meet of contexts to falsify θ^C. As the proof of Theorem 6 shows such a context can be obtained systematically from a counter model for θ. The simplest constraint model that refutes θ is the three-world model $M = (\{0, 1, 2\}, \sqsubseteq_i, \sqsubseteq_m, V, \emptyset)$ in which the accessibilities are such that $n \sqsubseteq_i m$ iff $n \leq m$ and $\sqsubseteq_m = \{(0,0), (1,1), (2,2), (1,2)\}$, and the valuation is $V(0) = V(1) = \emptyset$, $V(2) = \{\beta\}$. The following picture illustrates the situation and indicates the validity of subpropositions of θ showing that indeed $0 \not\models \theta$ in this model:

$$M \quad \begin{array}{ccccc} 0 & \sqsubseteq_i & 1 & \sqsubseteq_m & 2 \\ \bullet\text{-----------}\!\!\rightarrow\!\bullet & & \bullet\!\!-\!\!-\!\!-\!\!-\!\!-\!\!-\!\!\rightarrow\!\bullet \end{array}$$

$$
\begin{array}{lll}
\not\models (\bigcirc\beta \supset \beta) \vee & \models \bigcirc\beta & \models \beta \\
\quad \bigcirc\beta \supset (\beta \vee \bigcirc false) & \not\models \beta & \\
\not\models \bigcirc(\bigcirc\beta \supset \beta \vee & \not\models \bigcirc false & \\
\quad \bigcirc\beta \supset (\beta \vee \bigcirc false)) & \not\models \bigcirc\beta \supset \beta & \\
& \not\models \bigcirc\beta \supset (\beta \vee \bigcirc false) &
\end{array}
$$

Now we consider the semantical completion M^* of M, which is just M but with additional propositional variables $\alpha_0, \alpha_1, \alpha_2$ that are validated at worlds $0, 1, 2$ respectively. So, $V^*(0) = \{\alpha_0\}$, $V^*(1) = \{\alpha_1\}$, $V^*(2) = \{\alpha_2, \beta\}$. In pictures,

$$M^* \quad \begin{array}{ccccc} 0 & \sqsubseteq_i & 1 & \sqsubseteq_m & 2 \\ \bullet\text{-----------}\!\!\rightarrow\!\bullet & & \bullet\!\!-\!\!-\!\!-\!\!-\!\!-\!\!-\!\!\rightarrow\!\bullet \end{array}$$

$$
\begin{array}{lll}
\models \alpha_0 & \models \alpha_0 & \models \alpha_0 \\
& \models \alpha_1 & \models \alpha_1 \\
& & \models \alpha_2 \\
& & \models \beta
\end{array}
$$

We will use the new propositional variables $\alpha_0, \alpha_1, \alpha_2$, which represent the worlds $0, 1, 2$, respectively, to expand the meaning of $\bigcirc$ entirely in terms of propositions, following Lemma 7. Note that in M^*, $\alpha_0 \equiv true$ since α_0 is valid in every world of the model. The stable worlds in M^* are $Stab = \{0, 2\}$ since these have no proper modal successors. Their immediate successor sets are $iSucc(0) = \{1\}$ and $iSucc(2) = \emptyset$. Therefore, by Lemma 7 the following equivalence must hold in M^*, for arbitrary φ:

$$
\begin{aligned}
\bigcirc\varphi &\equiv \bigwedge_{w \in Stab} [\alpha_w, \bigvee_{w' \in iSucc(w)} \alpha_{w'}) \varphi \\
&= [\alpha_0, \bigvee_{w' \in iSucc(0)} \alpha_{w'}) \varphi \wedge [\alpha_2, \bigvee_{w' \in iSucc(2)} \alpha_{w'}) \varphi \\
&\equiv [\alpha_0, \alpha_1) \varphi \wedge [\alpha_2, \bigvee_{w' \in \emptyset} \alpha_{w'}) \varphi \\
&\equiv [true, \alpha_1) \varphi \wedge [\alpha_2, false) \varphi \\
&\equiv ([true, \alpha_1) \sqcap [\alpha_2, false)) \varphi.
\end{aligned}
$$

Because of this equivalence and the fact that $M^* \not\models \theta$ it must be the case that $M^* \not\models \theta^C$ where C is the constraint context $C =_{df} [true, \alpha_1) \sqcap [\alpha_2, false)$. The proposition θ is an example of a propositional scheme that requires a proper composition of two constraints, *viz.* an implicational $\alpha_2 \supset \cdot$ and a disjunctive one $\cdot \vee \alpha_1$, in order to be outed as a non-theorem of PLL. In general, so it turns out, the discriminative power of *all* standard contexts is needed in order to characterise PLL fully. We show this in the following section.

6   Finite Constraint Collections Are Inadequate

Neither the implicational nor the disjunctive contexts alone are sufficient to characterise PLL. The implicational constraints $\bigcirc = [\alpha, false)$ validate the scheme

$(\bigcirc\varphi \supset \bigcirc\psi) \supset \bigcirc(\varphi \supset \psi)$ and $\bigcirc(\bigcirc\varphi \supset \varphi)$, while the disjunctive ones $\bigcirc =$ $[true, \beta)$ validate $\bigcirc(\varphi \vee \psi) \supset (\bigcirc\varphi \vee \bigcirc\psi)$ and $\bigcirc\varphi \supset (\varphi \vee \bigcirc false)$. It is easy to see that basic constraints $\bigcirc = [\alpha, \beta)$ satisfy the scheme $\bigcirc(\bigcirc\varphi \supset \varphi \vee \bigcirc false)$. The general pattern that emerges is as follows. Let $p_1, p_2, \ldots$ be a list of distinct propositional variables. We define a sequence of propositional schemes χ_m such that χ_m has exactly the p_i, $i \leq m$ as its (free) variables:

$$\chi_0 =_{df} \bigcirc false$$
$$\chi_{m+1} =_{df} \bigcirc(\bigcirc p_{m+1} \supset (p_{m+1} \vee \chi_m)).$$

Then χ_m is a valid scheme for all standard constraints of depth at most m.

Lemma 8. *Let C be a standard constraint of depth n. Then,* IPL $\vdash \chi_m^C$ *for all* $m \geq n$.

Proof. We prove this by induction on the depth n of the constraint. For $n = 0$ the constraint is $C = [false, false)$. Since for all $m \geq 0$, χ_m is of the form $\bigcirc\theta$ we find $\chi_m^C = false \supset \theta \vee false$ which is equivalent to *true*. Now let $C = \bigcap_{i=0}^n [\varphi_i, \psi_i)$ be a constraint of depth $n + 1$. Then, $m \geq n + 1 \geq 1$ and

$$\chi_m^C = \bigwedge_{i=0}^{n} \varphi_i \supset ((\bigcirc p_m \supset (p_m \vee \chi_{m-1}))^C \vee \psi_i).$$

So, we have to show that for all $i = 0, \ldots, n$, IPL derives $\varphi_i \vdash (\bigcirc p_m \supset (p_m \vee \chi_{m-1}))^C \vee \psi_i$. We will actually show that we can derive $\varphi_i \vdash (\bigcirc p_m \supset (p_m \vee \chi_{m-1}))^C$, or, which amounts to the same thing, that the sequents

$$\varphi_i, \bigwedge_{j=0}^{n} \varphi_j \supset (p_m \vee \psi_j) \vdash p_m \vee \chi_{m-1}^C$$

are derivable in IPL. Since the assumption includes φ_i and the implication $\varphi_i \supset (p_m \vee \psi_i)$ it is enough to show that $\varphi_i, p_m \vee \psi_i \vdash p_m \vee \chi_{m-1}^C$, *i.e.* the two sequents

$$\varphi_i, p_m \vdash p_m \vee \chi_{m-1}^C$$
$$\varphi_i, \psi_i \vdash p_m \vee \chi_{m-1}^C.$$

The first obviously is derivable immediately. For the second we proceed as follows: Let $C_i = \bigcap_{j \neq i} [\varphi_j, \psi_j)$ be the reduced constraint of depth n where we have dropped the interval $[\varphi_i, \psi_i)$. Then it is not difficult to see that we have $\varphi_i, \psi_i \vdash \chi_{m-1}^C \equiv \chi_{m-1}^{C_i}$. This follows essentially from the equivalence $\varphi_i, \psi_i \vdash \varphi_i \supset (\theta \vee \psi_i) \equiv true$. Thus, to obtain the sequent $\varphi_i, \psi_i \vdash p_m \vee \chi_{m-1}^C$ it suffices to prove $\vdash \chi_{m-1}^{C_i}$. But this follows from the induction hypothesis since C_i is a strictly smaller constraint of depth n and $m - 1 \geq n$. $\quad\square$

Lemma 8 says that standard constraints of depth up to and including m satisfy the axiom χ_m. However, these "characteristic" schemes are not theorems of PLL as we shall now see.

Lemma 9. *For every $m \geq 0$ there is a constraint model M such that $M \not\models \chi_m$.*

Proof. For $m \geq 0$ consider the linear constraint model $M_m =_{df} (W, \sqsubseteq_i, \sqsubseteq_m, V, \emptyset)$, where $W = \{0, 1, \ldots, 2m\}$, $\sqsubseteq_i = \leq$,

$$\sqsubseteq_m = \{ (2k+1, 2k+2) \mid k = 0, \ldots m-1 \} \cup \{ (i, i) \mid i = 0, \ldots, 2m \},$$

and for all $k \leq m$, $V(2k) = V(2k+1) = \{p_{m-k+1}, p_{m-k+2}, \ldots, p_m\}$. So, *e.g.* $V(0) = V(1) = \{\}$ and $V(2) = V(3) = \{p_m\}$, $V(4) = V(5) = \{p_{m-1}, p_m\}$, *etc.* In pictures the model looks like this:

We claim that $M_m \not\models \chi_m$. We show this by induction on n. For $m = 0$ the model M_0 is the trivial non-fallible one-world model which obviously refutes $\chi_0 = \bigcirc false$. Now consider M_{m+1}. Recall that $\chi_{m+1} = \bigcirc(\bigcirc p_{m+1} \supset (p_{m+1} \vee \chi_m))$. We first observe that the suffix model $M_m(2)$ of M_m that starts with world 2 is precisely the same as M_{m-1}, if propositional variable p_{m+1} is ignored, whence by induction hypothesis $2 \not\models \chi_m$, so in particular $1 \not\models \chi_m$. Also, $1 \not\models p_{m+1}$, whence $1 \not\models p_{m+1} \vee \chi_m$. On the other hand, $2 \models p_{m+1}$ and since $1 \sqsubseteq_m 2$ we find $1 \models \bigcirc p_{m+1}$. This shows $0 \not\models \bigcirc p_{m+1} \supset (p_{m+1} \vee \chi_m)$. Finally, since $0 \sqsubseteq_m k$ implies $0 = k$, this implies $0 \not\models \bigcirc(\bigcirc p_{m+1} \supset (p_{m+1} \vee \chi_m))$ as desired.   □

The desired theorem is a direct consequence of Lemmas 8 and 9:

Corollary 10. *No finite subset of $\mathbb{S}$ is complete for* PLL.

Proof. Let $\mathbb{D} \subset \mathbb{S}$ be a finite subset of standard constraints. Then, there exists a number $m \geq 0$ such that all $D \in \mathbb{D}$ are of depth at most m. By Lemma 8 χ_m^D is a theorem of IPL for each $D \in \mathbb{D}$. On the other hand by Lemma 9 the proposition χ_m is not a theorem of PLL.   □

This brings to a satisfactory conclusion the programme of this paper in that we not only showed that the infinite Boolean algebra $\mathbb{S}$ of standard constraints provides a sound and complete interpretation of PLL, but also that no finite subset of $\mathbb{S}$ would suffice.

7   Final Remarks

This paper offers one solution to Curry's programme of internally characterising the modality $\bigcirc$ by provability in IPL. A parallel can be found in [15, 14] which give an abstract representation of nuclei—the algebraic counterparts of $\bigcirc$—on a complete Heyting algebra in terms of implicational and disjunctive nuclei. Let us expand on this a bit. Let $M^* = (W, \sqsubseteq_i, \sqsubseteq_m, V^*, F)$ be a semantically complete constraint model. We form the Alexandroff topology $\Upsilon M^* = (W^+, \subseteq, \cap, \bigcup)$ of subsets of W upward closed under $\sqsubseteq_i$ and containing F. That is, $U \in W^+$ iff $F \subseteq U$ and whenever $u \in U$ and $u \sqsubseteq_i v$ then $v \in U$. ΥM^* is a complete Heyting algebra (cHA). Each formula φ of IPL corresponds to an element $M^*(\varphi) =_{df} \{w \mid M^*, w \models \varphi\}$ of W^+ so that *true* corresponds to W, *false* corresponds to F, $\wedge$ corresponds to $\cap$ and $\vee$ to $\cup$. Also $M^*(\varphi \supset \psi) = M^*(\varphi) \Rightarrow M^*(\psi)$ where $U \Rightarrow V$ is the interior of $\overline{U} \cup V$, *i.e.* the largest upper closed subset of $(W \setminus U) \cup V$. The algebraic counterpart to $\bigcirc$ is a *nucleus* on ΥM^*, that is, a monotone operation j on W^+ satisfying $U \subseteq j(U)$, $j(j(U)) \subseteq j(U)$ and $j(U \cap V) = j(U) \cap j(V)$. Then $\sqsubseteq_m$ determines a specific nucleus j_{M^*} given by $j_{M^*}(U) =_{df} \{u \in W \mid \forall v. u \sqsubseteq_i v \Rightarrow \exists r. v \sqsubseteq_m r \ \& \ r \in U \cup F\}$. Remarkably, it turns out that the nuclei on an arbitrary cHA H themselves form a cHA $\mathcal{N}(H) = (N(H), \leq, \wedge, \rightarrow, \bigvee)$, where $N(H)$ is the set of nuclei on H and $\leq$, $\wedge$ are given pointwise and $\rightarrow$, $\bigvee$ hardly ever pointwise. In [15] and [14] it is shown that every nucleus on a cHA H can be expressed in the form $\bigvee_{i \in H} o(K_i) \wedge c(L_i)$ where $o(K)$ is the open nucleus sending x to $K \rightarrow x$ and $c(L)$ is the closed nucleus sending x to $x \vee L$. If M^* is finite, then two facts emerge. Firstly we may represent every element $U \in W^+$ syntactically by $\psi_U =_{df} \bigvee\{\alpha_u \mid u \text{ minimal in } U \text{ w.r.t. } \sqsubseteq_i\}$. Secondly, since ΥM^* is also finite, $\mathcal{N}(\Upsilon M^*)$ is a finite Boolean algebra [15]. In this case, we may also represent its join $\bigvee$ syntactically as follows. By DeMorgan's rules, we have

$$\bigvee_{i \in I} o(K_i) \wedge c(L_i) = \bigwedge_{A \subseteq I} \left(\bigvee_{i \in A} o(K_i) \right) \vee \left(\bigvee_{i \in I \setminus A} c(L_i) \right) \tag{1}$$

$$= \bigwedge_{A \subseteq I} o\left(\bigwedge_{i \in A} K_i \right) \vee c\left(\bigvee_{i \in I \setminus A} L_i \right) \tag{2}$$

where step (2) follows from the equivalences $(K_1 \supset \varphi) \vee (K_2 \supset \varphi) \equiv (K_1 \wedge K_2) \supset \varphi$ and $(\varphi \vee L_1) \vee (\varphi \vee L_2) \equiv \varphi \vee (L_1 \vee L_2)$. It remains to find a syntactic representation for joins of the form $o(K) \vee c(L)$. But [15, Lem. 2.1] tells us that $o(K) \vee j = o(K) \circ j$ for any nucleus j and so we may define $(o(K) \vee c(L)) \varphi$ to be $K \supset (\varphi \vee L) = [K, L] \varphi$. Since we can define $o(K)$ as $[K, false)$ and $c(L)$ as $[true, L]$ it makes sense to take as basic constraints those of the form $[K, L]$ as we have done in this paper.

The constraint algebra $\mathbb{S}$ is not the only collection of constraints that might be considered for PLL. In type theory other computational interpretations of $\bigcirc$ have been proposed. For instance a generalisation of double negation $\bigcirc p = (p \supset \alpha) \supset \alpha$, used for typing continuations, yields a sound semantics, too. It can be shown that the context $a(\alpha)(\varphi) = (\varphi \supset \alpha) \supset \alpha$ in general cannot be represented by a fixed standard constraint that does not depend on x. However, given a finite constraint model M^*, results of [15] show that any nucleus j in $\mathcal{N}(\Upsilon M^*)$ may be represented as a meet of nuclei of the form $\bigwedge\{a(x) \mid x \text{ stable}\}$ where x is stable if $j(x) = x$. Letting $Stab^+$ denote the set of elements U of W^+ satisfying

$j_{M^*}(U) = U$, this means that for finite M^*, $M^* \models \bigcirc \varphi \equiv \bigwedge_{U \in Stab^+} (\varphi \supset \psi_U) \supset \psi_U$ and so the class of constraints of the form $\bigwedge_{i \in I} a(K_i)$ provides another sound and complete interpretation of PLL. More work needs to be done to study the inherent structure of the class of all constraint contexts $C[x]$ expressible in IPL.

Acknowledgements

We would like to thank Harold Simmons for a series of email discussions on finite modal algebras, which gave us fruitful insights that are reflected in this work. We are also grateful to Andy Pitts who prompted this investigation by challenging us with the question if Curry contexts are complete for PLL. Finally thanks are due to the anonymous referees who made many good suggestions for improving this paper.

References

[1] P. Aczel. The Russell-Prawitz modality. *Mathematical Structures in Computer Science*, 11(4), 2001. Special issue: *Modalities in Type Theory* (Proceedings of the 1999 Workshop on Intuitionistic Modal Logic and Applications, Trento, Italy).

[2] N. Benton and V. de Paiva. Computational types from a logical perspective. *Journal of Functional Programming*, 8(2), March 1998.

[3] P. N. Benton. A unified approach to strictness analysis and optimising transformations. Technical Report 388, University of Cambridge Computer Laboratory, February 1996.

[4] H. B. Curry. The elimination theorem when modality is present. *Journal of Symbolic Logic*, 17:249–265, 1952.

[5] H. B. Curry. *A Theory of Formal Deducibility*, volume 6 of *Notre Dame Mathematical Lectures*. Notre Dame, Indiana, second edition, 1957.

[6] R. Davies and F. Pfenning. A modal analysis of staged computation. In Jr. Guy Steele, editor, *Proc. 23rd POPL*. ACM Press, January 1996.

[7] J. Despeyroux and P. Leleu. Recursion over objects of functional type. *Mathematical Structures in Computer Science*, 2001.

[8] J. Despeyroux, F. Pfenning, and C. Schürmann. Primitive recursion for higher-order abstract syntax. In *Typed Lambda Calculi and Applications*, pages 147–163. Springer, 1997. LNCS 1210.

[9] M. Fairtlough and M. Mendler. Propositional Lax Logic. *Information and Computation*, 137(1):1–33, August 1997.

[10] M.V.H. Fairtlough and M. Walton. Quantified Lax Logic. Technical Report CS-97-11, Department of Computer Science, The University of Sheffield, July 1997.

[11] R. I. Goldblatt. Grothendieck topology as geometric modality. *Zeitschrift für mathematische Logik und Grundlagen der Mathematik*, 27:495–529, 1981.

[12] J. Hatcliff and O. Danvy. A computational formalisation for partial evaluation. *Math. Struct. in Comp. Science*, 7:507–541, 1997.

[13] B. P. Hilken. Duality for intuitionistic modal algebras. UMCS-96-12-2, Manchester University, Department of Computer Science, 1996. To appear in Journal of Pure and Applied Algebra.

[14] P. T. Johnstone. *Stone Spaces*. Cambridge University Press, 1982.

[15] D. S. Macnab. Modal operators on Heyting algebras. *Algebra Universalis*, 12:5–29, 1981.

[16] M. Mendler. Constrained proofs: a logic for dealing with behavioural constraints in formal hardware verification. In G. Jones and M. Sheeran, editors, *Workshop on Designing Correct Circuits*. Springer, 1991.

[17] M. Mendler. *A Modal Logic for Handling Behavioural Constraints in Formal Hardware Verification*. PhD thesis, Department of Computer Science, University of Edinburgh, ECS-LFCS-93-255, March 1993.

[18] E. Moggi. Notions of computation and monads. *Information and Computation*, 93:55–92, 1991.

[19] D. A. Turner. Elementary strong functional programming. In P. H. Hartel and M. J. Plasmeijer, editors, *Functional Programming Languages in Education FPLE'95*, pages 1–13. Springer, 1995. LNCS 1022.

[20] P. Wadler. Comprehending monads. In *Conference on Lisp and Functional Programming*. ACM Press, June 1990.

[21] P. Wadler. Monads for functional programming. In *Lecture Notes for the Marktoberdorf Summer School on Program Design Calculi*. Springer Verlag, August 1992.

Constructive Reals in Coq: Axioms and Categoricity

Herman Geuvers and Milad Niqui

Department of Computer Science, University of Nijmegen, The Netherlands
{herman,milad}@cs.kun.nl

Abstract. We describe a construction of the real numbers carried out in the Coq proof assistant. The basis is a set of axioms for the constructive real numbers as used in the FTA (Fundamental Theorem of Algebra) project, carried out at Nijmegen University. The aim of this work is to show that these axioms can be satisfied, by constructing a model for them. Apart from that, we show the robustness of the set of axioms for constructive real numbers, by proving (in Coq) that any two models of it are isomorphic. Finally, we show that our axioms are equivalent to the set of axioms for constructive reals introduced by Bridges in [2]. The construction of the reals is done in the 'classical way': first the rational numbers are built and they are shown to be a (constructive) ordered field and then the constructive real numbers are introduced as the usual Cauchy completion of the rational numbers.

1 Introduction

The FTA project at the University of Nijmegen (see [10]) had as goal to formalise a constructive proof of the Fundamental Theorem of Algebra in the proof assistant Coq [8]. For reasons of modularity, it was decided not to start with a specific construction of the real numbers, but to work axiomatically. So, a list of axioms was defined, which together define the notion of a *real number structure*. To do this, Coq allows a nice modular approach, using *dependent labelled record types*, on which simple inheritance relations can be defined using *coercions*. So, a real number structure is defined as a Cauchy complete Archimedean ordered field, where an ordered field is again defined as a field extended with an ordering satisfying certain properties. Fields are defined in terms of rings, which are defined in terms of groups, etcetera. In this way there is an algebraic hierarchy of structures that extend each other and that inherit operations, relations and properties from each other. For the present exposition we do not describe the full hierarchy, but just the relevant nodes: constructive setoids, rings, fields, ordered fields and real number structures.

The outline of the paper is as follows. After describing the relevant part of the algebraic hierarchy, we give the construction of the rational numbers $\mathbb{Q}$ and show that they form a constructive ordered field. Then we define the reals as Cauchy sequences over $\mathbb{Q}$ and show that they form a real number structure. This roughly follows the standard texts [13, 1]. As a matter of fact, we proceed

P. Callaghan et al. (Eds.): TYPES 2000, LNCS 2277, pp. 79–95, 2002.

in a slightly more general fashion by showing that the Cauchy sequences over *any* Archimedean constructive ordered field form a real number structure. The notion of real number structure describes a collection of models and we may wonder whether the axioms characterise the reals completely. This is the case: any two real number structures are isomorphic. This property has been formally stated and proved inside Coq. Finally, we compare our axioms for the reals and prove (within Coq) that they are equivalent to the ones of [2].

The mathematics described in this paper is formalised in Coq. To make the paper of interest to constructive mathematicians in general, we have tried to keep Coq syntax to a minimum. Some features of (the type theory of) Coq that are important to the formalisation are: the propositions-as-types interpretation of logic (to formalise logical reasoning), record types and inductive types (to formalise mathematical structures and to have a type of natural numbers), dependent types (to formalise, e.g. the notion of subset) and the Axiom of Choice.

The present work gives a constructive version of what is already in the Coq standard library. This constructive version is required by the FTA project [9]. The Coq standard library contains classical real numbers, see [7]. They are axiomatised, based on a parameter R which is of type Type (and hence does not satisfy our axioms), together with relations and operations on R, which satisfy the *classical* axioms of real numbers. Other constructions of the reals in a type theoretic proof assistants are described in [12] (in the system LEGO), in [4] (in the system NuPrl) and [5] (in Coq). In [12] a real number is defined as a collection of arbitrary small nested intervals with rational endpoints. More precisely, a general construction for completing a metric space is given, with the metric space of the rationals as primary example. In [4] the set of rational numbers is defined, and then the type of reals is defined following Bishop's original construction [1], using regular Cauchy sequences. Finally the Cauchy completeness of reals is proved as a theorem in Nuprl. In [5], a construction of the reals in Coq is described, using infinite (lazy) streams. The work is especially of interest, as special care is taken to obtain an *efficient* implementation of the reals, with which one might actually compute. In [6], a (constructive) axiomatisation for these reals is given. The axioms are very similar to ours, with the distinguished exception that in [6] special attention is paid to the minimality of the proposed axioms. This has lead to a minimal set of 16 axioms, which is compared to our axioms (Section 2) and Bridges axioms [2]. A good classical construction of the reals can be found in [11], which is especially interesting because much attention is devoted to the optimisation of the operations and generating useful decision procedures.

2   Ordered Fields and Real Number Structures

We now present the definitions, in Coq, of a constructive ordered field and a real number structure. We present them as part of a constructive algebraic hierarchy, of which we present here only the part that is relevant for the purpose of understanding the real number axioms. See [9] for the full details.

Coq Definition 1. Record CSetoid   : Type :=
```
{ C                 :> Set;
  [=]               : C->C->Prop;
  [#]               : C->C->Prop;
  ap_irr            : (x:C)~(x [#] x);
  ap_sym            : (x,y:C)(x [#] y) -> (y [#] x);
  ap_cot            : (x,y:C)(x [#] y) -> (z:C)(x [#] z)\/(z [#] y);
  ap_tight          : (x,y:C)~(x [#] y) <-> (x [=] y)                }.
```

An important remark to make is that the above is *not* syntactically correct Coq code. The symbols [#] and [=] cannot be used as labels in a record structure. These symbols are only introduced *after* the record definition as syntactic sugared infix notation (via grammar rules). For readability we will throughout this paper use the syntactic sugared symbols already *in* the definition, as infix operations.

The notation C :> Set signifies that there is a coercion from constructive setoids to Set. This allows, in Coq, to use a term of type CSetoid (a constructive setoid) in places where a term of type Set (a set) is asked for: the type checker will insert C, the projection on the first field of the record.

The definition says that a constructive setoid (a term of type CSetoid) is a tuple $\langle C, =, \#, a_1, a_2, a_3, a_4 \rangle$, with C : Set, $=$ and $\#$ binary relations over C, a_1 a proof of the irreflexivity of $\#$, a_2 a proof of the symmetry of $\#$, a_3 a proof of the *cotransitivity* of $\#$ and a_4 saying that $\#$ is *tight* with respect to $=$. On real numbers the apartness is basic; a constructive setoid axiomatises its basic properties, of which the cotransitivity is the most noteworthy. Cotransitivity gives the positive meaning to the apartness.

It is also possible to do without equality altogether and define $=$ as the negation of $\#$. A reason for not doing so is that $=$ is very basic and the universal algebraic laws are all stated in terms of $=$. With respect to functions between constructive setoids, there are now several choices. Given the constructive setoids $S := \langle C, =, \#, a_1, a_2, a_3, a_4 \rangle$ and $S' := \langle C', =', \#', a_1', a_2', a_3', a_4' \rangle$, there are the terms of type $C \to C'$, which can be seen as algorithms that take a representation of an element of S and return a representation of an element in S'. Obviously, such an algorithm need not preserve the equality. For a *constructive setoid function*, we want more. First of all that it preserves the equality: we want to consider only those $f : C \to C'$ for which $\forall x, y : C . (x = y) \to f(x) =' f(y)$ holds. As we are in a constructive setting, we even want a bit more: we want a *constructive setoid function* to reflect the apartness. This is called *strong extensionality* and it is defined as (for f unary and g binary, respectively)

$$\forall x, y : C . f(x) \#' f(y) \to (x \# y)$$
$$\forall x_1, x_2, y_1, y_2 : C . g(x_1, x_2) \#' g(y_1, y_2) \to (x_1 \# y_1) \lor (x_2 \# y_2)$$

Strong extensionality of a function f says that f cannot distinguish (via $\#'$) two elements that were indistinguishable (via $\#$). Note that strong extensionality of binary functions arises naturally from the notion of binary product on setoids. Strong extensionality of f implies that it preserves the equality, so we don't

have to assume that separately. As an example of the power (and usefulness) of
the notion, we remark that strong extensionality of multiplication * (see below)
implies that $\forall x, y{:}C.(x * y)\#0 \rightarrow (x\#0 \wedge y\#0)$ holds.

```
Coq Definition  2. Record CRing      : Type :=
{ R                  :> CSetoid;
  zero               : R;
  add                : R->R->R;
  add_strext         : (x1,x2,y1,y2:R)((x1 [+] y1) [#] (x2 [+] y2)) ->
                                       (x1 [#] x2)\/(y1 [#] y2);
  add_assoc          : (x,y,z:R)((x [+] (y [+] z)) [=]
                                              ((x [+] y) [+] z));
  add_unit           : (x:R)((x [+] zero) [=] x);
  add_commut         : (x,y:R)((x [+] y) [=] (y [+] x));

  minus              : R->R;
  minus_strext       : (x,y:R)([--]x [#] [--]y) -> (x [#] y);
  minus_proof        : (x:R)((x [+] [--]x) [=] zero);
  one                : R;
  mult               : R->R->R;
  mult_strext        : (x1,x2,y1,y2:R) ((x1 [*] y1) [#] (x2 [*] y2))->
                                        (x1 [#] x2)\/(y1 [#] y2);
  mult_assoc         : (x,y,z:R)
                             ((x [*] (y [*] z)) [=] ((x [*] y) [*] z));
  mult_unit          : (x:R)((x [*] one) [=] x);
  mult_commut        : (x,y:R)((x [*] y) [=] (y [*] x));
  dist               : (x,y,z:R) ((x [*] (y [+] z)) [=]
                                       ((x [*] y) [+] (x [*] z)));
  non_triv           : (one [#] zero)                            }.
```

So, a constructive ring consists of a constructive setoid as a carrier, extended
with an addition function add, with infix notation [+], a unary minus function,
minus, with prefix notation [--] and a multiplication function mult, with infix
notation [*]. The axioms are the usual ones, apart from the fact that we assume
all functions to be strongly extensional.

A constructive field is a ring extended with a reciprocal function, which is a
partial function on the type F. The partiality is expressed by requiring both an
$x : F$ and a proof of $x\#0$ as input. In the formalisation the reciprocal is defined
as a function on the *subsetoid* of non-zeros of F. Given a predicate over a setoid,
say P : S -> Prop, the 'set' of elements satisfying P is defined by the record:

Coq Definition 3.
```
Record subsetoid [S : CSetoid; P : S -> Prop] : Set :=
  { elem  :> S;
    prf   : (P elem)
  }.
```

This subsetoid is then turned into a constructive setoid by inheriting the apartness and equality from those on S (i.e. ignoring the proof term).

To define fields we use the subsetoid of non-zeroes of a ring. Given R : CRing, we define (NonZeros R) as (subsetoid R [x:F](x [#] zero)). Constructive fields are then defined by the following record structure.

Coq Definition 4. Record CField   : Type :=
```
{ F                :> CRing;
  rcpcl            : (NonZeros F) -> (NonZeros F);
  rcpcl_strext     : (x,y:(NonZeros F))
                     ((rcpcl x) [#] (rcpcl y)) -> (x [#] y);
  rcpcl_proof      : (x:(NonZeros F))
                     ((nzinj x) [*] (nzinj (rcpcl x )) [=] one) }.
```

Here, nzinj : (NonZeros F) -> F is the function that maps a non-zero (of a ring) to the underlying element. So it is just the projection to the elem field for the subsetoid of non-zeros.

Coq Definition 5. Record COrdField : Type :=
```
{ OF               :> CField;
  [<]              : OF->OF->Prop;
  less_strext      : (x1,x2,y1,y2:OF) (x1 [<] y1) ->
                            (x2 [<] y2)\/(x1 [#] x2)\/(y1 [#] y2);
  less_trans       : (x,y,z:OF)(x [<] y) -> (y [<] z) -> (x [<] z);
  less_irr         : (x:OF)~(x [<] x);
  less_asym        : (x,y:OF)(x [<] y) -> ~(y [<] x);
  add_resp_less    : (x,y:OF)(x [<] y) ->
                            (z:OF)((x [+] z) [<] (y [+] z));
  times_resp_pos   : (x,y:OF)(zero [<] x) -> (zero [<] y)->
                            (zero [<] (x [*] y));
  less_conf_ap     : (x,y:F)(x [#] y) <-> ((x [<] y)\/(y [<] x))  }.
```

An ordered field is the union of a field with an ordered ring. Apart from the standard requirements for the ordering [<], we require it to be *strongly extensional*, which is asserted by less_strext. This is (just like strongly extensional functions) a positive way of saying that the relation cannot distinguish between elements that are indistinguishable. Note that it follows from the strong extensionality that order respects the equality ($x < y \wedge y = z \rightarrow x < z$ etcetera). The final axiom connects apartness with the ordering in the expected way and together with the cotransitivity of the apartness implies the important cotransitivity property of $<$, i.e. the property $x < y \rightarrow x < z \vee z < y$.

A *real number structure* is an ordered field that is *Archimedean* (every real is majorised by some natural number), and Cauchy complete, i.e. every Cauchy sequence has a limit. For the Archimedean property, we define, for F an arbitrary field, the function nreal : nat -> F, which maps 0 to zero and (S n) to ((nreal n) [+] one). If F is ordered, this is an injection. We want to define a Cauchy sequence over the ordered field F as an $s:\mathbb{N} \to F$ such that

$$\forall \varepsilon:F_{>0}.\exists N:\mathbb{N}.\forall m \geq N.(|s_m - s_N| < \varepsilon).$$

However, constructively, the absolute value function is not definable in an arbitrary ordered field, so it has to be replaced with something else. For this purpose the notion AbsSmall has been introduced.

Coq Definition 6.
AbsSmall [e,x:F]: Prop := ([--]e [<] x) /\ (x [<] e).

So AbsSmall e x expresses that, for a positive e, x is small with respect to e. Now the Cauchy property can be defined.

Coq Definition 7. Definition cauchy [g:nat->F]: Prop :=
    (e:F)(zero [<] e) ->
        (EX N:nat | (m:nat)(le N m) -> (AbsSmall e ((g m)[-](g N)))).

Note that we do not explicitly require Cauchy sequences to have a modulus of convergence. By the Axiom of Choice, every Cauchy sequence has a modulus of convergence and the Axiom of Choice is provable in our framework. Next, for a real number structure we assume a function lim that takes a Cauchy sequence and returns its limit.

Coq Definition 8. Record CReals : Type :=
{IR   :> CField;
lim   : (s:nat->IR)(cauchy s)->IR;
lim_proof   : (s:nat->F)(c:(cauchy s))
                        (e:F)(zero [<] e) ->
                            (EX N:nat | (m:nat)(le n m) ->
                                        (AbsSmall e (s m)[-](lim s c)));
arch_proof   : (x:F)(EX n:nat | (x [<] (nreal n)))   }.

2.1   On the Choice of Primitives: $\frac{1}{k}$ versus ε

The notion of Cauchy sequence (and similarly the notion of limit) is defined above via the 'ε-definition':

$$\forall \varepsilon{:}F_{>0}.\exists N{:}\mathbb{N}.\forall m \geq N.(|s_m - s_N| < \varepsilon).$$

Let us call such a sequence s an ε-Cauchy sequence. An alternative is the '$\frac{1}{k}$-definition':

$$\forall k{:}\mathbb{N}.\exists N{:}\mathbb{N}.\forall m \geq N.(|s_m - s_N| < \frac{1}{k+1})$$

and we will call this a $\frac{1}{k}$-Cauchy sequence. Any ε-Cauchy sequence is a $\frac{1}{k}$-Cauchy sequence. For the reverse implication we need the Archimedean property. (To find an $N \in \mathbb{N}$ such that $\forall m \geq N.|s_m - s_N| < \varepsilon$, we have to find $k, N \in \mathbb{N}$ such that $\forall m \geq N.|s_m - s_N| < \frac{1}{k}$ and $\frac{1}{k} < \varepsilon$. The inequality $\frac{1}{k} < \varepsilon$ is solved by applying the Archimedean property to $\frac{1}{\varepsilon} < k$.)

Similarly one can define the notion of limit via the ε-definition (the ε-limit) or via the $\frac{1}{k}$-definition (the $\frac{1}{k}$-limit). If s has an ε-limit, then s has a $\frac{1}{k}$-limit and the reverse holds by the Archimedean property.

So, in an Archimedean ordered field, the two notions are equivalent. In a non-Archimedean field (e.g. a non-standard model of the reals) they are not equivalent and one may wonder what the 'best' definition of Cauchy sequence is (and of limit). We feel that the question isn't very relevant, because for the analysis of non-standard reals, sequences of length ω are just too short. We have a slight preference for the ε-definitions for the following reasons.

(1) If one defines Cauchy-completeness as 'all $\frac{1}{k}$-Cauchy sequences have a $\frac{1}{k}$-limit', then $1, \frac{1}{2}, \frac{1}{3}, \ldots$ has a limit, which seems unnatural in non-standard reals (all infinitesimals are limit of this sequence). Also, limits are not unique anymore: if a and b are $\frac{1}{k}$-limit of s, then $\forall k.|a - b| < \frac{1}{k+1}$, but to conclude $a = b$ from this we need the Archimedean property.

(2) For constructing the reals out of the rationals, we consider Cauchy-sequences of rationals. The construction can be made more modular by considering Cauchy sequences over an arbitrary ordered field (even a non-Archimedean one). It turns out that $\frac{1}{k}$-Cauchy sequences over an ordered field do not necessarily form a field, whereas ε-Cauchy sequences do. See the discussion in Remark 4.9 for details.

As a final remark we want to point out that there is yet another (useful) definition of Cauchy completeness as 'all $\frac{1}{k}$-Cauchy sequences have an ε-limit'. Then the Archimedean property follows from Cauchy completeness: let $x > 0$ and to find an $n \in \mathbb{N}$ with $n > x$ consider the $\frac{1}{k}$-Cauchy sequence $1, \frac{1}{2}, \frac{1}{3}, \ldots$. This sequence has an ε-limit, and if we take ε to be $\frac{1}{x}$, we find an $N \in \mathbb{N}$ for which

$$\forall m \geq N. |\frac{1}{m+1} - \frac{1}{N+1}| < \frac{1}{x}.$$

Taking m to be $N + 1$, we conclude that $x < (N + 1)(N + 2)$ and we are done.

3   Rational Numbers

The job is to construct a real number structure, that is, a term of type CReals. It will be constructed from Cauchy sequences of rational numbers, so we need to have a construction of rational numbers in Coq. In the Coq standard library the rational numbers are not developed. So we have to construct them. Arithmetic is developed quite considerably in Coq: natural numbers are defined as an inductive type nat with constructors zero and successor, and relations and operations on them are all defined using induction or recursion. The inductive natural numbers are very useful for proving theorems about, but for actual computation this unary representation is very inefficient. For this reason, the integers are defined as the disjoint union of the one element type and two copies of the type of binary sequences (one denoting the positive numbers and one denoting the negative numbers). This yields a relatively fast (standard binary) implementation of the integers.

We represent an element of $\mathbb{Q}$ as a pair $\langle p, n \rangle$ of an integer and a natural number, denoting the rational $\frac{p}{n+1}$. This method is rather useful as it avoids carrying around the proof obligation corresponding to a non zero denominator.

In Coq, this is encoded using the Record constructor. We define the equality on $\mathbb{Q}$ in such a way that it corresponds to the intended interpretation of the pair $\langle p, n \rangle$: $\langle p, n \rangle =_{\mathbb{Q}} \langle q, m \rangle$ is defined as $p(m + 1) = q(n + 1)$. This means that we identify the two rationals $\frac{1}{2}$ and $\frac{2}{4}$. So unlike the case of natural numbers and integers, we deal with an inductive type for which the useful equality is not Leibniz equality. This is only because our way of representing rational numbers has redundancy in contrast to the definition of natural numbers and integers. Whenever we want to prove an equality on $\mathbb{Q}$, it boils down to proving an equality between integers, which we can verify using the arithmetics developed in the ZArith library. We define the *apartness* relation on $\mathbb{Q}$ to be the negation of the equality and we prove the required properties for it. The constants zero and one and the addition operation are defined in the expected way: $\langle p, n \rangle + \langle q, m \rangle :=$ $\langle p(m+1)+q(n+1), nm+n+m \rangle$. Similarly, the unary minus and the multiplication function are defined in the expected way. Then we prove that these operations are strongly extensional and satisfy the ring properties.

Defining the reciprocal is a bit more complicated, since we have to give a partial function taking nonzero elements of $\mathbb{Q}$ to nonzero elements of $\mathbb{Q}$. The idea is that, as $\langle p, n \rangle$ represents the rational $\frac{p}{n+1}$, the inverse of $\langle p, n \rangle$ should be $\langle n+1, p-1 \rangle$. But $p-1$ should be a natural, so we have to take $\langle -(n+1), -(p-1) \rangle$ if $p < 0$. (The case $p = 0$ does not occur, as we only consider the non-zeros of $\mathbb{Q}$.) So, the inverse on $\mathbb{Q}_{\#0}$ is defined as follows.

$$\frac{1}{\langle p, n \rangle} := \begin{cases} \langle n + 1, p - 1 \rangle & \text{if } p > 0 \\ \langle -(n + 1), -(p - 1) \rangle & \text{if } p < 0 \end{cases}$$

We prove that the inverse is strongly extensional and satisfies the field axiom. We have equipped our model of rational numbers with all the constructive field operations. In Coq terms, we have given a term of type CField. Now we define an order relation on $\mathbb{Q}$ in the natural way, using the order on $\mathbb{Z}$:

$$\langle p, n \rangle <_{\mathbb{Q}} \langle q, m \rangle := p * (m + 1) <_{\mathbb{Z}} q * (n + 1).$$

We prove that $<_{\mathbb{Q}}$ is strongly extensional, and that it satisfies the order axioms. It is also easy to prove that $\mathbb{Q}$ is Archimedean. So we wrap up what we did in this section, in the following theorem:

Theorem 3.1. $\mathbb{Q}$ *together with the operations and relations defined above forms an Archimedean constructive ordered field.*

The rational numbers are the 'simplest' example of a constructive ordered field: every constructive ordered field contains a copy of the rational numbers. An important property of rational numbers is that equality is decidable (which is, for example, not the case for the set of Cauchy sequences over $\mathbb{Q}$). Decidability of equality allows to define the maximum function. In general, this is not possible for constructive ordered fields. However it is possible for real number structures, see [10].

4   Real Numbers

Now we need to construct a real number structure out of the set of Cauchy sequences over the rationals. Unless stated otherwise, we will always talk about ε-Cauchy sequences (see Section 2.1). For reasons of readability, we will not use Coq notation, but ordinary mathematical notation. So, for example, we write $\mathbb{N}$ for the type of naturals (and $\mathbb{Q}$ for the type of rationals as we already did in the previous section). Instead of working with the concrete rationals $\mathbb{Q}$, we prove a general result about Archimedean constructive ordered fields.

Notation. Let F be a constructive ordered field. We denote the set of Cauchy sequences over F by $\mathrm{CauchySeq}_F$.

Theorem 4.1. *Let F be a constructive ordered field. If F is Archimedean then $\mathrm{CauchySeq}_F$ is a real number structure.*

The largest part of the formalisation consists of the proof of this theorem. Instantiating it with the term obtained in Theorem 3.1, we get a concrete term of the type 'real number structure', and our implementation is finished. We divide the proof of Theorem 4.1 into two separate parts, especially to highlight the use of the Archimedean property. We prove the following two results.

Theorem 4.2. *Let F be a constructive ordered field. Then $\mathrm{CauchySeq}_F$ is also a constructive ordered field.*

Theorem 4.3. *Let F be a constructive ordered field. If F is Archimedean then $\mathrm{CauchySeq}_F$ is Archimedean and Cauchy complete.*

For the proof of Theorem 4.2, we systematically define the constructive ordered field operations and relations on $\mathrm{CauchySeq}_F$. We introduce the following shorthands, corresponding to Coq Definition 6.

Notation. For x and ε elements of a constructive ordered field F, we denote $\mathrm{AbsSmall}(\varepsilon, x) := -\varepsilon < x \wedge x < \varepsilon$, and $\mathrm{AbsBig}(\varepsilon, x) := 0 < \varepsilon \wedge (x < -\varepsilon \vee \varepsilon < x)$.

The familiar real number properties that are stated using the absolute value function can all be stated in terms of these two predicates. We now define the relations on Cauchy sequences.

Definition 4.4 (Order). *Let $g, h : \mathrm{CauchySeq}_F$, we define $g < h$ as*

$$\exists \varepsilon : F_{>0}. \exists N : \mathbb{N}. \forall m \geq N. (\varepsilon < h_m - g_m).$$

An alternative definition of the ordering (that goes along naturally with the alternative definition of Cauchy sequence, as discussed in Section 2.1) can be given using $\frac{1}{k}$ instead of ε as follows: $g < h$ if $\exists k \in \mathbb{N}. \exists N : \mathbb{N}. \forall m \geq N. (\frac{1}{k+1} < h_m - g_m)$. If F is Archimedean, the $\frac{1}{k}$-order and the ε-order are the same, but in general only the first implies the second and not the other way around. If F is non-Archimedean, then in $\mathrm{CauchySeq}_F$ with the $\frac{1}{k}$-order, some distinct elements become identified, which may be undesirable.

Definition 4.5. *We define apartness # and equality = on Cauchy sequences in terms of order as follows.* $f \# g := f < g \vee g < f$, $f = g := \neg(f \# g)$.

We prove that # satisfies the axioms of apartness. Then we define some equivalent definitions for apartness and equality, that are useful in practice. We mention two of them.

Lemma 4.6. *Let* $f, g : \text{CauchySeq}_F$. $f \# g$ *is equivalent to*

$$\exists \varepsilon : F_{>0}.\exists N : \mathbb{N}.\forall m \geq N.\text{AbsBig}(\varepsilon, f_m - g_m).$$

$f = g$ *is equivalent to*

$$\forall \varepsilon : F_{>0}.\exists N : \mathbb{N}.\forall m \geq N.\text{AbsSmall}(\varepsilon, f_m - g_m).$$

The following lemma states another important property of Cauchy sequences that is used several times in our proofs. It gives us an upper-bound for a Cauchy sequence.

Lemma 4.7. *Let* $s : \text{CauchySeq}_F$. *Then* $\exists K : F_{\#0}.\forall m : \mathbb{N}.\text{AbsSmall}(K, s_m)$.

To turn the Cauchy sequences into a ring, we define zero and one as the sequences which are constant zero, respectively one. Similarly we define addition, minus and multiplication on Cauchy sequences via the pointwise addition, minus and multiplication. We prove that these operations are strongly extensional. In the case of multiplication, the proof of strong extensionality uses Lemma 4.7. Finally, we prove that we have constructed a constructive ring.

Now we have to define the reciprocal. Let s be a Cauchy sequence and suppose $s \# 0$. By Lemma 4.6, this means that, from a certain N onwards we know that we are a positive distance from zero. So, determine ε and N such that

$$\forall m \geq N.\text{AbsBig}(\varepsilon, s_m).$$

So for $m \geq N$, the elements s_m are all $\# 0$. Now we take the reciprocals of these elements to get the reciprocal of the Cauchy sequence x:

Definition 4.8 (Reciprocal). *Let the non-zero Cauchy sequence* s *be given and let* N *be such that for* $m \geq N$, $s_m \# 0$. *Define the sequence* s^{-1} *as follows.*

$$s_m^{-1} = \begin{cases} 0 & \text{if } m < N \\ \frac{1}{s_m} & \text{if } m \geq N \end{cases}$$

This definition uses the Axiom of Choice to assign an N to each s. We prove that the above sequence is a Cauchy sequence and is apart from zero. Then we show that the operation $(\_)^{-1}$ is strongly extensional and satisfies the field axiom for reciprocal.

Remark 4.9. [$\frac{1}{k}$-Cauchy sequences versus ε-Cauchy sequences] If we use $\frac{1}{k}$-Cauchy sequences, the reciprocal can not be defined without assuming the Archimedean property.

(1) If we use the ε-order (and hence the ε-apartness and ε-equality of Definitions 4.4, 4.5), one needs the Archimedean property to show that s^{-1} is a $\frac{1}{k}$-Cauchy sequence: we need to find $N \in \mathbb{N}$ such that $\forall m \geq N |\frac{1}{s_m} - \frac{1}{s_N}| < \frac{1}{k+1}$. This is equivalent to finding N such that $\forall m \geq N |\frac{s_m - s_N}{s_m s_N}| < \frac{1}{k+1}$. If we have a lower bound $\frac{1}{K_0}$ (for $K_0 \in \mathbb{N}$) on $|s_m s_{N_0}|$ (for all m from a certain N_0 onward), this N can be computed from the fact that s is a $\frac{1}{k}$-Cauchy sequence. The existence of the lower bound $\frac{1}{K_0}$ is equivalent to the Archimedean property.

(2) If we use the $\frac{1}{k}$-order (see the remark just after Definition 4.4) and also $\frac{1}{k}$-apartness and $\frac{1}{k}$-equality, then s^{-1} is a $\frac{1}{k}$-Cauchy sequence indeed. (The lower bound $\frac{1}{K_0}$ that was required above, is now derived from the fact that $s\#0$, which now yields: $\exists K \exists N \forall m \geq N |s_m| > \frac{1}{K}$.) But now we can not prove that $s^{-1}\#0$, for $s\#0$, because the reciprocal of an 'infinite element' is just 0 under the $\frac{1}{k}$-equality. (We need the Archimedean property.)

Finally we prove that the order as defined in Definition 4.4 satisfies the required axioms and then we have proved Theorem 4.2.

In order to prove Theorem 4.3, we assume that F is Archimedean and let s:CauchySeq$_F$. We use Lemma 4.7 to obtain a bound K:F for s. The Archimedean property for F gives us an n:$\mathbb{N}$ such that $K < n$ in F. It is now easy to prove that the constant sequence λx:$\mathbb{N}.n$ bounds s in CauchySeq$_F$.

To prove completeness, we use a kind of diagonalisation argument. Note, however, that if $s^1, s^2, \ldots$ are Cauchy sequences, then λi:$\mathbb{N}.s_i^i$ may not be the right choice for the limit of $\{s^i\}_{i=0}^{\infty}$ (it may not even be a Cauchy sequence). Instead we have to take λi:$\mathbb{N}.s_{\tau(i)}^i$, where $\tau(i)$ is a number N (depending on the sequence s^i) such that s^i is *sufficiently close* to its limit from N onwards. To formalise the notion of 'sufficiently close', we use $\frac{1}{i+1}$ as a bound.

Definition 4.10. *Given a sequence of Cauchy sequences $\{s^i\}_{i=0}^{\infty}$, we define a function $\tau : \mathbb{N} \to \mathbb{N}$ satisfying*

$$\forall m > \tau(i).\mathrm{AbsSmall}(\frac{1}{i+1}, s_m^i - s_{\tau(i)}^i).$$

The map $\tau(i)$ is defined using the Cauchy property for the sequence s^i, which gives us (for $\varepsilon = \frac{1}{i+1}$) an N such that the above holds. Due to the fact that Cauchy sequences are not represented using a *modulus of convergence*, but via an existential quantifier, we need the Axiom of Choice to define τ.

Definition 4.11. *Given the sequence of Cauchy sequences $\{s^i\}_{i=0}^{\infty}$, we define the limit sequence l by*

$$l := \lambda n.s_{\tau(n)}^n$$

To prove that l has the Cauchy property we need the Archimedean property: l is 'convergent with respect to $\frac{1}{k}$' by construction, but it needs to be 'convergent with respect to ε'. The proof that l is the limit of $\{s^i\}_{i=0}^{\infty}$ similarly uses the Archimedean property and moreover the fact that the canonical embedding of F into CauchySeq_F is dense.

This finishes the proof of Theorem 4.3 and hence the proof of Theorem 4.1. We already have a proof that $\mathbb{Q}$ is Archimedean and by applying Theorem 4.1 to this result we have proved the following theorem.

Theorem 4.12. $\text{CauchySeq}_{\mathbb{Q}}$, *the set of Cauchy sequences of rational numbers, together with the relations and operations defined on Cauchy sequences as in 4.4- 4.11, is a constructive real number structure.*

5   Homomorphisms and Isomorphisms

In the end of Section 3 we mentioned that every constructive ordered field F contains a 'copy' of $\mathbb{Q}$. In this and the following Section, we will extend this to real number structures: every real number structure contains a 'copy' of $\text{CauchySeq}_{\mathbb{Q}}$. The result is even stronger: every real number structure is essentially $\text{CauchySeq}_{\mathbb{Q}}$. More precisely, we prove that every two real number structures are isomorphic (and hence isomorphic to $\text{CauchySeq}_{\mathbb{Q}}$). To do this we define the notion of a morphism between two real number structures.

Definition 5.1. *Let R_1 and R_2 be two real number structures. We say that $\varphi : R_1 \longrightarrow R_2$ is a* homomorphism *from R_1 to R_2, if it has the following properties.*

1. *(strongly extensional)* $\forall x, y : R_1.\ \varphi(x) \# \varphi(y) \rightarrow x \# y$.
2. *(order preserving)* $\forall x, y : R_1.\ x < y \rightarrow \varphi(x) < \varphi(y)$.
3. *(addition preserving)* $\forall x, y : R_1.\ \varphi(x + y) = \varphi(x) + \varphi(y)$.
4. *(multiplication preserving)* $\forall x, y : R_1.\ \varphi(x * y) = \varphi(x) * \varphi(y)$.

The equality used in this definition is the setoid equality. (In the definition, the relations and operations are taken from the appropriate real number structures, R_1, resp. R_2, which is left implicit.) Note that a homomorphism between real number structures is just a homomorphism between the underlying ordered rings. The preservation of reciprocals ($\varphi(\frac{1_{R_1}}{x}) = \frac{1_{R_2}}{\varphi(x)}$) and limits ($\varphi(\lim\{g_i\}_{i=0}^{\infty}) = \lim\{\varphi(g_i)\}_{i=0}^{\infty}$) comes as a consequence. To state the latter we first have to show that φ maps Cauchy sequences over R_1 to Cauchy sequences over R_2. In the following, let φ be a homomorphism from R_1 to R_2.

Proposition 5.2. *1. (order reflecting)* $\forall x, y : R_1.\ \varphi(x) < \varphi(y) \rightarrow x < y$.
2. *(apartness preserving)* $\forall x, y : R_1.\ x \# y \rightarrow \varphi(x) \# \varphi(y)$.
3. *(zero preserving)* $\forall x : R_1.\ \varphi(0_{R_1}) = 0_{R_2}$.
4. *(minus preserving)* $\forall x : R_1.\ \varphi(-x) = -\varphi(x)$.
5. *(unit preserving)* $\forall x : R_1.\ \varphi(1_{R_1}) = 1_{R_2}$.
6. *(reciprocal preserving)* $\forall x : R_1.x \# 0_{R_1} \rightarrow (\varphi(x) \# 0_{R_2} \wedge \varphi(\frac{1_{R_1}}{x}) = \frac{1_{R_2}}{\varphi(x)})$.
7. *(Cauchy preserving)* *If $\{g_i\}_{i=0}^{\infty}$ is Cauchy, then $\{\varphi(g_i)\}_{i=0}^{\infty}$ is Cauchy.*

Proof (of 7). This follows from the following property of φ:

$$\forall \varepsilon : R_2 . \varepsilon > 0 \rightarrow \exists \delta : R_1 . \delta > 0 \wedge \varphi(\delta) < \varepsilon \tag{1}$$

(which again is a direct consequence of the Archimedean property for R_2). To find, for a given $\varepsilon > 0$, the $N \in \mathbb{N}$ such that $\forall m \geq N |\varphi(g_m) - \varphi(g_N)| < \varepsilon$, we take the $\delta : R_1$ that results from (1) and then we take $N \in \mathbb{N}$ (using the Cauchy property of $\{g_i\}_{i=0}^{\infty}$) to be such that $\forall m \geq N |g_m - g_N| < \delta$. Then $\forall m \geq N |\varphi(g_m) - \varphi(g_N)| < \varphi(\delta) < \varepsilon$. (N.B. With $\frac{1}{k}$-Cauchy sequences, we wouldn't need the Archimedean property in this proof, because the '$\frac{1}{k}$-analogue' of (1) is immediate) □

Lemma 5.3. *The homomorphism* $\varphi : R_1 \longrightarrow R_2$ *preserves limits, that is* $\forall g : \text{CauchySeq}_{R_1} . \varphi(\lim\{g_i\}_{i=0}^{\infty}) = \lim\{\varphi(g_i)\}_{i=0}^{\infty}$.

An isomorphism is defined as a pair of homomorphisms $\langle \varphi, \psi \rangle$ that are inverses to each other, i.e. $\forall x : R_2 . \varphi(\psi(x)) =_{R_2} x$ and $\forall x : R_1 . \psi(\varphi(x)) =_{R_1} x$. Note that to define a composition of homomorphisms, we do not just compose the maps but we also 'compose' the two homomorphism-proofs into a proof that the composed map is a homomorphism.

We now construct an isomorphism between any two real number structures. (An alternative is to show that any real number structure is isomorphic to CauchySeq$_\mathbb{Q}$, which amounts to the same technical work.) The basic idea is to define a map which is the 'identity' on rational numbers and then to extend this map in the canonical way to Cauchy sequences We first define for any real number structure R, the canonical injection $\mathbb{Q}2R$ of $\mathbb{Q}$ into R. (This injection is defined by first defining the injection nring of natural numbers into R and the injection zring of integers into R. Then $\mathbb{Q}2R$ is the map that takes the term $\langle p, q \rangle : \mathbb{Q}$ to $\frac{\text{zring } p}{\text{nring}(q+1)}$.) The map $\mathbb{Q}2R$ is strongly extensional and preserves addition, multiplication, negation, order, AbsSmall and the Cauchy property. Moreover, the image of $\mathbb{Q}$ under $\mathbb{Q}2R$ is dense in R.

If no confusion arises, we omit the injection nring and $\mathbb{Q}2R$ that map $\mathbb{N}$, respectively $\mathbb{Q}$ into R.

Theorem 5.4. *For any* $x : R$, *we can construct a sequence* $q_0, q_1, \ldots$ *of elements of* $\mathbb{Q}$ *such that* $x = \lim\{q_i\}_{i=0}^{\infty}$. *In other words, every real number is the limit of a Cauchy sequence of (images of) rational numbers.*

Proof. We construct (using Axiom of Choice), two Cauchy sequences of rationals $q_0, q_1, \ldots$ and $r_0, r_1, \ldots$ which both converge to x. Using the Archimedean property of R for x and $-x$, we obtain two natural numbers N_1 and N_2 such that $-N_2 < x < N_1$. Define q_0 as N_2 and r_0 as N_1 (both seen as elements of $\mathbb{Q}$). So $q_0 < x < r_0$. Now assume that we have constructed q_n and r_n. To get q_{n+1} and r_{n+1} we use an interval trisection argument (which is the constructive counterpart of interval bisection in classical analysis). Since $\frac{2q_n+r_n}{3} < \frac{q_n+2r_n}{3}$, we have (using cotransitivity of $<$) *either* $\mathbb{Q}2R(\frac{2q_n+r_n}{3}) < x$ *or* $x < \mathbb{Q}2R(\frac{q_n+2r_n}{3})$. In the first case we set $q_{n+1} := \frac{2q_n+r_n}{3}$, $r_{n+1} = r_n$ and in the second case we set $q_{n+1} := q_n$, $r_{n+1} := \frac{q_n+2r_n}{3}$. We prove following facts about q_i and r_i:

1. $\forall n{:}\mathbb{N}(r_n - q_n = (r_0 - q_0)(\tfrac{2}{3})^n)$,
2. $\forall m, n{:}\mathbb{N}(m \leq n \rightarrow q_m \leq q_n \wedge r_n \leq r_m)$,

Hence $\{q_i\}_{i=0}^{\infty}$ and $\{r_i\}_{i=0}^{\infty}$ are Cauchy sequences in $\mathbb{Q}$ and the injections of these sequences in R are Cauchy sequences in R. We also prove the following facts.

$$\forall n{:}\mathbb{N}(q_n < x < r_n),$$

$$\forall n{:}\mathbb{N}(\text{AbsSmall}((r_0 - q_0)(\tfrac{2}{3})^n, x - q_n)).$$

And it follows (using the Archimedean property) that x is the limit of $\{q_i\}_{i=0}^{\infty}$

$\square$

Definition 5.5. *For $x : R$, we define $G(x)$ as the following Cauchy sequence over $\mathbb{Q}$.*

$$G(x) := \lambda n.q_n,$$

where $\{q_i\}_{i=0}^{\infty}$ is the sequence constructed from x in the proof of Theorem 5.4.

We are now ready to construct an isomorphism $\langle \varphi, \psi \rangle$ between two real number structures R_1 and R_2. For clarity, we present a sequence $\{x_i\}_{i=0}^{\infty}$ as $\lambda n{:}\mathbb{N}.x_n$. We define $\varphi : R_1 \rightarrow R_2$ by composing the maps G, $\mathbb{Q}2R_2$ and lim as indicated in the following diagram. So, $\varphi(x)$ is the map $\lim(\lambda n{:}\mathbb{N}.\mathbb{Q}2R_2(G(x)_n))$ (for $x : R_1$). Similarly we define $\psi : R_2 \rightarrow R_1$.

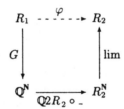

We have the following properties of lim and G.

- lim and G preserve $<$, $+$ and $*$.
- lim and G reflect $<$.

Using these properties we prove the following for φ and ψ.

Proposition 5.6.   *1. φ and ψ are inverses to each other.*

2. φ and ψ are strongly extensional and they preserve order, addition and multiplication.

Then, by Lemma 5.3, φ and ψ are homomorphisms, and since they are inverses to each other, they form an isomorphism from R_1 to R_2.

Theorem 5.7. *All real number structures are isomorphic.*

6   Cauchy Completion versus Order Completion

Classically, Dedekind cuts are an alternative way of constructing the reals out of the rationals. In this approach, the emphasis lies on the order completion of the rationals. However, the classical least upper bound principle, does not hold constructively: for a subset of $\mathbb{Q}$ to have a least upper bound, it is not enough to be bounded. But the principle can be modified, using a constructive version of boundedness. This approach is taken in the axiomatisation of the reals in [2]: the axioms for the reals are the same as ours except for the Cauchy completeness, which is replaced by a constructive version of the least upper bound principle. We will show that the two axiomatisations are equivalent. To introduce the axiom of [2], we need some notations and definitions. First note that in type theory, a subset is formalised via a predicate over a 'carrier type'. As we will not be using type theoretic notation, we just write $x \in S$ if we want to denote that x is in the subset S, leaving the carrier type and the precise encoding of subsets in type theory implicit.

Definition 6.1. *Let F be an Archimedean constructive ordered field. Let $S \subseteq F$, nonempty. An element $b : F$ is the* least upper bound *of S, if the following hold.*

– *b is an* upper bound *of S, that is, $\forall s \in S.s \leq b$*
– *for each $b' < b$ there exists $s : S$ such that $s > b'$,*

where $\leq$ is defined by $x \leq y := \forall z(x > z \rightarrow y > z)$ (which is equivalent to $\neg(y < x)$).

We will write $x \geq S$ if x is an *upper-bound* of the set S. It is easily shown that a least upper bound is unique, if it exists. The classical axiom says that all nonempty bounded subsets have a least upper-bound; constructively speaking, only the *weakly located subsets* do. The following definition is implicit in [2].

Definition 6.2. *A subset S of F is* weakly located *if (1) it is inhabited, (2) it has an upper bound and (3) for all $x, y : F$ with $x < y$, either $y \geq S$ or $\exists s \in S(s > x)$.*

The following axiom replaces the Cauchy completeness axiom. We show that the axiom holds for all real number structures as defined in Section 2.

Least Upper Bound Principle. Every weakly located subset has a least upper bound.

The argument in the proof of the next theorem is used in [3] to justify the least upper bound principle.

Theorem 6.3. *Let F be a real number structure, then F satisfies the least upper bound principle.*

Proof. Assume S is a weakly located subset of F. Let $s \in S$ and $b \geq S$. We construct a least upper bound for S by repeating the interval trisection argument that we also used to prove Theorem 5.4. So, we define Cauchy sequences $\{l_i\}_{i=0}^{\infty}$ and $\{r_i\}_{i=0}^{\infty}$ such that $\forall i(l_i < r_i)$, $\{l_i\}_{i=0}^{\infty}$ and $\{r_i\}_{i=0}^{\infty}$ have the same limit and $\forall i(r_i \geq S \land \exists s' \in S(s > l_i))$.

We start by taking $l_0 := s - 1$ and $r_0 := b + 1$. Now, given that we have l_n, r_n satisfying the requirements, we consider $x = \frac{2l_n + r_n}{3}$ and $y = \frac{l_n + 2r_n}{3}$. Then $x < y$ so by the weakly-locatedness of S we can distinguish cases:

- if $y \geq S$, take $l_{n+1} := l_n$ and $r_{n+1} := y$.
- if there is an $s' \in S$ such that $s' > x$, take $l_{n+1} := x$ and $r_{n+1} := r_n$.

It is easy to show that $\{l_i\}_{i=0}^{\infty}$ and $\{r_i\}_{i=0}^{\infty}$ are Cauchy sequences that satisfy the requirements above. Their limit is the least upper-bound of S. $\square$

The proof above is formalised in Coq. (It is very similar to the formalised proof of Theorem 5.4.) As one might expect, the converse of this theorem also holds: if a constructive ordered field satisfies the least upper bound principle, then it is Cauchy complete. A proof of this fact is given in [2]. So the two axiomatisations are equivalent. We have also formalised this proof inside Coq. In other words we have formally proved the following statement.

Theorem 6.4. *Let F be an Archimedean constructive ordered field that satisfies the least upper bound principle. Then every Cauchy sequence in F has a limit.*

Taking the least upper bound principle as an axiom originates from the construction of reals as subsets of $\mathbb{Q}$. This construction is given (constructively) in [13]. There, a Dedekind cut is defined as a nonempty bounded *located* (a modification of 6.2 phrased for sets of rational numbers) subset of $\mathbb{Q}$. This yields the canonical order completion of rational numbers. We want to emphasise that we *have not* constructed a model of the Dedekind real numbers as special subsets of $\mathbb{Q}$ in Coq. Such a formalisation requires a step by step construction of the operations and relations on located subsets of $\mathbb{Q}$, similar to what we have done for Cauchy sequences over $\mathbb{Q}$. Having proved all the required properties, Theorem 6.4 then yields a concrete term of type CReals out of the Dedekind real numbers.

7   Conclusion

The above results have all been formalised in Coq; the source files are available as part of the FTA project in [10]. Of course, the source files contain a lot of collateral lemmata which we haven't mentioned here. Most of them are not interesting from a mathematical point of view, and deal with microscopic details needed for the proofs. For the results in Section 3 we have mainly used the standard libraries Arith and ZArith of Coq. When doing arithmetic on integers we have used the tactic Ring of Coq. For the remaining results, the tactics Algebra and Rational of the FTA project have been thoroughly used. The whole formalisation uses the syntactic definitions and symbols introduced by the algebraic hierarchy of the FTA project [9]

As for the implementation, we have mentioned the main points in this paper, and this may be useful for possible implementations of constructive real numbers in other proof assistants. As we have already pointed out in the introduction, there are more constructions of the reals in type theory ([4, 12, 5]), implemented in various proof assistants with different aims. The added value of the present work is that we work from a set of axioms. We use our construction to show that the axioms can be satisfied (there is a model for them) and that the axiomatisation is categorical (any two models are isomorphic). This shows that Coq can serve very well as a logical framework in which both axiomatic and model-theoretic reasoning can be formalised.

Acknowledgements

We want to thank Venanzio Capretta, Freek Wiedijk and Jan Zwanenburg for the many fruitful discussions and their useful Coq suggestions. We thank Bas Spitters for the useful discussions on constructive analysis. We thank the referees for their insightful comments and useful suggestions.

References

[1] E. Bishop and D. Bridges. *Constructive Analysis*. Number 279 in Grundlehren der mathematischen Wissenschaften. Springer, Berlin, 1985.

[2] D. Bridges. Constructive mathematics: a foundation for computable analysis. *Theoretical Computer Science*, 219:95–109, 1999.

[3] D. Bridges and S. Reeves. Constructive mathematics in theory and programming practice. *Philosophia Mathematica*, 7, 1999.

[4] J. Chirimar and D. Howe. Implementing constructive real analysis. In J.P. Myers and M.J. O'Donnel, editors, *Constructivity in Computer Science*, number 613 in LNCS, pages 165–178, 1992.

[5] A. Ciaffaglione and P. Di Gianantonio. A coinductive approach to real numbers. In Th. Coquand, P. Dybjer, B. Nordström, and J. Smith, editors, *Types 1999 Workshop, Lökeberg, Sweden*, number 1956 in LNCS, pages 114–130, 2000.

[6] A. Ciaffaglione and P. Di Gianantonio. A tour with constructive real numbers. In *Types 2000 Workshop, Durham, UK*, 2001. This Volume.

[7] D. Delahaye and M. Mayero. Field: une procédure de décision pour les nombres réels en Coq. In *Proceedings of JFLA 2001*. INRIA, 2001.

[8] B. Barras et al. *The Coq Proof Assistant Reference Manual, Version 7.1*. INRIA, http://coq.inria.fr/doc/main.html, sep 2001.

[9] H. Geuvers, R. Pollack, F. Wiedijk, and J. Zwanenburg. The algebraic hierarchy of the FTA project. In *Calculemus 2001 Proc.*, pages 13–27, Siena, Italy, 2001.

[10] H. Geuvers, F. Wiedijk, J. Zwanenburg, R. Pollack, M. Niqui, and H. Barendregt. FTA project. http://www.cs.kun.nl/gi/projects/fta/, nov 2000.

[11] J. Harrison. *Theorem Proving with the Real Numbers*. Distinguished dissertations. Springer, London, 1998.

[12] C. Jones. Completing the rationals and metric spaces in LEGO. In G. Huet and G. Plotkin, editors, *Logical Environments*, pages 297–316. CUP, 1993.

[13] A. Troelstra and D. van Dalen. *Constructivism in Mathematics, vol I*, volume 121 of *Studies in Logic and The Foundation of Math*. North Holland, 1988. 342 pp.

A Constructive Proof of the Fundamental Theorem of Algebra without Using the Rationals

Herman Geuvers, Freek Wiedijk, and Jan Zwanenburg

Department of Computer Science, University of Nijmegen, The Netherlands
{herman,freek,janz}@cs.kun.nl

Abstract. In the FTA project in Nijmegen we have formalised a constructive proof of the Fundamental Theorem of Algebra. In the formalisation, we have first defined the (constructive) algebraic hierarchy of groups, rings, fields, etcetera. For the reals we have then defined the notion of *real number structure*, which is basically a Cauchy complete Archimedean ordered field. This boils down to axiomatising the constructive reals. The proof of FTA is then given from these axioms (so independent of a specific construction of the reals), where the complex numbers are defined as pairs of real numbers.
The proof of FTA that we have chosen to formalise is the one in the seminal book by Troelstra and van Dalen [17], originally due to Manfred Kneser [12]. The proof by Troelstra and van Dalen makes heavy use of the rational numbers (as suitable approximations of reals), which is quite common in constructive analysis, because equality on the rationals is decidable and equality on the reals isn't. In our case, this is not so convenient, because the axiomatisation of the reals doesn't 'contain' the rationals. Moreover, we found it rather unnatural to let a proof about the reals be mainly dealing with rationals. Therefore, our version of the FTA proof doesn't refer to the rational numbers. The proof described here is a faithful presentation of a fully formalised proof in the Coq system.

1 Introduction

The Fundamental Theorem of Algebra states that the field of complex numbers is algebraically closed. More explicitly, it says that

For every non-constant polynomial

$$f(z) = a_n z^n + a_{n-1} z^{n-1} + \ldots + a_1 z + a_0$$

with coefficients in $\mathbb{C}$, the equation $f(z) = 0$ has a solution.

This theorem has a long and illustrious history (see [6] or [11] for an overview). It was proved for the first time in Gauss's Ph.D. thesis from 1799. Many proofs of the Fundamental Theorem of Algebra are known, most of which have a constructive version.

The proof that we're presenting here was invented by Manfred Kneser [12] (inspired by a proof of his father, Hellmuth Kneser, in [11]), and is a constructive

P. Callaghan et al. (Eds.): TYPES 2000, LNCS 2277, pp. 96–111, 2002.
© Springer-Verlag Berlin Heidelberg 2002

version of the simple proof that derives a contradiction from the assumption that
the (non-constant) polynomial f is minimal at z_0 with $|f(z_0)| \neq 0$. We briefly
repeat the classical proof here. Let $f(z) = a_n z^n + a_{n-1} z^{n-1} + \ldots + a_1 z + a_0$ be
a non-constant polynomial.

First note that $|f(z)|$ must have a minimum somewhere, because $|f(z)| \to \infty$
if $|z| \to \infty$. We may assume the minimum to be reached for $z = 0$. (If the
minimum is reached for z_0, consider the polynomial $g(z) = f(z + z_0)$.) Now,
assume the minimum of $|f(z)|$ is not 0 (i.e. $f(0) \neq 0$). The function $f(z)$ has the
form

$$f(z) = a_0 + a_k z^k + O\left(z^{k+1}\right)$$

with $a_k \neq 0$. Because of this, $f(0) = a_0 \neq 0$ and we can take

$$z = \epsilon \sqrt[k]{-\frac{a_0}{a_k}}$$

with $\epsilon \in \mathbb{R}_{>0}$, and if ϵ is small enough, the part $O\left(z^{k+1}\right)$ will be negligible
compared to the rest, and we get a $z \neq 0$ for which

$$|f(z)| = a_0 + a_k (\epsilon \sqrt[k]{-\frac{a_0}{a_k}})^k$$

$$= a_0(1 - \epsilon^k)$$

$$< |f(0)|$$

So $|f(0)|$ is not the minimum and we have derived a contradiction.

By iterating this idea, one can try to construct a Cauchy sequence to a zero
of the polynomial. The main difficulty with this approach is that we have two
conflicting requirements for the choice of ϵ:

- if ϵ is chosen too small each time, we may not reach the zero in countably
  many steps (we will go down, but might not go down all the way to zero).
- if ϵ is not small enough, we are not allowed to ignore the $O\left(z^{k+1}\right)$ part.

The solution to this is that, instead of using the above representation (in which
the term $a_k z^k$ is the *smallest* power with a non-zero coefficient), in the constructive proof one just takes *some* appropriate k (not necessarily the smallest) and
writes $f(z)$ as

$$f(z) = a_0 + a_k(z - z_0)^k + \textit{the other terms}$$

That way one can make sure that not only $|f(z)| < |f(0)|$, but in fact $|f(z)| <
q|f(0)|$ for some fixed $q < 1$.

The FTA proof along these lines presented by Manfred Kneser in [12] is classical ('to improve readability'), but it is stated that it can be made constructive
without any serious problems. In [17], a constructive version of this proof is
given, using rational approximations to overcome the undecidability of equality
on the reals. Another constructive version of the Kneser proof is presented by
Schwichtenberg in [15], also using rational approximations, but along different
lines. The constructive version of FTA reads as follows.

For every polynomial

$$f(z) = a_n z^n + a_{n-1} z^{n-1} + \ldots + a_1 z + a_0$$

with coefficients in $\mathbb{C}$, such that $a_k \mathbin{\#} 0$ for some $k > 0$, the equation $f(z) = 0$ has a solution.

As the equality on $\mathbb{R}$ (and therefore on $\mathbb{C}$) is not decidable (we don't have $\forall x, y \in \mathbb{R}(x = y \lor x \neq y)$) we can't just write $f(z)$ as $a_n z^n + a_{n-1} z^{n-1} + \ldots + a_1 z + a_0$ with $a_n \neq 0$. Therefore, in constructive analysis, one works with the notion of apartness, usually denoted by $\#$, which is a 'positive inequality': $a \mathbin{\#} b$ if we positively know them to be distinct, i.e. we know a distance ϵ between them. Now, one can constructively find a root of f if we positively know some coefficient a_k ($k > 0$) to be distinct from 0. The proof of constructive FTA proceeds by first proving it for *monic* polynomials (i.e. where $a_n = 1$).

The original Kneser proof of FTA for monic polynomials makes use of an approximation of the polynomial with coefficients in $\mathbb{Q}$, because it needs to compare the size of various expressions (which is not decidable in $\mathbb{R}$). We found this unsatisfactory: the rational numbers don't seem to have anything to do with the Fundamental Theorem of Algebra! Also, in our Coq formalisation of Kneser's proof, we introduced the real numbers axiomatically (so *a priori* we didn't have $\mathbb{Q}$ in our formalisation), and it seemed silly to reconstruct the rational numbers inside our real numbers just to be able to formalise this proof. Therefore, instead of constructing $\mathbb{Q}$, we modified the proof so that it no longer referred to the rationals. The result is presented here.

The main idea behind the modification of the proof is that we introduce 'fuzziness' in the comparisons. The proof will contains a 'fuzziness parameter' ϵ Instead of having to decide whether

$$x < y \lor x = y \lor x > y,$$

all we need to establish is whether

$$x < y + \epsilon \lor x > y - \epsilon$$

(which we might write as

$$x \leq_\epsilon y \lor x \geq_\epsilon y$$

using a relation $\leq_\epsilon$). Constructively we have *cotransitivity* of the order relation

$$x < y \Rightarrow x < z \lor z < y$$

from which it follows that the disjunction with the ϵ's is decidable.

Apart from not needing $\mathbb{Q}$, another difference between the proof presented here and the proof in [17] is that we have avoided using Vandermonde determinants. In the original proof, this is used to prove FTA from FTA for monic polynomials. We prove this implication directly, using some polynomial arithmetic. Therefore there's no use of linear algebra in the proof anymore.

We have formalised the proof presented here using the Coq system: this was known as the FTA project [7]. In the formalisation, we treat the real numbers axiomatically. More precisely, the reals form a part of a *constructive algebraic hierarchy*, which consists (among other things) of the abstract notions of rings, fields and ordered fields. See [8] for details. The base level of this hierarchy consists of the notion of *constructive setoid*, which is basically a pair of a type and an *apartness* relation over the type. (For constructive reals, 'being apart' is more basic then 'being equal', so we start from apartness.) In this hierarchy, a *real number structure* is defined as a Cauchy complete Archimedean ordered field. In the FTA project, the Fundamental Theorem of Algebra was proven for any real number structure, so as a matter of fact the theorem was proven from the axioms for the constructive reals. Also it was shown that real number structures exist by actually constructing one. Details on this construction can be found in [9], where also other axiomatisations are discussed and it is shown that any two real number structures are isomorphic.

The whole formalisation turned out to be 930K of Coq source code, which includes the construction of the real numbers by Milad Niqui, see [9]. The parts that directly correspond to the mathematics in this paper is about 65K of Coq source. The final lemma that was proved in the formalisation was, in Coq syntax:

```
(f:(cpoly_cring CC))(nonConst ? f) -> (EX z | f!z [=] Zero)
```

The plan of the paper is as follows: for an overview we first present the root-finding algorithm that's implicit in Kneser's proof (for simplicity we give the classical version of that algorithm). After that we give the full constructive Kneser proof, which contains a correctness proof of the algorithm.

2   The Kneser Algorithm, Classically

Let
$$f(z) = z^n + a_{n-1}z^{n-1} + \ldots + a_1 z + a_0$$
be a monic polynomial over the complex numbers of degree $n > 0$. Let be given an arbitrary complex number z_0. We are going to describe an algorithm that computes a Cauchy sequence
$$z_0, z_1, z_2, \ldots$$
that converges to a zero of this polynomial.

Suppose that z_i has already been established. ¿From this we have to determine the next term in the sequence, z_{i+1}. There are two possibilities:

- In the case that $f(z_i) = 0$ we already are at a zero, and so we will take $z_{i+1} = z_i$.
- In the case that $f(z_i) \neq 0$ we consider the polynomial f_{z_i}, defined by[1] $f_{z_i}(z) \equiv f(z + z_i)$, find an appropriate offset δ_i and then take $z_{i+1} = z_i + \delta_i$.

---

[1] The shift from f to f_{z_i} corresponds to the step in the classical FTA proof (see Section 1) where the polynomial is shifted so that the alleged minimum is reached in 0

So in the second case we have the polynomial

$$f_{z_i}(z) = b_n z^n + b_{n-1} z^{n-1} + \ldots + b_1 z + b_0$$

(the coefficients b_k really depend on z_i, but we won't write this dependency to keep the formulas simple), with $b_n = 1$ and $b_0 = f_{z_i}(0) = f(z_i) \neq 0$, and we have to determine δ_i.

First, we will determine $|\delta_i|$. Define

$$r_0 = \min_{k \in \{1, \ldots, n\}, b_k \neq 0} \sqrt[k]{|b_0|/|b_k|}$$

and from this define a sequence of radii

$$r_0 > r_1 > r_2 > \ldots$$

by

$$r_j = 3^{-j} r_0$$

(so every radius is $\frac{1}{3}$ of the previous one).

Now for each j let k_j be the element of $\{1, \ldots, n\}$ such that

$$\left| b_{k_j} \right| r_j^{k_j}$$

is maximal (if there are more elements of $\{1, \ldots, n\}$ for which the maximum is attained, then take the least one). This will give a decreasing sequence[2]

$$k_0 \geq k_1 \geq k_2 \geq \ldots$$

Take the least $j > 0$ for which

$$k_{j-1} = k_j = k_{j+1}$$

and let $r = r_j$ and $k = k_j$. We will define δ_i such that $|\delta_i| = r$, and such that $b_k \delta_i^k$ points opposite to b_0 in the complex plane. This means that we take

$$\delta_i = r \sqrt[k]{-\frac{b_0/b_k}{|b_0/b_k|}}$$

and $z_{i+1} = z_i + \delta_i$. This concludes the description of the classical version of the Kneser algorithm.

Note that this last step introduces ambiguity, because there are k different complex roots. So the sequence

$$z_0, z_1, z_2, \ldots$$

---

[2] That this sequence is decreasing is seen by the following argument: if $\left| b_{k_j} \right| r_j^{k_j}$ is the maximum among $\{|b_1| r_j, \ldots, |b_n| r_j^n\}$, then $\left| b_{k_j} \right| r_{j+1}^{k_j} > |b_i| r_{j+1}^i$ for all $i > k_j$, because $|b_i| r_{j+1}^i = \frac{1}{3^i} |b_i| r_j^i \leq \frac{1}{3^i} \left| b_{k_j} \right| r_j^{k_j} < \frac{1}{3^{k_j}} \left| b_{k_j} \right| r_j^{k_j} = \left| b_{k_j} \right| r_{j+1}^{k_j}$.

really is a path in an infinite tree which this algorithm computes. Of course, following different paths in this tree one might find different zeroes.

The correctness of the algorithm is a consequence of the following properties of the choice for δ_i (and r). (These properties and the correctness will be proved in detail in the next Section.)

$$|f_{z_i}(\delta_i)| < q\,|f_{z_i}(0)| \text{ for some fixed } q < 1,$$
$$r^n < |f_{z_i}(0)|.$$

The first inequality says that $|f(z_{i+1})| < q\,|f(z_i)|$, so the f-values of the sequence $z_0, z_1, z_2, \ldots$ converge to 0. The second inequality says that $|z_{i+1} - z_i|^n = |\delta_i|^n = r^n < |f(z_i)|$, so the sequence $z_0, z_1, z_2, \ldots$ converges.

3   The Kneser Proof, Constructively

We will now present our variation on Kneser's proof of the Fundamental Theorem of Algebra. This variant of the proof doesn't make use of $\mathbb{Q}$, unlike the proof from [17] that it was based on. In the proof we have isolated the parts that are about the reals and the parts that really need the complex numbers.

The only essential property of the complex numbers that is used is the existence of k-th roots, which can be proved independently of FTA. The most well-known proof of this fact proceeds by first moving to a polar coordinate representation of $\mathbb{C}$. As we have chosen $\mathbb{R}^2$ as a representation of $\mathbb{C}$, this is not an easy proof to formalise. (One would first have to define the arctan function and establish an isomorphism between the two representations.) Therefore we have chosen a different proof, which appears e.g. in [5], and [13] and is basically constructive. Here, the existence of k-th roots in $\mathbb{C}$ is derived directly from the existence of square roots in $\mathbb{C}$ and the fact that all polynomials over $\mathbb{R}$ of odd degree have a root. The proof of these properties have all been completely formalised in Coq. Note here that the intermediate value theorem (which implies directly that all polynomials over $\mathbb{R}$ of odd degree have a root) is not valid constructively. However, the intermediate value theorem can be proved for polynomials. (In our formalisation we have followed the proof of [17], using Lemma 6 for a substantial shortcut.)

The proof of FTA goes through three lemmas, which in the Coq formalisation have been called 'the Key Lemma', 'the Main Lemma' and 'the Kneser Lemma'. The presentation that we give here directly corresponds to the way it was formalised in Coq.

We first state an auxiliary lemma, that says that constructively it's possible to find the maximum of a sequence of numbers 'up to ϵ':

Lemma 1. *For $n > 0$, $\epsilon > 0$ and $c_1, \ldots, c_n \in \mathbb{R}$, there is a k such that for all $i \in \{1, \ldots, n\}$:*

$$c_k > c_i - \epsilon$$

The proof is a straightforward induction using the cotransitivity of the $<$ relation: to determine the 'maximum up to ϵ' of $c_1, \ldots, c_{n+1}$, first determine (induction hypothesis) the 'maximum up to $\epsilon/2$' of $c_1, \ldots, c_n$, say c_k and then choose c_k if $c_{n+1} < c_k + \epsilon$ and c_{n+1} if $c_{n+1} > c_k - \epsilon/2$. The latter choice can be made because of cotransitivity of $<$.

We now state the Key Lemma:

Lemma 2 (Key Lemma). *For every $n > 0$, $\epsilon > 0$ and $a_0 > \epsilon$, $a_1 \ldots, a_{n-1} \geq 0$, $a_n = 1$, there exist $r_0 > 0$ and $k_j \in \{1, \ldots, n\}$ with $k_0 \geq k_1 \geq k_2 \geq \ldots$ such that*

$$a_{k_0} r_0{}^{k_0} = a_0 - \epsilon$$

and for all $j \in \mathbb{N}$, if we define $r_j = 3^{-j} r_0$, for all $i \in \{1, \ldots, n\}$ it holds that

$$a_{k_j} r_j{}^{k_j} > a_i r_j{}^i - \epsilon$$

This lemma corresponds directly to the part of the algorithm from the previous section that establishes r_0 and the sequence $k_0 \geq k_1 \geq k_2 \geq \ldots$ (what is called $|b_i|$ there, is called a_i here, because that way the Key Lemma doesn't need to refer to complex numbers). The choice for r_0 in the classical situation as $r_0 = \min_{k \in \{1, \ldots, n\}, b_k \neq 0} \sqrt[k]{|b_0|/|b_k|}$ is is here represented by choosing r_0 such that $|b_0| = \max_{k \in \{1, \ldots, n\}} |b_k| r_0^k$.

The real difference with the classical situation is that 'taking the maximum' during the selection of the k_j is just 'up to ϵ': a term $a_i r_j{}^i$ different from $a_{k_j} r_j{}^{k_j}$ may actually be the biggest, but it may not exceed the selected one by more than ϵ.

We will now prove the Key Lemma:

Proof. We first select k_0 and r_0. This is done by taking initial values for k_0 and r_0 and then considering in turn for i the values $n - 1$ down to 1, preserving the following invariant:

$$a_{k_0} r_0{}^{k_0} = a_0 - \epsilon,$$
$$a_{k_0} r_0{}^{k_0} > a_l r_0{}^l - \epsilon \text{ for all } l \in \{i, \ldots, n\}.$$

Start with the initial values $k_0 = n$ and $r_0 = \sqrt[n]{a_0 - \epsilon}$. Then, at each i (from $n - 1$ down to 1) we update the values of k_0 and r_0 as follows.

- If $a_i r_0^i < a_0$, do nothing. The invariant trivially remains to hold.
- If $a_i r_0^i > a_0 - \epsilon$, set k_0 to i and r_0 to $\sqrt[i]{(a_0 - \epsilon)/a_i}$ (in which case r_0 will decrease). The first part of the invariant trivially remains to hold. For the second part: $a_{k_0} r_0{}^{k_0} = a_0 - \epsilon$, which is larger than each of the $a_l r_0{}^l - \epsilon$ (by the invariant for the previous choice of i and the fact that r_0 has decreased).

After this, k_0 and r_0 have the appropriate values.

To get k_{j+1} from k_j, let $k = k_j$, $r = 3^{-j} r_0$ and apply Lemma 1 with $\epsilon/2$ to the sequence

$$a_1(r/3), a_2(r/3)^2, \ldots, a_k(r/3)^k$$

to get $k' = k_{j+1}$. (So $k_j \geq k_{j+1}$.) Then for $i \leq k$ the inequality $a_{k'}(r/3)^{k'} > a_i(r/3)^i - \epsilon$ follows directly, while for $i > k$ we have:

$$a_k(r/3)^k = 3^{-k}a_k r^k > 3^{-k}\left(a_i r^i - \epsilon\right) = 3^{-k}a_i r^i - 3^{-k}\epsilon > a_i(r/3)^i - \epsilon/2$$

and so:

$$a_{k'}(r/3)^{k'} > a_k(r/3)^k - \epsilon/2 > a_i(r/3)^i - \epsilon$$

□

We will now state and prove the Main Lemma, which isolates the part of the proof that's about the real numbers from the part that involves the complex numbers.

Lemma 3 (Main Lemma). *For every $n > 0$, $\epsilon > 0$, $a_0 > \epsilon$, $a_1 \ldots, a_{n-1} \geq 0$, $a_n = 1$, there exists an $r > 0$ and a $k \in \{1, \ldots, n\}$ that satisfy the inequalities*

$$r^n < a_0$$
$$3^{-2n^2}a_0 - 2\epsilon < a_k r^k < a_0$$

and have the property

$$\sum_{i=1}^{k-1} a_i r^i + \sum_{i=k+1}^{n} a_i r^i < (1 - 3^{-n})a_k r^k + 3^n\epsilon$$

The Main Lemma corresponds to the choice for r and k in the description of the classical algorithm. The first condition states that r cannot be too large in comparison to the previous value: this corresponds to the property $r^n < |f_{z_i}(0)|$, mentioned in the discussion of the classical algorithm. The second condition is to make sure that, if we let $b_k \delta_i^k$ point in the opposite direction of b_0, then $|b_0 + b_k \delta_i^k|$ gets sufficiently smaller. (The Main Lemma is about reals, but it will be applied by taking $a_i = |b_i|$, where the b_i are the coefficients of the polynomial.) Moreover, the second and the third condition together make sure that the sum of the remaining terms $a_i r^i$ is negligible.

We will now prove the Main Lemma.

Proof. Apply the Key Lemma to get sequences $k_0, k_1, k_2 \ldots$ and $r_0, r_1, r_2 \ldots$ Because the sequence k_j is non-increasing in the finite set $\{1, \ldots, n\}$ there exists a (smallest) $j < 2n$ with

$$k_{j-1} = k_j = k_{j+1}$$

Take $k = k_j$ and $r = r_j$.

Because $r = 3^{-j}r_0 \leq r_0$, for all i we have $a_i r^i \leq a_i r_0^i < a_{k_0} r_0^{k_0} + \epsilon = a_0$. Of this statement $r^n < a_0$ and $a_k r^k < a_0$ are special cases. From $a_{k_0} r_0^{k_0} = 3^{-jk_0}a_{k_0}r_0^{k_0} \geq 3^{-jn}a_{k_0}r_0^{k_0} = 3^{-jn}(a_0 - \epsilon) \geq 3^{-jn}a_0 - \epsilon$ it follows that $a_k r^k > a_{k_0}r^{k_0} - \epsilon \geq 3^{-jn}a_0 - 2\epsilon > 3^{-2n^2}a_0 - 2\epsilon$.

¿From $k = k_{j+1}$ we get that for all $i \in \{1, \ldots, n\}$

$$a_k(r/3)^k > a_i(r/3)^i - \epsilon$$

and from that it follows that

$$a_i r^i < 3^{i-k} a_k r^k + 3^i \epsilon$$

and therefore

$$\sum_{i=1}^{k-1} a_i r^i < \left(\sum_{i=1}^{k-1} 3^{i-k}\right) a_k r^k + \left(\sum_{i=1}^{k-1} 3^i\right)\epsilon$$

$$= \frac{1}{2}(1 - 3^{1-k}) a_k r^k + \frac{1}{2}(3^k - 3)\epsilon$$

$$< \frac{1}{2}(1 - 3^{-n}) a_k r^k + \frac{1}{2} 3^n \epsilon$$

In exactly the same way we get from $k = k_{j-1}$ that

$$a_i r^i < 3^{k-i} a_k r^k + 3^{-i} \epsilon$$

and so

$$\sum_{i=k+1}^{n} a_i r^i < \frac{1}{2}(1 - 3^{-n}) a_k r^k + \frac{1}{2} 3^n \epsilon$$

Together this gives

$$\sum_{i=1}^{k-1} a_i r^i + \sum_{i=k+1}^{n} a_i r^i < (1 - 3^{-n}) a_k r^k + 3^n \epsilon$$

□

We now state and prove the 'Kneser Lemma'. This lemma states that we can find what was called δ_i in the previous Section: an appropriate vector that moves us sufficiently closer to a zero. In the classical version of the Kneser proof, one distinguishes cases according to $f(0) = 0$ or $f(0) \neq 0$. In the first case we are done, while in the second case one finds a $z \in \mathbb{C}$ such that

$$|z|^n < |f(0)|$$

and

$$|f(z)| < q|f(0)|$$

(where $q < 1$ is some fixed multiplication factor that only depends on the degree of the polynomial).

However, we don't know $f(0) = 0 \vee f(0) \neq 0$ constructively. Therefore, we here have a $c > 0$ that takes the role of $|f(0)|$. This c can get arbitrary close to $|f(0)|$ from above. Here is the constructive version of the Kneser Lemma:

Lemma 4 (Kneser Lemma). *For every $n > 0$ there is a q with $0 < q < 1$, such that for all monic polynomials $f(z)$ of degree n over the complex numbers, and for all $c > 0$ such that*

$$|f(0)| < c$$

there exists a $z \in \mathbb{C}$ such that

$$|z|^n < c$$

and

$$|f(z)| < qc$$

Proof. First of all, we give the factor q explicitly:

$$q = 1 - 3^{-2n^2 - n}$$

We now show how to find z.

Write the polynomial $f(z)$ as

$$f(z) = b_n z^n + b_{n-1} z^{n-1} + \ldots + b_1 z + b_0$$

Because $f(z)$ is monic we have that $b_n = 1$, Also, we have that $b_0 = f(0)$, so the condition about c states that $|b_0| < c$. As $qc > 0$ we can make the following case distinction

$$|f(0)| < qc \quad \vee \quad |f(0)| > 0.$$

In the first case we are done by taking $z := 0$. In the second case we proceed as follows. Define $a_i = |b_i|$ for $i \in \{0, \ldots, n\}$ and choose an $\epsilon > 0$ such that

$$2\epsilon < 3^{-2n^2} a_0 \tag{1}$$

$$(3^n + 1)\epsilon < q(c - a_0) \tag{2}$$

Then $\epsilon < a_0$ and we apply the Main Lemma (Lemma 3) to $a_0, \ldots, a_n$ to obtain $r > 0$ and $k \in \{1, \ldots, n\}$ satisfying

$$r^n < a_0$$
$$3^{-2n^2} a_0 - 2\epsilon < a_k r^k < a_0$$
$$\sum_{i=1}^{k-1} a_i r^i + \sum_{i=k+1}^{n} a_i r^i < (1 - 3^{-n}) a_k r^k + 3^n \epsilon$$

Finally take

$$z = r \sqrt[k]{-\frac{b_0/b_k}{a_0/a_k}}$$

(This makes use of inequality (1) $2\epsilon < 3^{-2n^2} a_0$, because we need to know that $a_k > 0$.) Then because $|b_0| = a_0$ and $|b_k| = a_k$ we have

$$|z| = r$$

From this, we get

$$|z|^n = r^n < a_0 < c$$

For the second property of z we start by computing $|b_0 + b_k z^k|$:

$$|b_0 + b_k z^k| = \left| b_0 + b_k r^k \left(-\frac{b_0/b_k}{a_0/a_k} \right) \right|$$

$$= \left| \frac{b_0}{a_0} (a_0 - a_k r^k) \right|$$

$$= |a_0 - a_k r^k|$$

$$= a_0 - a_k r^k$$

(Using the inequality $a_k r^k < a_0$.)

By the triangle inequality for the complex numbers, we then get

$$\left|\sum_{i=0}^{n} b_i z^i\right| \leq \left|b_0 + b_k z^k\right| + \sum_{i=1}^{k-1} a_i r^i + \sum_{i=k+1}^{n} a_i r^i$$

$$< a_0 - a_k r^k + (1 - 3^{-n})a_k r^k + 3^n \epsilon$$

$$= a_0 - 3^{-n} a_k r^k + 3^n \epsilon$$

$$< a_0 - 3^{-n}(3^{-2n^2} a_0 - 2\epsilon) + 3^n \epsilon$$

$$= (1 - 3^{-2n^2-n})a_0 + 3^n \epsilon + 3^{-n} 2\epsilon$$

$$< qa_0 + 3^n \epsilon + \epsilon$$

$$< qc,$$

where the final inequality follows from (2) $(3^n + 1)\epsilon < q(c - a_0)$.   □

Next we prove the special case of the Fundamental Theorem of Algebra for Monic Polynomials:

Lemma 5 (Fundamental Theorem of Algebra for Monic Polynomials).
For every monic polynomial $f(z)$ of degree $n > 0$ over the complex numbers, there exists $z \in \mathbb{C}$ such that $f(z) = 0$.

Proof. Take any $c > 0$ with $c > |f(0)|$. We construct a sequence $z_i \in \mathbb{C}$ such that for all i

$$|f(z_i)| < q^i c \tag{3}$$

$$|z_{i+1} - z_i| < (q^i c)^{1/n} \tag{4}$$

where $q < 1$ is given by the Kneser Lemma 4. This sequence is constructed by iteratively applying the Kneser Lemma to $f_{z_i}(z) \equiv f(z + z_i)$ to find $z_{i+1} - z_i$. The required properties of z_i then follow directly from the properties in the Kneser Lemma, by induction on i.

Because of 4, the z_i form a Cauchy sequence:

$$|z_{m+i} - z_m| \leq |z_{m+i} - z_{m+i-1}| + \ldots + |z_{m+1} - z_m|$$

$$< (q^{(m+i-1)/n} + q^{(m+i-2)/n} + \ldots + q^{m/n})c^{1/n}$$

$$= \frac{q^{m/n} - q^{(m+i)/n}}{1 - q^{1/n}} c^{1/n}$$

$$= q^{m/n} \frac{1 - q^{i/n}}{1 - q^{1/n}} c^{1/n}$$

$$< q^{m/n} \frac{c^{1/n}}{1 - q^{1/n}}.$$

By choosing m sufficiently large (n is fixed), this last expression can be made arbitrarily small.

Then, because z_i is a Cauchy sequence, the limit $z = \lim_{i \to \infty} z_i$ exists and by continuity of f one has

$$|f(z)| = \lim_{i \to \infty} |f(z_i)| \leq \lim_{i \to \infty} q^i c = 0$$

so $f(z) = 0$.   □

Finally we prove the full Fundamental Theorem of Algebra. A polynomial is called *non-constant* if for some $k > 0$ one of its coefficients a_k is apart from zero. We denote this by $f \# 0$. This a_k doesn't necessarily need to be the head coefficient a_n of the polynomial. In fact the head coefficient a_n might be zero (we can't know this), so proving the full Fundamental Theorem of Algebra is not as easy as just dividing by a_n and then applying Lemma 5.

We need one more important property stating, in a sense, the opposite of the Fundamental Theorem of Algebra: instead of showing that there is an argument for which the polynomial is zero, it shows that there is an argument for which the polynomial is *apart* from zero. This fact comes as an immediate corollary of the following lemma.

Lemma 6. *Given a polynomial f of degree at most n and $n + 1$ distinct points $z_0, z_1, \ldots, z_n \in \mathbb{C}$, $f(z_i) \# 0$ for at least one of the z_i.*

Proof. Write $f(z)$ in the form

$$f(z) = a_n z^n + a_{n-1} z^{n-1} + \ldots + a_1 z + a_0$$

which means that $f(z)$ has at most degree n. Then for any $n + 1$ different $z_0, z_1, \ldots, z_n \in \mathbb{C}$ one can write

$$f(z) = \sum_{i=0}^{n} f(z_i) \frac{(z - z_0) \cdots (z - z_{i-1})(z - z_{i+1}) \cdots (z - z_n)}{(z_i - z_0) \cdots (z_i - z_{i-1})(z_i - z_{i+1}) \cdots (z_i - z_n)}$$

because both sides have at most degree n, and coincide on $n + 1$ points (and hence they are equal). This means that we can write $f(z)$ in the form

$$f(z) = \sum_{i=0}^{n} f(z_i) f_i(z)$$

for some $n + 1$ polynomials f_i. Because this sum is $\# 0$, there is some $i \in \{0, \ldots, n\}$ for which the polynomial $f(z_i) f_i \# 0$ and therefore for this i we have that $f(z_i) \# 0$.   □

Corollary 1. *For every polynomial $f \# 0$ over the complex numbers, there exists $z \in \mathbb{C}$ such that $f(z) \# 0$.*

Theorem 1 (Fundamental Theorem of Algebra). *For every non-constant polynomial $f(z)$ over the complex numbers, there exists $z \in \mathbb{C}$ such that $f(z) = 0$.*

Proof. We write

$$f(z) = a_n z^n + a_{n-1} z^{n-1} + \ldots + a_1 z + a_0$$

Because a_n might be zero, we call n the *length* of f instead of calling it the degree of f. We'll prove the theorem with induction on this length n.

With Corollary 1 find a $z_0 \in \mathbb{C}$ such that $f(z_0) \# 0$. Then if we define $f_{z_0}(z) \equiv f(z + z_0)$, it is sufficient to find a zero of f_{z_0}, because if z is a zero of f_{z_0} then $z + z_0$ will be a zero of f. So all we need to prove is that f_{z_0} has a zero.

We write

$$f_{z_0}(z) = b_n z^n + b_{n-1} z^{n-1} + \ldots + b_1 z + b_0$$

with $b_0 = f_{z_0}(0) = f(z_0) \# 0$. We define the *reverse* $f_{z_0}{}^{rev}(z)$ of this polynomial to be the polynomial

$$b_0 z^n + b_1 z^{n-1} + \ldots + b_{n-1} z + b_n$$

so with the coefficients in the opposite order. This reverse operation has the property that the reversal of a product is the product of the reversals: $(gh)^{rev} = g^{rev} h^{rev}$.

Now $f_{z_0}{}^{rev}(z)/b_0$ is monic, so by Lemma 5 it has a zero c, and so it can be written as $(z - c)g(z)$. Because, as we noted, reversals commute with products, this implies that the original f_{z_0} can be written as

$$f_{z_0}(z) = (c_1 z + c_0)h(z)$$

where $h(z)$ is a lower length polynomial of the form

$$h(z) = d_{n-1} z^{n-1} + \ldots + d_1 z + d_0$$

Because f_{z_0} is non-constant, we have $b_i \# 0$ for some $i > 0$. And because

$$b_i = c_0 d_i + c_1 d_{i-1}$$

we find that either $c_0 d_i \# 0$ or $c_1 d_{i-1} \# 0$.

- In the case that $c_0 d_i \# 0$, we get $d_i \# 0$ and therefore $h(z)$ is non-constant, has a zero by induction, and this zero will also be a zero of f_{z_0}.
- In the case that $c_1 d_{i-1} \# 0$, we get $c_1 \# 0$ and then $-c_0/c_1$ will be a zero of f_{z_0}.

□

4 Convergence Speed of the Kneser Algorithm

The Kneser proof (and the algorithm that is implicit in it) as presented in this paper differs from the Kneser proof from [17] in an important respect. In this paper we define the sequence

$$r_0 > r_1 > r_2 > \ldots$$

(and the matching sequence $k_0 \geq k_1 \geq k_2 \geq \ldots$) to start at zero. In [17] the r_j sequence starts at minus one

$$r_{-1} > r_0 > r_1 > r_2 > \ldots$$

Each r_j is three times as small as the previous one, so in the other variant of the proof the search for an appropriate r starts at a radius that is three times as big as the radius r_0. To distinguish the two proofs we'll call the proof that is in this paper the *slow* variant of the proof and the one where the sequences r_j and k_j start at -1 the *fast* variant of the proof.

In Coq we formalised the slow variant of the proof. It is a simpler proof and we wanted to finish the formalisation as fast as possible. Also in Coq it's easier to formalise a sequence starting at 0 than a sequence starting at -1. (One could shift the sequence by one but that would complicate the formulas in various places.)

The fast variant of the proof has the advantage that the corresponding algorithm behaves like Newton-Raphson when the algorithm gets close to the zero of the polynomial. The algorithm from the slow variant of the proof converges slower, because close to a zero it only takes one third of a Newton-Raphson step. In the slow variant of the proof, close to a zero we get $k_0 = k_1 = k_2 = \ldots = 1$, which means that $j = 1$ and so $r = r_1 = \frac{1}{3}r_0$, where r_0 is the Newton-Raphson distance. Note that close to the zero, the value of the polynomial will then be multiplied with approximately a factor of $2/3$ at each step, which is much better than the 'worst case' factor of $q = 1 - 3^{-2n^2 - n}$ which appears in the proof. As an example of the behaviour of the algorithm from the slow variant of the proof we calculate $\sqrt{2} \approx 1.41421$ by finding a root of $z^2 - 2$, starting from $z_0 = 1$:

$z_0 =$	1	$= 1$
$z_1 =$	$7/6$	≈ 1.16667
$z_2 =$	$317/252$	≈ 1.25794
$z_3 =$	$629453/479304$	≈ 1.31326
$z_4 =$	$2440520044877/1810196044272$	≈ 1.34821
$z_5 =$	$\ldots$	≈ 1.37075
$z_6 =$	$\ldots$	≈ 1.38547

In this sequence the Kneser algorithm takes $\log(1/10)/\log(2/3) \approx 5.7$ steps to gain one decimal of precision.

In the fast variant of the proof we get, close to a zero $k_{-1} = k_0 = k_1 = k_2 = \ldots = 1$, which means that then $j = 0$ and so $r = r_0$. In the case of $\sqrt{2}$ this gives

$z_0 =$	1	$= 1$
$z_1 =$	$3/2$	≈ 1.5
$z_2 =$	$17/12$	$\approx 1.4166666666666666666666667$
$z_3 =$	$577/408$	$\approx 1.4142156862745098039215686$
$z_4 =$	$665857/470832$	$\approx 1.4142135623746899106262956$
$z_5 =$	$886731088897/627013566048$	$\approx 1.4142135623730950488016896$
$z_6 =$	$\ldots$	$\approx 1.4142135623730950488016887$

This is the same sequence that the Newton-Raphson algorithm calculates. This particular sequence consists of continued fraction approximations of $\sqrt{2}$ and the correct number of decimals doubles with every step. Note that the Kneser algorithm of the fast variant of the proof only coincides with Newton-Raphson close to the zero. With Newton-Raphson not all start values lead to a convergent sequence, but with the Kneser algorithm it does.

To change the proof in this paper to the fast variant (where the sequences r_j and k_j start at -1), only the proof of the Key Lemma needs to be modified. Apart from going from k_j to k_{j+1} we will also need to go from k_0 to k_{-1}. To be able to do that, the k_0 and r_0 will need to satisfy a stricter restriction than before. It needs to satisfy

$$a_{k_0} r_0^{k_0} > a_i r_0^i - \epsilon'$$

where

$$\epsilon' = 3^{-n} \epsilon$$

To find such k_0 and r_0 one proceeds like before, but this time distinguishing between

$$a_i r_0^i < a_0 - \epsilon + \epsilon'$$

and

$$a_i r_0^i > a_0 - \epsilon$$

at every iteration. Then to get k_{-1} from k_0 one applies Lemma 1 to the sequence

$$a_{k_0}(3r_0)^{k_0}, \ldots, a_n(3r_0)^n$$

with a reasoning similar to the k_{j+1} from k_j case.

5   Brief Overview of Other Constructive Proofs

The first constructive proof of FTA (for monic polynomials) is from Weyl [18], where the winding number is used to simultaneously find all zeros of a (monic) polynomial. A similar but more abstract proof, also using the winding number, occurs in [1], where FTA is proved for arbitrary non-constant polynomials. Based on Weyl's approach, [10] presents an implementation of an algorithm for the simultaneous determination of the zeros of a polynomial.

In [2], Brouwer and De Loor give a constructive proof of FTA for monic polynomials by first proving it for polynomials with rational complex coefficients (which have the advantage that equality is decidable) and then make the transition (viewing a complex number as the limit of a series of rational complex numbers) to general monic polynomials over $\mathbb{C}$. This proof – and also Weyl's and other FTA proofs – are discussed and compared in [14].

Brouwer [3] was the first to generalise the constructive FTA proof to arbitrary non-constant polynomials (where we just know *some* coefficient to be apart from 0). In [16] it is shown that, for general non-constant polynomials, there is a continuous map from the coefficients to the set of zeros.

Acknowledgements

We thank the referees for their very valuable comments, which led us to improve part of the proof. We thank Henk Barendregt, Randy Pollack for inspiring discussions and their valuable comments. Thanks to Helmut Schwichtenberg for the enlightening and stimulating discussions on the FTA proof of Kneser and for providing us with a copy of [15]. We thank Bas Spitters and Wim Veldman for various discussions on the constructive aspects of analysis.

References

1. E. Bishop and D. Bridges, *Constructive Analysis*, Number 279 in Grundlehren der mathematischen Wissenschaften. Springer, 1985.
2. L.E.J. Brouwer and B. de Loor, Intuitionistischer Beweis des Fundamentalsatzes der Algebra, in *Proceedings of the KNAW*, 27, pp. 186–188, 1924.
3. L.E.J. Brouwer, Intuitionistische Ergänzung des Fundamentalsatzes der Algebra, in *Proceedings of the KNAW*, 27, pp. 631–634, 1924.
4. B. Dejon and P. Henrici, Editors, *Constructive Aspects of the Fundamental Theorem of Algebra*, Proceedings of a symposium at IBM Research Lab, Zürich-Rüschlikon, June 5-7, 1967, Wiley-Interscience, London.
5. H.-D. Ebbinghaus et al. (eds.), *Numbers*, Springer, 1991, 395 pp.
6. B. Fine and G. Rosenberger, *The Fundamental Theorem of Algebra*, Undergraduate Texts in Mathematics, Springer, 1997, xii+208 pp.
7. H. Geuvers, F. Wiedijk, J. Zwanenburg, R. Pollack, M. Niqui, H. Barendregt, FTA project, `http://www.cs.kun.nl/gi/projects/fta/`.
8. H. Geuvers, R. Pollack, F. Wiedijk, and J. Zwanenburg. The algebraic hierarchy of the FTA project, in *Calculemus 2001 workshop proceedings*, pp. 13–27, Siena, 2001.
9. H. Geuvers, M. Niqui, Constructive Reals in Coq: Axioms and Categoricity, *Types 2000 Workshop, Durham, UK*, this volume.
10. P. Henrici and I. Gargantini, Uniformly convergent algorithms for the simultaneous approximation of all zeros of a polynomial, in [4], pp. 77–113.
11. H. Kneser, Der Fundamentalsatz der Algebra und der Intuitionismus, *Math. Zeitschrift*, 46, 1940, pp. 287–302.
12. M. Kneser, Ergänzung zu einer Arbeit von Hellmuth Kneser über den Fundamentalsatz der Algebra, *Math. Zeitschrift*, 177, 1981, pp. 285–287.
13. J.E. Littlewood, Every polynomial has a root, *Journal of the London Math. Soc.* 16, 1941, pp. 95–98.
14. B. de Loor, *Die Hoofstelling van die Algebra van Intuïtionistiese standpunt*, Ph.D. Thesis, Univ. of Amsterdam, Netherlands, Feb. 1925, pp. 63 (South-African).
15. Helmut Schwichtenberg, Ein konstruktiver Beweis des Fundamentalsatzes, Appendix A (pp. 91–96) of Algebra, Lecture notes, Mathematisches Institut der Universität München 1998, `http://www.mathematik.uni-muenchen.de/schwicht/lectures/algebra/ws98/skript.ps`
16. E. Specker, The Fundamental Theorem of Algebra in Recursive Analysis, in [4], pp. 321–329.
17. A. Troelstra and D. van Dalen, *Constructivism in Mathematics*, vols. 121 and 123 in Studies in Logic and The Found. of Math., North-Holland, 1988.
18. H. Weyl, Randbemerkungen zu Hauptproblemen der Mathematik, *Math. Zeitschrift*, 20, 1924, pp. 131–150.

A Kripke-Style Model for the Admissibility of Structural Rules*

(Extended Abstract)

Healfdene Goguen

AT&T Labs
180 Park Ave., Florham Park, NJ 07932, USA
hhg@att.com

Abstract. This paper demonstrates the admissibility of the basic structural rules of dependent type theory, such as weakening, substitution and splitting lemmas, through a Kripke-style term model.

1 Introduction

There are two common approaches to understanding the structural rules in type theory. The first approach, arising from the study of syntax, such as the presentations of type systems such as PTS [7] and the Edinburgh LF [10], gives a minimal set of rules and shows the admissibility of the structural properties by induction on the derivations of the system. The second approach, arising from the study of semantics, and characterized by Streicher's approach to soundness for the Calculus of Constructions [20], takes the structural properties as given and simply includes rules of inference for them.

The syntactic presentation has many benefits over the semantic when studying metatheoretic properties of type theory. In particular, there are few rules, and a carefully chosen collection of inference rules can have strong inversion principles, making the proofs of results such as subject reduction much easier. Furthermore, syntactic presentations often use conversion to capture the notion of equality, which removes some of the complexity of the metatheory: reduction and conversion can be studied without reference to the typing judgment. These properties of the presentation encourage studying syntactic presentations of type systems with a sequence of inductions.

In contrast, dependent systems presented with judgmental equality have an interdependence of judgments between typing and equality. This makes demonstrating certain syntactic properties sequentially impractical. Two such classes of properties are the presupposition lemmas, for example that if $\Gamma \vdash M : A$ then $\Gamma \vdash A$, and the splitting lemmas, for example that if $\Gamma \vdash M = N : A$ then $\Gamma \vdash M : A$. The difficulty is exemplified by the congruence rule for application,

---

* Most of the technical work reported in this paper was carried out at the Laboratory for Foundations in Computer Science, University of Edinburgh, in the first half of 1998. The work was presented in Munich in May 1998.

P. Callaghan et al. (Eds.): TYPES 2000, LNCS 2277, pp. 112–124, 2002.

that $\Gamma \vdash M_1 = N_1 : \Pi x : A_1.A_2$ and $\Gamma \vdash M_2 = N_2 : A_1$ imply $\Gamma \vdash M_1(M_2) = N_1(N_2) : [M_2/x]A_2$. In order to know that $\Gamma \vdash N_1(N_2) : [M_2/x]A_2$ we need to know that $\Gamma \vdash [N_2/x]A_2 = [M_2/x]A_2$. A similar but simpler problem arises for the application rule, where $\Gamma \vdash [M_2/x]A_2$ must follow from the well-formedness of $\Pi x : A_1.A_2$ and M_2.

We study these problems in the context of Martin-Löf's Logical Framework [17]. We resolve them by giving a Kripke-style inductive definition at the level of kinds: $\Delta \models \Pi x : A_1.A_2 = \Pi x : B_1.B_2$ is well-formed if $\Delta, \Delta' \models [M_2/x]A_2 = [N_2/x]B_2$ is well-formed for every valid context Δ, Δ' and every $\Delta, \Delta' \models M_2 = N_2 : A_1$. The definition of term equality, $\Delta \models M = N : A$, is simply that $\Delta, \Delta' \vdash^= M = N : A$ for every context extension $\Delta, \Delta' \geq \Delta$. These definitions are relative to a minimal equational system $\vdash^=$ with judgments only for equality. Soundness is proved as usual using parallel substitutions $\Delta \models \gamma = \delta : \Gamma$.

This understanding of the syntactic properties seems more in keeping with the meaning explanation of type theory, since each rule of inference is justified by referring to an understanding of the meaning of the language, instead of being understood as an isolated property established by induction.

We show the admissibility of these properties with respect to a very weak system, motivated by the atomic weakening rule for PTS [16, 21]. Our system has, for example, the following weak rule for variables:[1]

$$Var \; \frac{\Gamma_0, x{:}A, \Gamma_1 \vdash^= A \text{ kind}}{\Gamma_0, x{:}A, \Gamma_1 \vdash^= x \; : \; A}$$

A simpler result to prove would be that if $\Gamma \vdash^- M : A$ and $\Delta \models \gamma : \Gamma$ then $\Delta \models M\gamma : A\gamma$, where $\vdash^-$ is a presentation of the Logical Framework with a minimal set of rules, where $M\gamma$ and $A\gamma$ denote the parallel substitution of γ in M and A, and where $\Gamma \models M : A$ is defined as $\Gamma, \Gamma' \vdash^- M : A$ for all valid contexts Γ, Γ'. This follows by a straightforward induction, and weakening and substitution are consequences. For our purposes in this article, this simpler method does not allow us to exploit the full power of logical relations to show the presupposition and splitting properties, but it should be appropriate for the metatheory of type theories for which logical relations cannot be constructed, such as arbitrary PTS.

Although we refer to our construction as a term model, the uniform rather than inductive definition of term equality is very weak. The observation that weakening and substitution can be proved without any inductive definitions justifies this simplification: the Kripke-style construction is only used for structural properties requiring a strong inversion property for kinds. Framing our work in the context of a general model theory for type theory [11, 19, 20] would clarify the precise sense in which our construction can be considered a model.

Our technique should work for the Martin-Löf Logical Framework with constants and equalities representing inductive types. Extending it to the Calculus

---

[1] This rule is in fact strictly weaker than the atomic weakening rule in PTS, because there can be no intervening uses of judgmental equality.

of Constructions would require the standard technique of interpreting type operators as set-theoretic functions, similar to the logical-relation interpretations used for strong normalization proofs.

The proofs given here should also extend to η-equality. We did not include η because it in turn suggests studying Strengthening, which is a structural rule that cannot be shown admissible by the model construction in this article. However, while Strengthening cannot be shown admissible, it should also have no effect on our model to include both it and η-equality as rules of inference in each of the systems presented here.

2   The Logical Framework

In this section we introduce the language of terms, kinds and contexts, and we introduce the judgments and rules of inference for the Full Logical Framework, the Minimal Logical Framework, and the Minimal Equational Logical Framework.

2.1   Terms, Kinds, and Contexts

The language of types, terms and contexts for the Logical Framework is defined by the following grammar:

$$A, B, C \in K ::= \text{Type} \mid \text{El}(M) \mid \Pi x{:}A.B$$
$$M, N, P \in T ::= x \mid \lambda x{:}A.M \mid M(N)$$
$$\Gamma, \Delta, \Phi \in C ::= () \mid \Gamma, x{:}A$$

Substitution, $[N/x]M$, is defined as usual for terms, with the obvious extension to kinds and contexts. We identify terms, kinds and contexts up to α-equivalence.

Parallel substitution, $M\gamma$, for γ a list of pairs of variables and terms, is the simultaneous replacement of each of the variables by its corresponding value. We write () for the empty parallel substitution and $\gamma[x \leftarrow N]$ for the extension of γ with value N for x. Parallel substitution extends beneath binders in the following way:

$$(\lambda x : A_1.M_0)\gamma = \lambda y : A_1\gamma.(M_0(\gamma[x \leftarrow y])) \quad y \text{ fresh in } M_0$$

and similarly for Π. We use standard properties of substitution, such as that $([N/x]M)\gamma \equiv M(\gamma[x \leftarrow N\gamma])$, without further comment.

2.2   The Full and Minimal Logical Frameworks

We introduce two variants on the Logical Framework, which share the same basic judgment forms: the Full Logical Framework and the Minimal Logical Framework. The Full Logical Framework has five judgment forms:

- $\Gamma \vdash$ ok, meaning that Γ is a valid context of assumptions,
- $\Gamma \vdash A$ kind, meaning that A is a type under assumptions Γ,
- $\Gamma \vdash A = B$, meaning that A and B are types and are equal under the assumptions Γ,
- $\Gamma \vdash M : A$, meaning that A is a type and that M has a canonical form that is in A, under assumptions Γ, and
- $\Gamma \vdash M = N : A$, meaning that A is a type, that M and N have canonical forms that are in A, and that those canonical forms are equal in A, under assumptions Γ.

The Full Logical Framework is defined by the standard rules of inference for term formation, as given in Figure 1, as well as the structural rules given in Figure 2.

The Minimal Logical Framework, with judgments written $\Gamma \vdash^- J$ with J ranging over the above judgment forms, is defined by the inference rules only of Figure 1. There is a trivial inclusion from the Minimal Logical Framework into the Full Logical Framework.

2.3   Minimal Equational Logical Framework

The judgments of the Minimal Equational Logical Framework are $\Gamma \vdash^= $ ok, $\Gamma \vdash^= A = B$, and $\Gamma \vdash^= M = N : A$. The rules of inference for the minimal equational system are given in Figure 3, where we write $\Gamma \vdash^= A$ kind for $\Gamma \vdash^= A = B$ for any B, and $\Gamma \vdash^= M : A$ for $\Gamma \vdash^= M = N : A$ for any N. The meaning of the system is given by Completeness, Proposition 1.

Lemma 1. *If $\Gamma, \Gamma' \vdash^= J$ then there is a not necessarily strict subderivation of $\Gamma \vdash^= $ ok.*

Proposition 1 (Completeness).

1. *If $\Gamma \vdash^= $ ok then $\Gamma \vdash^- $ ok.*
2. *If $\Gamma \vdash^= A = B$ then $\Gamma \vdash^- A$ kind, $\Gamma \vdash^- B$ kind, and $\Gamma \vdash^- A = B$.*
3. *If $\Gamma \vdash^= M = N : A$ then $\Gamma \vdash^- M : A$, $\Gamma \vdash^- N : A$, $\Gamma \vdash^- M = N : A$ and $\Gamma \vdash^- A$ kind.*

Proof. By induction on derivations, using Lemma 1 for *Var*.

3   Soundness

We now define the Kripke-style model and prove Soundness of the Logical Framework for it. Soundness of the Logical Framework for the Minimal Equational Logical Framework follows as a direct consequence.

3.1 The Model

Definition 1 $(\Delta \models M = N : A)$. $\Delta \models M = N : A$ if $\Delta \vdash^= $ ok and $\Delta, \Delta' \vdash^= M = N : A$ for all extensions Δ' such that $\Delta, \Delta' \vdash^= $ ok.

Definition 2 $(\Delta \models A = B)$. $\Delta \models A = B$ is the least relation such that:

- $\Delta \models \mathrm{Type} = \mathrm{Type}$ if $\Delta \vdash^= $ ok.
- $\Delta \models \mathrm{El}(M) = \mathrm{El}(N)$ if $\Delta \models M = N : \mathrm{Type}$.
- $\Delta \models \Pi x : A_1.A_2 = \Pi x : B_1.B_2$ if $\Delta \vdash^= $ ok and if, for all Δ' such that $\Delta, \Delta' \models $ ok, we have $\Delta, \Delta' \models A_1 = B_1$, and if for all M and N such that $\Delta, \Delta' \models M = N : A_1$ we have $\Delta, \Delta' \models [M/x]A_2 = [N/x]B_2$.

Definition 3 $(\Delta \models \gamma = \delta : \Gamma)$. $\Delta \models \gamma = \delta : \Gamma$ is the least relation such that:

- $\Delta \models () = () : ()$ if $\Delta \vdash^= $ ok.
- $\Delta \models \gamma[x \leftarrow M] = \delta[x \leftarrow N] : \Gamma, x : A$ if $\Delta \vdash^= $ ok and if, for all Δ' such that $\Delta, \Delta' \vdash^= $ ok, $\Delta, \Delta' \models \gamma = \delta : \Gamma$, $\Delta, \Delta' \models A\gamma = A\delta$, and $\Delta, \Delta' \models M = N : A\gamma$.

Definition 4 $(\Delta \models \Gamma = \Gamma')$. $\Delta \models \Gamma = \Gamma'$ is the least relation such that:

- $\Delta \models () = ()$ if $\Delta \vdash^= $ ok.
- $\Delta \models \Gamma, x : A = \Gamma', x : B$ if $\Delta \vdash^= $ ok and if, for all Δ' such that $\Delta, \Delta' \vdash^= $ ok, $\Delta, \Delta' \models \Gamma = \Gamma'$, and for all γ and δ such that $\Delta, \Delta' \models \gamma = \delta : \Gamma$, we have $\Delta, \Delta' \models A\gamma = B\delta$.

3.2 Properties of the Model

Lemma 2 (Monotonicity). *Suppose* $\Delta, \Delta' \vdash^= $ ok. *Then:*

1. If $\Delta \models M = N : A$ then $\Delta, \Delta' \models M = N : A$.
2. If $\Delta \models A = B$ then $\Delta, \Delta' \models A = B$.
3. If $\Delta \models \Gamma = \Gamma'$ then $\Delta, \Delta' \models \Gamma = \Gamma'$.
4. If $\Delta \models \gamma = \delta : \Gamma$ then $\Delta, \Delta' \models \gamma = \delta : \Gamma$.

Proof. By induction on the relevant definition.

Lemma 3.

1. If $\Delta \models A = B$ and $\Delta, \Delta' \vdash $ ok then $\Delta, \Delta' \vdash^= A = B$.
2. If $\Delta \models A = B$ then $\Delta, x : A \models x = x : A$ and $\Delta, x : B \models x = x : B$.

Proof. By simultaneous induction on proofs that $\Delta \models A = B$.

We consider the Π rule for case 1. Suppose $\Delta, \Delta' \vdash^= $ ok. Then $\Delta, \Delta' \models A_1 = B_1$ by definition, so $\Delta, \Delta', y : A_1 \models y = y : A_1$ and $\Delta, \Delta', y : B_1 \models y = y : B_1$ by the induction hypothesis case 2, and $\Delta, \Delta' \vdash^= A_1 = B_1$ by the induction hypothesis case 1. $\Delta, \Delta', y : A_1 \models [y/x]A_2 = [y/x]B_2$ by definition, and so $\Delta, \Delta', y : A_1 \vdash^= [y/x]A_2 = [y/x]B_2$ by the induction hypothesis case 1. Finally, $\Delta, \Delta', y : B_1 \models [y/x]A_2 = [y/x]B_2$, so by induction hypothesis case 1 $\Delta, \Delta', y : B_1 \vdash^= [y/x]A_2 = [y/x]B_2$ and by *KSym* $\Delta, \Delta', y : B_1 \vdash^= [y/x]B_2 = [y/x]A_2$.

Definition 5 (Partial Equivalence Relation). *A binary relation R is a partial equivalence relation if it is symmetric and transitive.*

The relations we defined above are all partial equivalence relations.

Lemma 4.

1. *$\Delta \models M = N : A$ is a partial equivalence relation.*
2. *$\Delta \models A = B$ is a partial equivalence relation.*
3. *$\Delta \models \gamma = \delta : \Gamma$ is a partial equivalence relation.*
4. *$\Delta \models \Gamma = \Delta$ is a partial equivalence relation.*

Proof. By induction on the left-hand side of each statement: for example, $\Delta \models A = B$ implies $\Delta \models B = A$ follows by induction on derivations that $\Delta \models A = B$.

Lemma 5. *If $\Delta \models M = N : A$ and $\Delta \models A = B$ then $\Delta \models M = N : B$.*

Proof. Suppose $\Delta, \Delta' \vdash^= ok$ for some Δ'. Then $\Delta, \Delta' \vdash^= M = N : A$ by definition and $\Delta, \Delta' \vdash^= A = B$ by Lemma 3, so $\Delta, \Delta' \vdash^= M = N : B$ by =-R and $\Delta \models M = N : B$ by definition.

We now prove lemmas that correspond to the operations performed by the structural rules.

Lemma 6 (Substitution). *Suppose $\Delta \models \Gamma_0, x : A, \Gamma_1 = \Gamma'_0, x : B, \Gamma'_1$, and let M and N be such that for all Δ', if $\Delta, \Delta' \models \gamma_0 = \delta_0 : \Gamma_0$ then $\Delta, \Delta' \models M\gamma_0 = N\delta_0 : A\gamma_0$. Then:*

1. *$\Delta \models \Gamma_0, [M/x]\Gamma_1 = \Gamma'_0, [N/x]\Gamma'_1$, and*
2. *for any Δ', if $\Delta, \Delta' \models \gamma_0\gamma_1 = \delta_0\delta_1 : \Gamma_0, [M/x]\Gamma_1$ then $\Delta, \Delta' \models \gamma_0[x \leftarrow M\gamma_0]\gamma_1 = \delta_0[x \leftarrow N\delta_0]\delta_1 : \Gamma_0, x : A, \Gamma_1$.*

Proof. By induction on derivations that $\Delta \models \Gamma_0, x : A, \Gamma_1 = \Gamma'_0, x : B, \Gamma'_1$.

Lemma 7 (Thinning). *Suppose that $\Delta \models \Gamma_0, \Gamma_1 = \Gamma'_0, \Gamma'_1$, and let A and B be such that for all Δ', if $\Delta, \Delta' \models \gamma_0 = \delta_0 : \Gamma_0$ then $\Delta, \Delta' \models A\gamma_0 = B\delta_0$. Then:*

1. *$\Delta \models \Gamma_0, x : A, \Gamma_1 = \Gamma'_0, x : B, \Gamma'_1$, and*
2. *for any Δ', if $\Delta, \Delta' \models \gamma_0[x \leftarrow M]\gamma_1 = \delta_0[x \leftarrow N]\delta_1 : \Gamma_0, x : A, \Gamma_1$ then $\Delta, \Delta' \models \gamma_0\gamma_1 = \delta_0\delta_1 : \Gamma_0, \Gamma_1$.*

Proof. By induction on derivations that $\Delta \models \Gamma_0, \Gamma_1 = \Gamma'_0, \Gamma'_1$.

Lemma 8 (Context Replacement). *Suppose that $\Delta \models \Gamma_0, x : A, \Gamma_1 = \Gamma'_0, x : A', \Gamma'_1$, and that for any Δ', if $\Delta, \Delta' \models \gamma_0 = \delta_0 : \Gamma_0$ then $\Delta, \Delta' \models A\gamma_0 = B\delta_0$. Then:*

1. *$\Delta \models \Gamma_0, x : B, \Gamma_1 = \Gamma'_0, x : A', \Gamma'_1$, and*
2. *for any Δ', if $\Delta, \Delta' \models \gamma_0[x \leftarrow M]\gamma_1 = \delta_0[x \leftarrow N]\delta_1 : \Gamma_0, x : B, \Gamma_1$ then $\Delta, \Delta' \models \gamma_0[x \leftarrow M]\gamma_1 = \delta_0[x \leftarrow N]\delta_1 : \Gamma_0, x : A, \Gamma_1$.*

Proof. By induction on derivations that $\Delta \models \Gamma_0, x : A, \Gamma_1 = \Gamma'_0, x : A', \Gamma'_1$.

3.3 Soundness

Finally, we are ready to prove Soundness.

Theorem 1 (Soundness). *Suppose $\Gamma \vdash J$ and $\Delta \vdash^= $ ok. Then $\Delta \models \Gamma = \Gamma$, and if $\Delta \models \gamma = \delta : \Gamma$ then:*

1. *If $\Gamma \vdash A$ kind then $\Delta \models A\gamma = A\delta$.*
2. *If $\Gamma \vdash A = B$ then $\Delta \models A\gamma = B\delta$.*
3. *If $\Gamma \vdash M : A$ then $\Delta \models A\gamma = A\delta$ and $\Delta \models M\gamma = M\delta : A\gamma$.*
4. *If $\Gamma \vdash M = N : A$ then $\Delta \models A\gamma = A\delta$ and $\Delta \models M\gamma = N\delta : A\gamma$.*

Proof. By induction on derivations. We consider several cases.

- λ. By the induction hypothesis $\Delta \models \Gamma, x : A_1 = \Gamma, x : A_1$, so $\Delta \models \Gamma = \Gamma$ by inversion. Next, suppose $\Delta \models \gamma = \delta : \Gamma$ and $\Delta, \Delta' \vdash^= $ ok. Then $\Delta, \Delta' \models A_1\gamma = A_1\delta$ by Monotonicity and the induction hypothesis, and so $\Delta, \Delta', y : A_1\gamma \models y = y : A_1\gamma$ by Lemma 3. Therefore, $\Delta, \Delta', y : A_1\gamma \models \gamma[x \leftarrow y] = \delta[x \leftarrow y] : \Gamma, x : A_1$ by definition and Monotonicity, so $\Delta, \Delta', y : A_1\gamma \models M_0(\gamma[x \leftarrow y]) = M_0(\delta[x \leftarrow y]) : A_2(\gamma[x \leftarrow y])$ by the induction hypothesis. Similarly, $\Delta, \Delta', y : A_1\delta \models M_0(\gamma[x \leftarrow y]) = M_0(\delta[x \leftarrow y]) : A_2(\gamma[x \leftarrow y])$, again using Lemma 3, Monotonicity and the induction hypothesis, so $\Delta, \Delta', y : A_1\delta \models M_0(\delta[x \leftarrow y]) : A_2(\gamma[x \leftarrow y])$ by Lemma 4. Hence, $\Delta, \Delta' \vdash^= (\lambda x : A_1.M_0)\gamma \equiv \lambda y : A_1\gamma.M_0(\gamma[x \leftarrow y]) = \lambda y : A_1\delta.M_0(\delta[x \leftarrow y]) \equiv (\lambda x : A_1.M_0)\delta : \Pi x : A_1\gamma.A_2(\gamma[x \leftarrow y]) \equiv (\Pi x : A_1.A_2)\gamma$ by λ-*Eq*, so $\Delta \models (\lambda x : A_1.M_0)\gamma = (\lambda x : A_1.M_0)\delta : (\Pi x : A_1.A_2)\gamma$.
  Finally, we need to show $\Delta \models (\Pi x : A_1.A_2)\gamma = (\Pi x : A_1.A_2)\delta$, which follows directly by definition and the induction hypothesis.
- El-*Eq*. By the induction hypothesis $\Delta \models M\gamma = N\delta : $ Type. Suppose $\Delta, \Delta' \vdash^= $ ok, then $\Delta, \Delta' \vdash^= M\gamma = N\delta : $ Type by definition, so $\Delta, \Delta' \vdash^= \text{El}(M)\gamma \equiv \text{El}(M\gamma) = \text{El}(N\delta) \equiv \text{El}(N)\delta$ by El-*Eq* for $\vdash^=$.
- *App-Eq*. By the induction hypothesis $\Delta \models M_1\gamma = N_1\delta : (\Pi x : A_1.A_2)\gamma$ and $\Delta \models (\Pi x : A_1.A_2)\gamma = (\Pi x : A_1.A_2)\delta$, and also $\Delta \models M_2\gamma = N_2\delta : A_1\gamma$ and $\Delta \models A_1\gamma = A_1\delta$. By Lemma 4 we know $\Delta \models (\Pi x : A_1.A_2)\gamma = (\Pi x : A_1.A_2)\gamma$. By definition and Lemma 3 $\Delta, \Delta' \vdash^= ([M_2/x]A_2)\gamma \equiv A_2(\gamma[x \leftarrow M_2\gamma]) = A_2(\gamma[x \leftarrow N_2\delta])$ for any Δ' such that $\Delta, \Delta' \models$ ok, and so $\Delta, \Delta' \vdash^= (M_1M_2)\gamma = (N_1N_2)\delta : A_2(\gamma[x \leftarrow M_2\gamma]) \equiv ([M_2/x]A_2)\gamma$ by *App-Eq*. Hence, by definition $\Delta \models (M_1M_2)\gamma = (N_1N_2)\delta : ([M_2/x]A_2)\gamma$.
  Furthermore, $\Delta \models \delta = \delta : \Gamma$ by Lemma 4, so $\Delta \models M_2\delta = N_2\delta : A_1\delta$ implies $\Delta \models M_2\delta = N_2\delta : A_1\gamma$ by Lemma 5, and so $\Delta \models M_2\gamma = M_2\delta : A_1\gamma$. Hence, $\Delta \models ([M_2/x]A_2)\gamma \equiv A_2(\gamma[x \leftarrow M_2\gamma]) = A_2(\delta[x \leftarrow M_2\delta]) \equiv ([M_2/x]A_2)\delta$ by definition.
- Context Replacement. We consider the rule for the typing judgment. By the induction hypothesis $\Delta \models \Gamma, x : A, \Gamma' = \Gamma, x : A, \Gamma'$. Also, if $\Delta, \Delta' \vdash^= $ ok and $\Delta, \Delta' \models \gamma_0 = \delta_0 : \Gamma$ for any γ_0 and δ_0 then by the induction hypothesis $\Delta, \Delta' \models A\gamma_0 = B\delta_0$. Therefore, by Lemma 8 $\Delta \models \Gamma, x : B, \Gamma' = \Gamma, x : A, \Gamma'$ and if $\Delta \models \gamma = \delta : \Gamma, x : B, \Gamma'$ then $\Delta \models \gamma = \delta : \Gamma, x : A, \Gamma'$. Hence, $\Delta \models \Gamma, x : B, \Gamma' = \Gamma, x : B, \Gamma'$ by Lemma 4, and $\Delta \models M\gamma = M\delta : C\gamma$ and $\Delta \models C\gamma = C\delta$ by the induction hypothesis.

Lemma 9. *If* $() \models \Gamma = \Gamma$ *then* $\Gamma \models \mathrm{id}_\Gamma = \mathrm{id}_\Gamma : \Gamma.$

Proof. By induction on derivations that $() \models \Gamma = \Gamma$, using Monotonicity and Lemma 3 Case 2 for the extension case.

Corollary 1.

1. *If* $\Gamma \vdash$ ok *then* $\Gamma \vdash^= $ ok.
2. *If* $\Gamma \vdash A$ *kind then* $\Gamma \vdash^= A = A.$
3. *If* $\Gamma \vdash A = B$ *then* $\Gamma \vdash^= A = B.$
4. *If* $\Gamma \vdash M : A$ *then* $\Gamma \vdash^= M = M : A.$
5. *If* $\Gamma \vdash M = N : A$ *then* $\Gamma \vdash^= M = N : A.$

Proof. We consider Case 3. We know that $() \vdash^= $ ok, so $() \models \Gamma = \Gamma$ by Soundness. Hence, $\Gamma \models \mathrm{id}_\Gamma = \mathrm{id}_\Gamma : \Gamma$ by Lemma 9, and so $\Gamma \models A \equiv A\mathrm{id}_\Gamma = B\mathrm{id}_\Gamma \equiv B$, again by Soundness. Therefore, $\Gamma \vdash^= A = B$ by Lemma 3, since $\Gamma \vdash^= $ ok follows trivially from $\Gamma \models \mathrm{id}_\Gamma = \mathrm{id}_\Gamma$.

Cases 4 and 5 follow using Definition 1.

Therefore, combining this result with Proposition 1, the equivalence between the Full Logical Framework $\vdash$ and the Minimal Equational Logical Framework $\vdash^=$ has been established.

4   Related Work

After writing this article, we became aware of three related articles in LICS 1999. Two of the three articles [5, 12] discuss pre-sheaves as semantics for syntax with variable binding operations. Certainly the present article could also be framed in categorical language, but this would involve heavier mathematical machinery to handle dependent types. As our aim in this article was to study syntactic properties, we use the simpler mechanism of Kripke-style models.

Fiore, Plotkin and Turi [5] show that pre-sheaves over the category of variables and renamings are an appropriate initial algebra semantics for syntax with variable binding operations. The present article demonstrates that only weakenings are necessary in the base category.[2] Hofmann [12] uses a similar structure to model higher-order abstract syntax as presented in [4]. Gabbay and Pitts [6] introduce a new set-forming operation of name-abstraction and show that it models α-equivalence classes of variable-binding syntax.

The traditional approach to the study of structural rules has a long history. Several good modern presentations are Luo [15], Geuvers and Nederpelt [7], and Pollack's formal treatment in LEGO [18].

Streicher's thesis [20] is the starting point for a great deal of work on the syntax and semantics of dependent type theory. In particular, it gives a very careful treatment of the rules of inference of the Calculus of Constructions, and the first thorough treatment of the proof of soundness for a categorical semantics.

---

[2] The author was aware of this on finishing [8], as Leleu [14] discovered independently.

Kripke-style term models [3], originally developed to prove strong normalization for the Calculus of Constructions, are constructed from well-formed terms with additional properties for saturated sets. The idea of using these models to show the admissibility of structural rules is implicit in Goguen's thesis [8], and is fully developed by Compagnoni and Goguen for a calculus of higher-order subtyping [2].

Goguen and McKinna [9] show the closure of PTSs under context morphisms using only one induction. The results of weakening and substitution follow as direct corollaries. The proof is via the class of renamings, as subsequently used by [5]. The renamings were needed to allow substitutions to pass under binders: if a term N with binders is substituted for a variable underneath another binder, then it may be necessary to change the parameters representing bound variables in N.

The current paper demonstrates that renamings are not necessary, because the induction hypothesis for the usual proof of substitution is too weak: if we instead require the term substituted to be closed under weakening then the proof goes through directly.

Jones, Luo and Soloviev [13] discuss the problem of splitting lemmas in the context of their system for coercive subtyping. They prove the splitting lemmas, presupposition lemmas, and admissibility of substitution through an intermediate system with weaker rules, together with an algorithm on derivations to implement the operations of splitting and the presupposition lemmas.

5   Conclusions

We have demonstrated the equivalence between three versions of the Logical Framework: the Full Logical Framework $\vdash$ corresponding to a semantic presentation; the Minimal Logical Framework $\vdash^-$ corresponding to a syntactic presentation; and the Minimal Equational Logical Framework $\vdash^=$, a system introduced to reduce bookkeeping in the proof of equivalence. The equivalence of these three systems demonstrates the admissibility of the structural rules of Figure 2 for the Minimal Logical Framework.

Acknowledgments

I would like to thank Prof. Schwichtenberg for inviting me to present this work to his group in Munich, James McKinna for many stimulating discussions, and the anonymous referees for helpful comments. I would also like to thank my wife Adriana Compagnoni both for helpful technical discussions and for her encouragement to write this article.

This work was partially financed by the British funding body EPSRC.

References

1. *14th Annual Symposium on Logic in Computer Science*, Trento, Italy, 1999. IEEE.
2. Adriana Compagnoni and Healfdene Goguen. Typed operational semantics for higher order subtyping. Technical Report ECS-LFCS-97-361, University of Edinburgh, July 1997. Submitted to *Information and Computation*.
3. Thierry Coquand and Jean Gallier. A proof of strong normalization for the theory of constructions using a Kripke-like interpretation. In *Workshop on Logical Frameworks-Preliminary Proceedings*, 1990.
4. Joëlle Despeyroux, Amy Felty, and André Hirchowitz. Higher-order abstract syntax in Coq. In *Proceedings of the International Conference on Typed Lambda Calculi and Applications*, volume 902 of *Lecture Notes in Computer Science*. Springer-Verlag, April 1995.
5. Marcelo Fiore, Gordon Plotkin, and Daniele Turi. Abstract syntax and variable binding. In *14th Annual Symposium on Logic in Computer Science* [1].
6. Murdoch Gabbay and Andrew Pitts. A new approach to abstract syntax involving binders. In *14th Annual Symposium on Logic in Computer Science* [1].
7. Herman Geuvers and Mark-Jan Nederhof. A modular proof of strong normalization for the calculus of constructions. *Journal of Functional Programming*, 1(2):155–189, April 1991.
8. Healfdene Goguen. *A Typed Operational Semantics for Type Theory*. PhD thesis, University of Edinburgh, August 1994.
9. Healfdene Goguen and James McKinna. Candidates for substitution. Technical Report ECS-LFCS-97-358, University of Edinburgh, May 1997.
10. Robert Harper, Furio Honsell, and Gordon D. Plotkin. A framework for defining logics. *Journal of the Association for Computing Machinery*, 40(1):143–184, 1992.
11. Martin Hofmann. On the interpretation of type theory in locally cartesian closed categories. In Jerzy Tiuryn and Leszek Pacholski, editors, *Proceedings of CSL*, volume 933. Springer-Verlag, 1994.
12. Martin Hofmann. Semantical analysis of higher-order abstract syntax. In *14th Annual Symposium on Logic in Computer Science* [1].
13. Alex Jones, Zhaohui Luo, and Sergei Soloviev. Some algorithmic and proof-theoretical aspects of coercive subtyping. In Eduardo Giménez and Christine Paulin-Mohring, editors, *Proceedings of TYPES'96*, volume 1512 of *Lecture Notes in Computer Science*, 1996.
14. Pierre Leleu. Metatheoretic results for a modal lambda calculus. Technical Report RR-3361, INRIA, 1998.
15. Zhaohui Luo. *An Extended Calculus of Constructions*. PhD thesis, University of Edinburgh, November 1990.
16. James McKinna and Robert Pollack. Pure type systems formalized. In M. Bezem and J. F. Groote, editors, *Proceedings of the International Conference on Typed Lambda Calculi and Applications*, pages 289–305. Springer-Verlag, LNCS 664, March 1993.
17. Bengt Nordström, Kent Petersson, and Jan Smith. *Programming in Martin-Löf's Type Theory: An Introduction*. Oxford University Press, 1990.
18. Robert Pollack. *The Theory of LEGO: A Proof Checker for the Extended Calculus of Constructions*. PhD thesis, University of Edinburgh, April 1995.
19. David Pym. Functorial Kripke models of the lambdaPi-calculus, I: Type theory and internal logic. Draft.
20. Thomas Streicher. *Semantics of Type Theory: Correctness, Completeness and Independence Results*. Birkhäuser, 1991.
21. L. van Benthem Jutting, James McKinna, and Robert Pollack. Typechecking in pure type systems. In Henk Barendregt and Tobias Nipkow, editors, *Types for Proofs and Programming*. Springer-Verlag, 1993.

Appendix

Valid Contexts

$$(Emp) \ \frac{}{() \vdash \text{ok}} \qquad (Weak) \ \frac{\Gamma \vdash A \text{ kind} \quad x \notin dom(\Gamma)}{\Gamma, x{:}A \vdash \text{ok}}$$

Types

$$(Type) \ \frac{\Gamma \vdash \text{ok}}{\Gamma \vdash \text{Type kind}} \qquad (El) \ \frac{\Gamma \vdash M \ : \ \text{Type}}{\Gamma \vdash \text{El}(M) \text{ kind}} \qquad (\Pi) \ \frac{\Gamma, x{:}A_1 \vdash A_2 \text{ kind}}{\Gamma \vdash \Pi x{:}A_1.A_2 \text{ kind}}$$

Type Equality

$$(KRefl) \ \frac{\Gamma \vdash A \text{ kind}}{\Gamma \vdash A = A} \qquad\qquad (KSym) \ \frac{\Gamma \vdash A = B}{\Gamma \vdash B = A}$$

$$(KTrans) \qquad \frac{\Gamma \vdash A = B \quad \Gamma \vdash B = C}{\Gamma \vdash A = C}$$

$$(El\text{-}Eq) \ \frac{\Gamma \vdash M = N \ : \ \text{Type}}{\Gamma \vdash \text{El}(M) = \text{El}(N)} \qquad (\Pi\text{-}Eq) \ \frac{\Gamma \vdash A_1 = B_1 \quad \Gamma, x{:}A_1 \vdash A_2 = B_2}{\Gamma \vdash \Pi x{:}A_1.A_2 = \Pi x{:}B_1.B_2}$$

Terms

$$(Var) \ \frac{\Gamma_0, x{:}A, \Gamma_1 \vdash \text{ok}}{\Gamma_0, x{:}A, \Gamma_1 \vdash x \ : \ A} \qquad (Eq) \ \frac{\Gamma \vdash M \ : \ A \quad \Gamma \vdash A = B}{\Gamma \vdash M \ : \ B}$$

$$(\lambda) \ \frac{\Gamma, x{:}A_1 \vdash M_0 \ : \ A_2}{\Gamma \vdash \lambda x{:}A_1.M_0 \ : \ \Pi x{:}A_1.A_2} \qquad (App) \ \frac{\Gamma \vdash M_1 \ : \ \Pi x{:}A_1.A_2 \quad \Gamma \vdash M_2 \ : \ A_1}{\Gamma \vdash M_1(M_2) \ : \ [M_2/x]A_2}$$

Term Equality

$$(Refl) \ \frac{\Gamma \vdash M \ : \ A}{\Gamma \vdash M = M \ : \ A} \qquad (Sym) \ \frac{\Gamma \vdash M = N \ : \ A}{\Gamma \vdash N = M \ : \ A}$$

$$(Trans) \qquad \frac{\Gamma \vdash M = N \ : \ A \quad \Gamma \vdash N = P \ : \ A}{\Gamma \vdash M = P \ : \ A}$$

$$(=R) \qquad \frac{\Gamma \vdash M = N \ : \ A \quad \Gamma \vdash A = B}{\Gamma \vdash M = N \ : \ B}$$

$$(\lambda\text{-}Eq) \qquad \frac{\Gamma \vdash A_1 = B_1 \quad \Gamma, x{:}A_1 \vdash M_0 = N_0 \ : \ A_2}{\Gamma \vdash \lambda x{:}A_1.M_0 = \lambda x{:}B_1.N_0 \ : \ \Pi x{:}A_1.A_2}$$

$$(App\text{-}Eq) \ \frac{\Gamma \vdash M_1 = N_1 \ : \ \Pi x{:}A_1.A_2 \quad \Gamma \vdash M_2 = N_2 \ : \ A_1}{\Gamma \vdash M_1(M_2) = N_1(N_2) \ : \ [M_2/x]A_2}$$

$$(\beta) \qquad \frac{\Gamma, x{:}A_1 \vdash M_0 \ : \ A_2 \quad \Gamma \vdash M_2 \ : \ A_1}{\Gamma \vdash (\lambda x{:}A_1.M_0)(M_2) = [M_2/x]M_0 \ : \ [M_2/x]A_2}$$

Fig. 1. Basic Rules of Inference for the Logical Framework

Substitution Rules

$$\frac{\Gamma, x{:}A, \Gamma' \vdash \text{ok} \quad \Gamma \vdash P : A}{\Gamma, [P/x]\Gamma' \vdash \text{ok}}$$

$$\frac{\Gamma, x{:}A, \Gamma' \vdash B \text{ kind} \quad \Gamma \vdash P : A}{\Gamma, [P/x]\Gamma' \vdash [P/x]B \text{ kind}} \qquad \frac{\Gamma, x{:}A, \Gamma' \vdash B \text{ kind} \quad \Gamma \vdash N = P : A}{\Gamma, [N/x]\Gamma' \vdash [N/x]B = [P/x]B}$$

$$\frac{\Gamma, x{:}A, \Gamma' \vdash M : B \quad \Gamma \vdash P : A}{\Gamma, [P/x]\Gamma' \vdash [P/x]M : [P/x]B} \qquad \frac{\Gamma, x{:}A, \Gamma' \vdash M : B \quad \Gamma \vdash N = P : A}{\Gamma, [N/x]\Gamma' \vdash [N/x]M = [P/x]M : [N/x]B}$$

$$\frac{\Gamma, x{:}A, \Gamma' \vdash B = C \quad \Gamma \vdash P : A}{\Gamma, [P/x]\Gamma' \vdash [P/x]B = [P/x]C} \qquad \frac{\Gamma, x{:}A, \Gamma' \vdash M = N : B \quad \Gamma \vdash P : A}{\Gamma, [P/x]\Gamma' \vdash [P/x]M = [P/x]N : [P/x]B}$$

Thinning

$$\frac{\Gamma, \Gamma' \vdash \text{ok} \quad \Gamma \vdash A \text{ kind} \quad x \notin FV(\Gamma, \Gamma')}{\Gamma, x{:}A, \Gamma' \vdash \text{ok}}$$

$$\frac{\Gamma, \Gamma' \vdash B \text{ kind} \quad \Gamma \vdash A \text{ kind} \quad x \notin FV(\Gamma, \Gamma')}{\Gamma, x{:}A, \Gamma' \vdash B \text{ kind}} \qquad \frac{\Gamma, \Gamma' \vdash M : B \quad \Gamma \vdash A \text{ kind} \quad x \notin FV(\Gamma, \Gamma')}{\Gamma, x{:}A, \Gamma' \vdash M : B}$$

$$\frac{\Gamma, \Gamma' \vdash B = C \quad \Gamma \vdash A \text{ kind} \quad x \notin FV(\Gamma, \Gamma')}{\Gamma, x{:}A, \Gamma' \vdash B = C}$$

$$\frac{\Gamma, \Gamma' \vdash M = N : B \quad \Gamma \vdash A \text{ kind} \quad x \notin FV(\Gamma, \Gamma')}{\Gamma, x{:}A, \Gamma' \vdash M = N : B}$$

Context Replacement

$$\frac{\Gamma, x{:}A, \Gamma' \vdash \text{ok} \quad \Gamma \vdash A = B}{\Gamma, x{:}B, \Gamma' \vdash \text{ok}}$$

$$\frac{\Gamma, x{:}A, \Gamma' \vdash C \text{ kind} \quad \Gamma \vdash A = B}{\Gamma, x{:}B, \Gamma' \vdash C \text{ kind}} \qquad \frac{\Gamma, x{:}A, \Gamma' \vdash M : C \quad \Gamma \vdash A = B}{\Gamma, x{:}B, \Gamma' \vdash M : C}$$

$$\frac{\Gamma, x{:}A, \Gamma' \vdash C = D \quad \Gamma \vdash A = B}{\Gamma, x{:}B, \Gamma' \vdash C = D} \qquad \frac{\Gamma, x{:}A, \Gamma' \vdash M = N : C \quad \Gamma \vdash A = B}{\Gamma, x{:}B, \Gamma' \vdash M = N : C}$$

Presuppositions

$$\frac{\Gamma \vdash A \text{ kind}}{\Gamma \vdash \text{ok}} \quad \frac{\Gamma \vdash A = B}{\Gamma \vdash A \text{ kind}} \quad \frac{\Gamma \vdash M : A}{\Gamma \vdash A \text{ kind}} \quad \frac{\Gamma \vdash M = N : A}{\Gamma \vdash M : A} \quad \frac{\Gamma, x{:}A \vdash \text{ok}}{\Gamma \vdash \text{ok}}$$

Fig. 2. Structural Rules for the Logical Framework

Valid Context

$$(Emp) \ \frac{}{() \vdash^= \text{ok}} \qquad (Weak) \ \frac{\Gamma \vdash^= A \ \text{kind} \quad x \notin dom(\Gamma)}{\Gamma, x{:}A \vdash^= \text{ok}}$$

Type Equality

$$(KSym) \ \frac{\Gamma \vdash^= A = B}{\Gamma \vdash^= B = A} \qquad (KTrans) \ \frac{\Gamma \vdash^= A = B \quad \Gamma \vdash^= B = C}{\Gamma \vdash^= A = C}$$

$$(Type\text{-}Eq) \ \frac{\Gamma \vdash^= \text{ok}}{\Gamma \vdash^= \text{Type} = \text{Type}} \qquad (El\text{-}Eq) \ \frac{\Gamma \vdash^= M = N : \text{Type}}{\Gamma \vdash^= El(M) = El(N)}$$

$$(\Pi\text{-}Eq) \ \frac{\Gamma \vdash^= A_1 = B_1 \quad \Gamma, x{:}A_1 \vdash^= A_2 = B_2 \quad \Gamma, x{:}B_1 \vdash^= B_2 \ \text{kind}}{\Gamma \vdash^= \Pi x{:}A_1.A_2 = \Pi x{:}B_1.B_2}$$

Term Equality

$$(Sym) \ \frac{\Gamma \vdash^= M = N : A}{\Gamma \vdash^= N = M : A} \qquad (Trans) \ \frac{\Gamma \vdash^= M = N : A \quad \Gamma \vdash^= N = P : A}{\Gamma \vdash^= M = P : A}$$

$$(=R) \ \frac{\Gamma \vdash^= M = N : A \quad \Gamma \vdash^= A = B}{\Gamma \vdash^= M = N : B} \qquad (Var) \ \frac{\Gamma_0, x : A, \Gamma_1 \vdash^= A \ \text{kind}}{\Gamma_0, x : A, \Gamma_1 \vdash^= x = x : A}$$

$$(\lambda\text{-}Eq) \ \frac{\Gamma \vdash^= A_1 = B_1 \quad \Gamma, x{:}A_1 \vdash^= M_0 = N_0 : A_2 \quad \Gamma, x{:}B_1 \vdash^= N_0 : A_2}{\Gamma \vdash^= \lambda x{:}A_1.M_0 = \lambda x{:}B_1.N_0 : \Pi x{:}A_1.A_2}$$

$$(App\text{-}Eq) \ \frac{\begin{array}{c}\Gamma \vdash^= M_1 = N_1 : \Pi x{:}A_1.A_2 \\ \Gamma \vdash^= M_2 = N_2 : A_1\end{array} \quad \Gamma \vdash^= [M_2/x]A_2 = [N_2/x]A_2}{\Gamma \vdash^= M_1(M_2) = N_1(N_2) : [M_2/x]A_2}$$

$$(\beta) \ \frac{\begin{array}{c}\Gamma, x{:}A_1 \vdash^= M_0 : A_2 \\ \Gamma \vdash^= M_2 : A_1\end{array} \quad \Gamma \vdash^= [M_2/x]M_0 : [M_2/x]A_2}{\Gamma \vdash^= (\lambda x{:}A_1.M_0)(M_2) = [M_2/x]M_0 : [M_2/x]A_2}$$

Fig. 3. Rules for Minimal Equational LF

Towards Limit Computable Mathematics

Susumu Hayashi[1] and Masahiro Nakata[2]

[1] Department of Computer and Systems Engineering
Kobe University, Rokko, Nada, Kobe, Japan
shayashi@kobe-u.ac.jp
http://kurt.cla.kobe-u.ac.jp/~hayashi/
[2] Graduate School of Science and Technology
Kobe University, Rokko, Nada, Kobe, Japan

Abstract. The notion of Limit-Computable Mathematics (LCM) will be introduced. LCM is a fragment of classical mathematics in which the law of excluded middle is restricted to Δ_2^0-formulas. We can give an accountable computational interpretation to the proofs of LCM. The computational content of LCM-proofs is given by Gold's limiting recursive functions, which is the fundamental notion of learning theory. LCM is expected to be a right means for "Proof Animation", which was introduced by the first author [10]. LCM is related not only to learning theory and recursion theory, but also to many areas in mathematics and computer science such as computational algebra, computability theories in analysis, reverse mathematics, and many others.

1 What Is LCM?

LCM (Limit-Computable Mathematics) is constructive mathematics augmented with some classical principles which are semi-executable by a notion of *computation in the limit* initiated by Gold [8]. It is remarkable that the same notion of limit had implicitly appeared in Hilbert's works of invariant theory in the late 1880's [11]. That notion of limit was beyond the algorithmic tendency of algebra of the time and was even called "theology". We will quote the "theology" from Hilbert's own words in his article in which was proved proved his now famous finite basis theorem for forms. (A form is a homogeneous polynomial.) Hilbert formulated it in a more concrete way than the contemporary formulation [12]:
If an infinite sequence of forms in the n variables x_1, x_2, ..., x_n is given, say F_1, F_2, F_3, ..., then there is always a number m such that every form in the sequence can be expressed as

$$F = A_1 F_1 + A_2 F_2 + \cdots + A_m F_m,$$

where A_1, A_2, ..., A_m are appropriate forms in the same n variables.

Hilbert proved this theorem by mathematical induction on n. However, Gordan was critical of the fact that Hilbert had not given any clear ordering of all forms along which a recursive method of proof is performed. It was due to the fact that Hilbert's proof was not constructive. He used non-constructive

P. Callaghan et al. (Eds.): TYPES 2000, LNCS 2277, pp. 125–144, 2002.
© Springer-Verlag Berlin Heidelberg 2002

arguments repeatedly in his proof, which was based on the mathematical induction. A typical argument appears in the base case $n = 1$. He argued as follows (p.144, [12]):

> In the simplest case $n = 1$ every form in the given sequence consists of only a single term of the form cx^r, where c denotes a constant. Let $c_1 x^{r_1}$ be the first form in the given sequence whose coefficient is $\neq 0$. We now look for the next form in the sequence whose degree is $< r_1$; let this form be $c_2 x^{r_2}$. We now look for the next form in the sequence whose degree is $< r_2$; let this form be $c_3 x^{r_3}$. Continuing in this way, after at most r_1 steps, we come to a form F_m of the given sequence which is followed by no form of lower order. Since every form in the sequence is divisible by this form F_m, m is a number with the property required by our theorem.

In the argument above, the minimum number principle(MNP)

$$\forall f : \mathbf{N} \to \mathbf{N}.\exists m.\forall n.f(m) \leq f(n)$$

is used to specify the m in F_m. Thus it is non-constructive. Non-constructivity of this argument is propagated throughout the proof in the induction steps. It is one of the first fully non-constructive proofs in algebra. This is why Gordan, who considered the matter in algorithmic ways, blamed it.

However, there is something strange in Hilbert's arguments. Anyone educated by modern mathematics would prove it in much fewer lines: "Let r_m be the least number among $r_1, r_2, \cdots$. Then F_m divides all the forms in the sequence, and hence m is a number with the property required by our theorem." However, Hilbert's "proof" is different from this standard modern proof using MNP by reductio ad absurdum. What then did he do? We may guess that he tried to give a "computational meaning" to the principle MNP.

Surprisingly, his arguments are practically the same as the arguments by Gold[8] and Berardi[3]. Gold's limit recursive functions are the central notion of algorithmic learning theory [8]. The values of a function $f(x)$ is called learnable if there is another recursive function $g(x, t)$ such that $\exists N.\forall t \geq N.f(x) = g(x, t)$. Consider t is a discrete time parameter. The the sequence $g(x, 1), g(x, 2), \cdots$ represents the process of an evolution of guesses on the value $f(x)$ by an entity. The entity guessing the value $f(x)$ may guess wrong values in the early stages of the evolution. However, it eventually converges to think the right value. It reaches the right value after the time N as above. Similarity of this guessing process to Hilbert's arguments is obvious. In Gold's words, Hilbert proved that the value m of the index of invariants system is learnable for the case $n = 1$. Furthermore, a careful inspection of Hilbert's proof shows that the solution of the finite basis theorem is learnable.

The relationship to Berardi's work is more complicated, but it is the one that we are most interested in. Berardi[3] tries to give understandable computational contents to classical proofs. Although there were several different approaches to the extended Curry-Howard isomorphism for classical logic, they shared a common defect. The algorithm or programs associated with classical proofs are extremely complicated, inefficient and illegible. Berardi tried to find out legible computational interpretations and investigated some candidates. One of them was an approximation theory of classical proofs [3]. According to his theory, one can associate a legible interpretation of a canonical proof of the minimum number principle. Remarkably, his interpretation was nearly the same as that of Gold and of Hilbert [3]. LCM was motivated by such a similarity. The background of the idea, however, comes from a practical issue in formal proof developments. We will illustrate the motivation in the next section. After illustrating the motivation, the notion of LCM is introduced in section 3, a relation to the arithmetical hierarchy is given in section 4, the scope of LCM is briefly examined in section 5, a relationship to computability analysis will be pointed out in section 6 and some open problems and related works are presented in section 7.

2   Proof Animation: Testing Proofs

The formal development of proofs is becoming more and more practical due to advancements of hardware and software. However, it is still very costly. There are several reasons why the task is so heavy. One of them, and probably the most serious one, is a mismatch between the formalized notion and the informal notion. In order to solve this problem, the first author introduced the notion of *proof animation* [10].

2.1   Debugging Proofs by Tests

The good old notion of "proofs as programs" reads "if a program is extracted from a checked proof, then it does not have bugs." The notion of proof animation is its contrapositive, that is, "if a program extracted from a proof has a bug, then the proof is not correct." The objective of the latter is not correct programs but checked proofs. Here proofs mean partial proofs under development and so they may have bugs in subgoals. Suppose we are going to prove a goal by an induction rule. We may apply an induction tactic with a wrong invariant. Even a finished and checked proof may have a bug, since formalizations of statements themselves can be wrong. On such considerations, the author introduced the notion of "proof animation" or "testing proofs". The idea is finding bugs in proofs by testing programs associated to proofs. See [10] for detailed discussions.

Proof animation is illustrated in [10] by a logic puzzle. Here we present a variant. We prove the following proposition:

Proposition 1 (false!). *If* $a_1, \cdots, a_n, \cdots$ *is a progression of 0 or 2, then the infinite product* $\prod_{i=1}^{\infty} a_i$ *converges.*

Of course, this is false, but we "prove" it. If an element in the progression is 0, then the product converges to 0. Thus, we may assume that all elements of the progression are 2. Thus $\prod_{i=1}^{n} a_i = 2^n$. We will prove that $2^n = 1$ by induction on n. Then the conclusion is obvious.

For the base case $n = 0$, $2^0 = 1$ holds. For the induction step, take n_0, n_1 such that $n + 1 = n_0 + n_1$ and $n + 1 > n_0, n_1$. Then, by the induction hypotheses $2^{n_0} = 1$ and $2^{n_1} = 1$ hold. Thus,

$$2^{n+1} = 2^{n_0} \cdot 2^{n_1} = 1 \cdot 1 = 1.$$

Q.E.D.

We will show how we can find the error of this (partial) proof by proof animation. An incorrect proof is similar to a program with bugs. In the case of a program, a program may produce a wrong result on correct inputs. Then, we *trace* the execution of the program to see where and when the correct inputs turn to a wrong result. In our case, there is an example for which the assumptions hold but the result is wrong. The progression without any zeros $a_0 = 2$, $a_1 = 2, a_2 = 2, \cdots$ is such an example. Thus, a false inference must be detected by this example. The proof is carried out by an application of the law of the excluded middle (LEM) that a zero exists in the progression or not. The progression above is the "not" case and the proof of this case uses the false equation $2^n = 1$. Here a bug resides. We may stop the animation here having found this wrong equation, but we now wish to locate the *exact* point where a wrong inference comes in.

We animate the proof by the example "n=5". The proof runs as follows:

1. 5 may be decomposed into $5 = 2 + 3$, and thus $2^5 = 2^2 \cdot 2^3 = 1$. $2^5 = 1$ is reduced to $2^2 = 1$ and $2^3 = 1$.
2. We go on to the case $n = 2$. By $2^2 = 2^1 \cdot 2^1$, the fault is reduced to $2^1 = 1$.
3. We go on to $2^1 = 1$. This is the case $n = 1$. To proceed, we must find n_1 and n_2 such that $1 = n_0 + n_1$ and $1 > n_0, n_1$. However, there is no such pair of natural numbers. We have found a bug.

We have found that the fault sneaked in at the subgoal in the induction step:

$$\exists a, b.(n + 1 = a + b \wedge n + 1 > a, b). \tag{1}$$

Note that this is a kind of error we may often make in proofs by induction, and recursive (or loop) programs. Proof animation is not a new technology. Mathematicians customarily analyze proofs as above by hands. Our aim is to *mechanize* it. Why not mechanize proof animation on proofs which are formalized and stored in a computer?

Actually, as pointed out in [10], proof animation of constructive proofs is possible by "proofs as programs" notion. For example, the proof of the lemma $2^n = 1$ above is constructive modulo the false subgoal (1). Since the subgoal (1) is existential, we assume two undefined functions $A(n)$ and $B(n)$, which return respectively the values a and b in the subgoal, and another undefined function $C(n)$ realizing the body of the subgoal $(n + 1 = a + b \wedge n + 1 > a, b)$. Since these are undefined, we have to supply values for them manually, when they are called in actual executions. Note that we have to choose the return values satisfying the condition given by the subgoal, that is, $n + 1 = A(n) + B(n)$ and $n + 1 > A(n), B(n)$.

$2^n = 1$ was proved by induction. The following function $g(n)$ defined by the corresponding recursion is associated with the proof:

$$g(0) = \text{base}$$
$$g(n+1) = \phi(x, g(A(n)), g(B(n))).$$

We trace the execution of g, A and B. Namely, we trace the theorem and the subgoal. By $g(5) = \phi(4, g(A(4)), g(B(4)))$, we have to choose values of $A(4)$ and $B(4)$. Note that these values must satisfy the conditions mentioned above and, if vales are once chosen, then they must be recorded and consistently reused on other calls on the same actual parameters. Suppose that we choose the values: $A(4) = 2$ and $B(4) = 3$. Then, the next step is the computation of $g(2)$.

$$g(2) = \phi(4, g(A(1)), g(B(1))).$$

Let us choose $A(1) = 1$ and $B(1) = 1$. Then,

$$g(1) = \phi(4, g(A(0)), g(B(0))).$$

We cannot choose values of $A(0)$ and $B(0)$ satisfying the condition $1 = A(0) + B(0)$ and $1 > A(0)$, $B(0)$. Here, we have found a bug. Note that the trace exactly corresponds to the informal proof animation as described above.

2.2 How to Animate Classical Proofs?

We have shown how constructive proofs are animated. However, majority of proofs are not constructive but classical. How can we animate them? Note that the proof of the false proposition above is classical. Nonetheless, we were able to animate it. We do not need to decide if zero appears or not for all progressions to find the bug in the proof, since error is found by tests on particular examples. It is enough to decide if zero appears or not only for the examples. In the example above, it was clear that we could find errors only on the example $2, 2, \cdots$. Thus, we did not need any essential use of the law of excluded middle. Similarly, the Curry-Howard isomorphism for constructive proofs is often adequate to animate simple classical proofs.

Another way to animate classical proofs by the Curry-Howard isomorphism is to use a "man-machine interaction." We may regard a classical proof as a constructive proof with some unproven lemmas, all of which happen to be instances of LEM. In the animation example above, such an unproven lemma (subgoal) was used, which was incidentally false, and when such a lemma is "called" in the proof, then *we guessed* realizers for the lemma. Similarly, on each call of LEM, we may guess its realizer. Many proofs with classical reasoning can be animated by such ad hoc ways.

However, systematic method animating classical proofs is still desirable. There are some proofs for which ad hoc animations illustrated above are difficult. For example, Hilbert's proof of finite basis theorem uses LEM repeatedly in an induction proof. Animating it by "man-machine interaction", the machine would ask you plenty of successive queries and you would lose good comprehension of the proof in such a busy job.

We thus need a method for smooth and natural animation of such classical proofs. Since proof animation is a means to understand proofs through computations, such proof execution methods or interpretation must meet the following two criteria:

1. computational contents (programs) associated with proofs are legible;
2. association between proofs and programs is legible.

We call a proof execution method with these criteria *accountable*.

Unfortunately, almost all proof execution methods for classical logic satisfy neither of the two criteria. The unique exception as far as we know is Berardi's theory. At least for some typical examples, e.g., the minimum number principle as mentioned above, his theory meets the first criterion. However, it is because he analyzed his interpretation in clever ways. No straightforward and mechanical method to associate the limit algorithm to his proof is known. This means that even Berardi's approximation theory does not satisfy the second criterion.

We now believe that there is no truly accountable proof execution methods for *full* classical mathematics. Thus we seek for an alternative approach. We try to find a *fragment F* of classical mathematics such that proof execution for *F* is accountable and significant part of mathematics can be carried out in *F*.

3   Limit-Computable Mathematics

The consideration explained above led the first author to the idea of *Limit-Computable Mathematics* (LCM). The idea is simple. LCM *is a fragment of classical mathematics whose Brouwer-Heyting-Kolmogorov-Kleene-interpretation is realized by limiting recursive functions rather than recursive functions.* Since, it is realized by limit-computable (limiting recursive) functions, it is called Limit-Computable Mathematics. Putting it on more formally, LCM is a fragment of mathematics whose realizability interpretation is realized by limit-computable functions. To explain the idea, a definite realizability interpretation for LCM and its formal theory are given below.

3.1   A Realizability and Formal System

A formal semantics of LCM is first given according to Kleene realizability interpretation by "taking the limit" of basic recursive function theory. We then give a formal theory which is sound for this interpretation. First, we will give a realizability interpretation of first order arithmetic. It is essentially Kleene's original interpretation except that the realizers are replaced with limit-computable functions. Basic Recursive Function Theory (BRFT) is an abstract definition of systems of recursive functions introduced by Strong and Wagner [20, 22]. A BRFT is a structure over a set A with at least two elements. Let F be a set of partial functions on A. For every $n \geq 0$, F_n is the set of n-ary partial functions in F. Then, the axioms of BRFT are as follows:

1. The constant functions and the projections are contained in F.
2. $\exists \Psi \in F_4.\ \forall abcx \in A.\ \Psi(a,b,c,x) \simeq \begin{cases} b \cdots & x = a \\ c \cdots & x \neq a \end{cases}$.
3. F is closed under composition.
4. A partial function $\Phi_m \in F_{m+1}$ is fixed for each $m > 0$, where $F_m = \{\lambda x_1 \cdots x_m.\Phi_m(x, x_1, \cdots, x_m) | x \in A\}$.
5. There is a function $S_n^m \in F_{m+1}$ for each $m, n > 0$ which satisfies the following:
   (a) $S_n^m(x, x_1, \cdots, x_m)$ is defined for all $x, x_1, \cdots, x_m$;
   (b) $\Phi_n(S_n^m(x, x_1, \cdots, x_m), y_1, \cdots, y_n) \simeq \Phi_{m+n}(x, x_1, \cdots, x_m, y_1, \cdots, y_n)$.

The equality $\simeq$ is the one of Logic of Partial Terms (LPT) in [2, 9]. Ψ is a call-by-value conditional function over equality test. Φ is a formalization of Kleene's bracket in recursion theory. It resembles a universal function in the sense of LISP. S_n^m is a formalization of s-m-n function of recursion theory. It can be regarded as a currying operator taking any code x of a function f with $m + n$ formal arguments and the first m actual arguments $x_1, \cdots, x_m$, and returning a code of the function $\lambda y_1, \cdots, y_n.f(x_1, \cdots, x_m, y_1, \cdots, y_n)$ with n formal arguments.

If the domain A of BRFT is the set of natural numbers, it is called ω-BRFT. We define **PRF** as a set of all partial recursive functions. Every ω-BRFT F with the successor function contains the **PRF**, since the recursion theorem holds for BRFT. Thus, we can define Kleene's recursive realizability interpretation of first order arithmetic on any ω-BRFT with the successor function, and Heyting Arithmetic HA is sound with respect to this interpretation. See [16] for details.

We now define the limit of any ω-BRFT F. We define the limit of a partial function f of F by

$$\lim_t f(x_1, \cdots, x_n, t) \simeq y \iff \exists a \forall b \geq a.f(x_1, \cdots, x_n, b) \simeq y.$$

f is called a *guessing (partial) function* of the limit. Set

$$\mathbf{Lim}(F)_n = \{\lim_t f(x_1, \cdots, x_n, t) | f \in F_{n+1}\},$$

$$\mathbf{Lim}(F) = \cup_n \mathbf{Lim}(F)_n$$

$\mathbf{Lim}(F)$ is called *the limit of F*. It is proved that $\mathbf{Lim}(F)$ is again an ω-BRFT in [16]. Then $\mathbf{Lim}(F)$ realizes **HA** plus some *restricted classical principles*.

Let us introduce these restricted classical principles. A Π_n^0-*formula* is a formula of the form $\forall x_1 \exists x_2 \cdots Q x_n.A$, where A is a formula for a recursive relation. Σ_n^0-*formula* is defined similarly.

- Σ_n^0-LEM is $A \vee \neg A$ for any Σ_n^0-formula A. LEM stands for Law of Excluded Middle. Π_n^0-LEM is defined similarly.
- Δ_n^0-LEM is $\forall x.(A \leftrightarrow B) \Rightarrow A \vee \neg A$, where x is the sequence of all free variable occurring in A and B, and A is a Σ_n^0-formula and B is a Π_n^0-formula
- Σ_n^0-DNE is $\neg\neg A \Rightarrow A$ for any Σ_n^0-formula A. DNE stands for Double Negation Elimination.

$\mathbf{Lim}(F)$ realizes Σ_2^0-DNE and it implies Σ_1^0-LEM, Π_1^0-LEM, Δ_2^0-LEM in $\mathbf{HA}$ (see [16]). Σ_1^0-LEM is the principle Hilbert used repeatedly in his invariant theory. $\mathbf{Lim}(F)$ is adequate to realize the theory and so his solution of Gordan's theorem produces a limit-algorithm computing the finite full invariant system from any finite system of forms. Of course, MNP is realized. The false Proposition 1 for proof animation example is also realized even if we consider limiting recursive functions relative to the *input* progression $a_0, a_1, \cdots$.

To formalize these facts, we introduced a formal theory $\mathbf{HA}+\Sigma_2^0$-DNE dubbed as $\mathbf{HAL}$ in [16]. It is the first formal theory for LCM. Although it is unsatisfactory with many respects for applications in our mind, it seems to be a fine theory to start with. In principle, the arguments on Hilbert's invariant theory of 1890 discussed above are formalized in $\mathbf{HAL}$. (This does not mean that a proof checker based on $\mathbf{HAL}$ is adequate to formalize them on computers. We need more expressive languages.) Formalization of MNP is a child's play. To formalize the incomplete proof of Proposition 1, we introduce a free function variable α and consider $\mathbf{HAL}(\alpha)$, which has the relativized scheme $\Sigma_2^{0,\alpha}$-DNE

$$\neg\neg\exists x\forall y.A(x, y, \alpha) \Rightarrow \exists x\forall y.A(x, y, \alpha),$$

where $A(x, y, \alpha)$ is recursive in α. By representing the infinite progression by α, it is straightforward to formalize the proof *using the false subgoal*.

The following theorem and corollary are shown in [16].

Theorem 1 (Soundness). *Let F be an ω-BRFT with successor function. If $\mathbf{HAL} \vdash A$ and $FV(A) = \{u_1, \cdots, u_n\}$, then there is an n-ary limiting partial function $f \in \mathbf{Lim}(F)_n$ such that $f(u) \in \mathbf{N}$ and $\mathbf{Lim}(F) \models f(u) \mathbf{r} A(u)$ hold for every $u = u_1, \cdots, u_n \in \mathbf{N}^n$. This also holds for $\mathbf{HAL}(\alpha)$ by interpreting α as a function of F.*

Corollary 1 (Program Extraction). *Suppose $\mathbf{HAL} \vdash \exists y.A$, and let $x_1, \cdots, x_n$ be the free variables of $\exists y.A$. Then there is an n-ary limiting recursive function f such that $\mathbf{HAL} \vdash A[f(x_1, \cdots, x_n)/y]$ holds. Furthermore, if A is a recursive formula, then f can be recursive. The result also holds for $\mathbf{HAL}(\alpha)$, if we take f to be a limiting recursive function in α.*

$\mathbf{HAL}(\alpha)$-part of the corollary is proved by taking F as the class of partial recursive functions relative to α.

3.2   Proof Animation by LCM

In the ad hoc animation of Proposition 1, we animated not the entire classical proof, but only the constructive subproof for the case that zero does not appear in the progression. It is sufficient for practical animation and we believe it is the right way. Nonetheless, we consider an animation of the entire proof just to show how limiting recursive functions animate such a classical proof.

The classical reasoning used in the proposition is a Σ_1^0-LEM on $\exists i.a_i = 0$. If $\exists i.a_i = 0$ holds, the infinite product actually converges and nothing is wrong.

If not, we get into the wrong part of the proof. We cannot decide which case actually happens, since there is no algorithm to decide if $\exists i. a_i = 0$ or not. This axiom is realized by a limiting recursive function relative to $\{a_n\}$. The realizer *guesses* as follows.

1. At the beginning, the realizer guesses that the answer is "there is no zero in it".
2. It successively checks if every a_i is non-zero as it guesses. As far as the guess is correct, it keeps the first guess.
3. However, if it finds a counterexample, it changes its mind. The new guess is "there is an i such that $a_i = 0$."
4. Once the new guess is obtained, the realizer never changes the guess. It will keep guessing that "there is an i such that $a_i = 0$."

Of course, it may stop when it finds a counterexample, since the guess never changes. With this realizer, the entire process of the animation of the proposition goes as follows.

1. Check the progression $a_0, a_1 \cdots$ successively if a_i is zero.
2. Assume that we have just checked a_n and so far there is no zero. Then, we get into the subproof for "there is no zero", and apply the proof animation of the proof $\prod_{i=1}^{n} a_i = 1$.
3. However, if we find $a_n = 0$, we get into the subproof of $\prod_{i=1}^{\infty} a_i = 0$.

This time, an animation is an infinite computation process. In a sense, it is a computation of a stream of guesses. If a_1 of the input progression is zero, we progress forever without finding any error in the proof. This is obvious, since the proposition is correct for this particular input. If we choose any progression such that $a_1 \neq 0$, the animation gets into the subproof of $2^1 = 1$ and we can find the error just as we did above.

Although it is more natural to do the proof animation of the false proof by the ad hoc way, this illustrates that the limiting computation serves as an accountable proof animation of LCM. For a more intricate case like Hilbert's invariant theory, it seems practically impossible to do ad hoc proof animation only with animation of constructive proofs. On the other hand, it is not difficult to read LCM-animation process of theorems and lemmas of Hilbert's invariant theory, e.g. his finite basis theorem from their original texts. Thus, we expect that limiting computational realizability serves a good basis for animation of LCM.

Proof animation of an LCM-theorem is an infinite computational process whose algorithm is automatically extracted from an LCM-proof of the theorem. We run the algorithm on some examples and watch out what happens. The process is infinite and we may not get the final value of the realizer of the theorem in a reasonable time. However, we do not care, since our objective is to find something not conformable to our intuition or actual inconsistency in the computation. If we find one, we stop. If we cannot find any, we will be more confident with the proof, although it does not mean the (incomplete) proof is really correct. Only the degree of correctness is increased, as it does not check the proof but *experiment* the proof.

4   LCM and Arithmetical Hierarchy

The class of limiting recursive functions is known to coincide with that of Δ_2^0-functions in the arithmetical hierarchy. In constructive mathematics, functions are computable, that is, they are Δ_1^0-functions. In this sense, constructive mathematics is Δ_1^0-mathematics. In the same sense, LCM is Δ_2^0-mathematics. Thus LCM naturally has deep relationships to the arithmetical hierarchy.

4.1   LCM as Jump

A total function is limiting recursive if and only if its graph is a Δ_2^0-set. Similarly, a set of natural numbers is a Δ_2^0-set if and only if its characteristic function is a limiting recursive function. In general, a set of natural numbers is a Δ_{n+1}^0-set if and only if its characteristic function is definable in the form $\lim_{t_1} \lim_{t_2} \cdots \lim_{t_n} f(t_1, t_2, \cdots, t_n, x)$, where f is a recursive function (Shoenfield's limit lemma; see [17]).

A set of natural numbers is Δ_{n+1}^0 if and only if it is recursive in a Σ_n^0-set or Π_n^0-set. Since negation is a recursive operator, the latter two conditions are the same. These observations tell us that Gold's limit operation embodies a single quantification modulo recursiveness. Since recursiveness is identified with constructiveness, we may regard LCM as mathematics with LEM of formulas having a single arithmetical quantification modulo constructive reasoning.

¿From this point of view, relativized LCM such as $\mathbf{HAL}(\alpha)$ may be characterized as *a logic of single-layered arithmetical quantification* or *a logic of jump*. Jump is a recursion theoretic operator on recursion theoretic degrees representing a single arithmetic quantification [17]. A degree (Turing-degree or T-degree) is a coset of the equivalence relation "mutually recursive in", that is, two sets (or predicates) have the same degree if and only if they are mutually recursive in. For example, $\forall y.\phi(x, y)$ and $\exists y.\neg\phi(x, y)$ have the same degree. The universal $\Sigma_1^{0,A}$-set $A' = \{x\}^A(x) \downarrow$ is called the jump of A. The jump operator $\mathbf{a}'$ on degrees is induced by this jump operator A' on sets. $\mathbf{a}'$ is called the jump of a degree $\mathbf{a}$. Realizers of theorems of $\mathbf{HAL}(\alpha)$ are recursive in α' by regarding α as its graph. LCM embodies a jump in this sense.

The Δ_n^0-sets are closed under boolean operations, but not under quantifications. The soundness and the extraction theorem for $\mathbf{HAL}$ are obviously extended to Δ_n^0-mathematics such as $\mathbf{HA}(\alpha) + \Delta_n^{0,\alpha}$-DNE. These results show that the mathematics based on Δ_n^0-functions has a good first order logic, although Δ_n^0-sets are not closed under quantification. This is because the logic is constructive except semi-classical principles like Δ_n^0-DNE, and constructive rules of quantification are recursive in the sense of Kleene's realizability.

A principle of numerical omniscience (NOS)

$$\forall x \in \mathbf{N}.(A(x) \vee \neg A(x)) \Rightarrow \forall x \in \mathbf{N}.A(x) \vee \exists x \in \mathbf{N}.\neg A(x),$$

which is clearly related to the discussion above, has been considered by several authors, [7, 14]. Iterated applications of this scheme implies Σ_n^0-LEM in $\mathbf{HA}$

and the system turns to **PA**. Namely, NOS maintains that realizers are closed under the jump operator and makes the levels of hierarchy indiscernible. Thus, it is *level-senseless*. On the other hand, LCM sensitively discerns levels of the arithmetical hierarchy and is *level-sensitive*.

A level-sensitive logic with the full arithmetical hierarchy might be obtained introducing disjunctions $\vee^{(n)}$ and existence quantifiers $\exists^{(n)}$ realized by Δ^0_{n-1}-functions, and adding the following stratified axioms of NOS ($\text{NOS}^{(k)}$):

$$\forall x \in \mathbf{N}.(A(x) \vee^{(n)} \neg A(x)) \Rightarrow \forall x \in \mathbf{N}.A(x) \vee^{(n+1)} \exists^{(n)} x \in \mathbf{N}.\neg A(x).$$

Note that constructive mathematics is a logic of $\vee^{(0)}$ and $\exists^{(0)}$, and LCM is a logic of $\vee^{(1)}$ and $\exists^{(1)}$. The minimum number principle is stated as

$$\forall f : \mathbf{N} \to^{(k)} \mathbf{N}.\exists^{(k+1)} m.\forall n \geq m.f(m) \leq f(n),$$

where $f : \mathbf{N} \to^{(k)} \mathbf{N}$ means that $\forall x : \mathbf{N}.\exists^{(k)} y.f(x) = y$.

We may regard $\vee^{(n)}$ and $\exists^{(n)}$ realized by $\Delta^{0,\alpha}_{n-1}$-functions relativized to any input stream α. Then the minimum principle above with $k = 0$ maintains that the value m attaining the minimum value of the stream is obtained limiting recursively relative to the stream. LCM would be formalized more naturally by a logic with variables for input streams α, β, $\cdots$ as well as $\vee$, $\exists$ realized by functions recursive in the streams and $\vee'$, $\exists'$ realized by functions limiting recursive in the streams. It will not be difficult to formalize such a logic or type theory, and it might resemble the type theory with informative and non-informative quantifiers. Note that $\mathbf{HAL}(\alpha)$ can be regarded as a fragment of such a logic, only with $\vee'$ and $\exists'$.

An important observation of LCM is that a great number of basic classical theorems, such as Hilbert finite basis theorem, MNP, and the maximum value theorem for continuous functions over $[0, 1]$, etc, are provable by a "single jump" rather than multi-jumps with iterated usages of NOS or NOS_k. Of course, we may use an output of one of these theorems as an input of another theorem. For example, the output of the maximum value theorem is a real number and we may apply MNP to find the minimal digit of a decimal composition of the number. This corresponds to input the output of NOS_k to NOS_{k+1}, and so the final outputs are elevated by two levels. Then we cannot directly animate the proof by limiting recursive realizers. However, it is not necessary to animate such a composition of proofs at once. We may animate two components, that is, the maximum number theorem and MNP, separately. In the same vein, proofs with NOS may be decomposed into LCM-proofs and each of them can be animated. It will be an interesting problem to find a method deciding if a proof with NOS is decomposable into LCM-proofs, and to decompose it, when it is possible.

4.2   Isolated Limits Do not Elevate Levels

Although repeated applications of NOS is prohibited, nested applications of LEM are allowed in LCM. Assume that we use a Σ^0_1-LEM $\forall x.A(x) \vee \exists x.\neg A(x)$ for a proof by cases. In the second case, we assume there is x_0 such that $\neg A(x_0)$ and may use another Σ_1^0-LEM $\forall y.B(x_0, y) \vee \exists y.\neg B(x_0, y)$. Note that x_0 is obtained by limiting recursion, say $x_0 = \lim_t \phi(t)$. Using x_0, the value by the second application of the restricted LEM, e.g, the value y_0 for y is obtained as $\lim_s \psi(x_0, s)$. We thus have

$$y_0 = \lim_s \psi(\lim_t \phi(t), s) = \lim_s \lim_t \psi(\phi(t), s).$$

On the surface, y_0 is obtained by nested limits for a Σ_3^0-function, due to Shoenfield's lemma. However, it turns out that this is equivalent to $\lim_t \psi(\phi(t), t)$ (see [16]). This is because ϕ does not depend on s. The nested limits in $\lim_s \psi(\lim_t \phi(t), s)$ are *isolated* in the sense of Takeuti [21]. Namely, any bound variable of a limit does not appear as a free variable in the scope of another limit. On the other hand, the limits in $\lim_s \psi(\lim_t \phi(t, s), s)$ are not isolated, since s appears in the scope of $\lim_t$. It is easy to see the following theorem.

Theorem 2. *(1) Assume that T is a first order term built from basic functions with limit and application. Let $\lim_{t_1} \phi_1(t_1), \lim_{t_2} \phi_2(t_2), \cdots, \lim_{t_n} \phi_n(t_n)$ be the occurrences of all limits in it. Let T' be the term obtained from T by replacing these limits with $\phi_1(t), \phi_2(t), \cdots, \phi_n(t)$ respectively. If the limits in T are isolated, then, T is equivalent to $\lim_t T'$.[1] (2) If the limits of T_1 and T_2 are isolated, then the limits in $T_1[T_2/x]$ are isolated. (3) A term with only isolated limits can be obtained by repeated substitutions of terms with at most one occurrence of limit.*

By regarding the substitution $T_1[T_2/x]$ as a composition of T_1 and T_2, the property (2) corresponds to the condition (3) of BRFT. Plugging a limiting process into another does not elevate levels of the arithmetical hierarchy. Constructions associated with ordinary logical rules are plugging of such processes except "λ-abstraction" corresponding to $\forall$-introduction and $\Rightarrow$-introduction. An abstraction (in BRFT) is an introduction of a limiting function, and so its value is definite from the beginning and the limit operation is not necessary. For this reason, LCM conserves the level of noncomputability at Δ_2^0.

4.3   A Hierarchy of Semi-classical Principles

Up to the equivalence relation "mutually recursive in", a Π_1^0-set is equivalent to a Σ_1^0-set. However, Markov's principle for recursive predicate, or Δ_1^0-DNE in our notation, is used to prove that Π_1^0-LEM and Σ_1^0-LEM are constructively equivalent. Thus the hierarchy of semi-classical principles of LCM are more subtle than the standard arithmetical hierarchy. Kohlenbach [15] and Hayashi have shown that there is a hierarchy of provability of semi-classical principles in **HA** as shown below.

---

[1] Gold's limit commutes with any "finite" construction of terms. We may even define "finite construction" by means of this computation as the notion of finite representability of the category theory.

Theorem 3. *In the following figure 1, the arrow → denotes the provability in* **HA** *and the dashed arrow↛ represents the unprovability in* **HA***. Note that for each axiom in the diagram, f and g are recursive functions.*

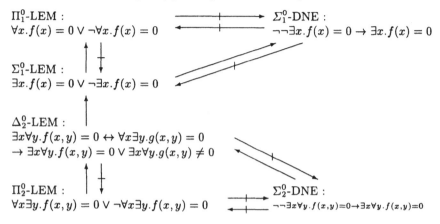

Π_1^0-LEM :
$\forall x.f(x) = 0 \vee \neg\forall x.f(x) = 0$

Σ_1^0-DNE :
$\neg\neg\exists x.f(x) = 0 \to \exists x.f(x) = 0$

Σ_1^0-LEM :
$\exists x.f(x) = 0 \vee \neg\exists x.f(x) = 0$

Δ_2^0-LEM :
$\exists x\forall y.f(x,y) = 0 \leftrightarrow \forall x\exists y.g(x,y) = 0$
$\to \exists x\forall y.f(x,y) = 0 \vee \exists x\forall y.g(x,y) \neq 0$

Π_2^0-LEM :
$\forall x\exists y.f(x,y) = 0 \vee \neg\forall x\exists y.f(x,y) = 0$

Σ_2^0-DNE :
$\neg\neg\exists x\forall y.f(x,y)=0 \to \exists x\forall y.f(x,y)=0$

Fig. 1. The hierarchy

Although we suspect that Δ_2^0-LEM would not be derivable from Σ_1^0-LEM, it is still an open problem. We conjecture that similar hierarchies exist at higher levels but it has not been solved either. Furthermore, we do not know a complete picture of the hierarchy even at the first and the second levels. Note that we have considered only prenex formulas. Under constructive logic, there are many of constructively non-equivalent variations which are classically equivalent to a prenex normal form. We therefore do not know how the full hierarchy looks like.

Furthermore, there is another interesting problem. *Boolean hierarchy* is a hierarchy of limiting recursive functions [17]. A guessing process is a progression of *change of mind* of an entity. The entity changes mind several times but eventually stops to do so. Then, its guess has a limit. For Σ_1^0-LEM $\forall x.A(x) \vee \exists x.\neg A(x)$, the number of changes of mind is at most once. Note that if the entity finds a counterexample $\neg A(x_0)$, it never changes the guess. However, some problems are known to need more changes. Boolean hierarchy is a hierarchy by the number of necessary changes. Is there a counterpart of boolean hierarchy as the proof theoretical hierarchy in Theorem 3?

4.4   Proof Theory of LCM

To investigate hierarchies mentioned above and reverse mathematics of constructivity in the next section, we will need proof theoretical investigation of LCM. It seems that many metamathematical results can be transferred from constructive mathematics to LCM. For example, the derived Markov's rule for Σ_1^0-formula, that is for Σ_1^0-formula A, if $\neg\neg A$ is provable in **HA**, then so is A, can be extended to Σ_{n+1}^0-formulas for **HA** $+ \Sigma_n^0$-LEM just by redoing a well-known proof. This was conjectured by Berardi for the case of $n = 1$, and extended and proved by Hayashi.

5   What Is the Reach? Reverse Mathematics of Constructivity

Reflecting Hilbert's invariant theory, we conjecture that many theorems of the 19th and the early 20th centuries mathematics will be in the range of LCM. The first author and Hajime Ishihara conjectured that many theorems, which are not provable in Bishop style constructive mathematics but whose "approximation versions" are provable, would be provable in LCM. Typical examples are the maximum and the minimum value theorems. It maintains that any continuous function f over $[0, 1]$ attains the maximum value m at a point p in the interval. The existence of m is constructively provable, but the existence of p is not. However, p can be proved to exist in LCM in the following sense:

1. we can define a $\Pi_2^{0, \{p_m(x)\}_m, e_f}$-predicate $P(n, a)$,
2. $\forall n \in \mathbf{N}. \exists! a \in \mathbf{Q}. P(n, a)$ is provable,
3. the sequence $\{a_n\}_n$ defined by $P(n, a_n)$ is a Cauchy sequence of rationals and $f(a_n)$ converges to m.

The set of rationals, $\mathbf{Q}$, is supposed to be identified with a subset of $\mathbf{N}$. We assume that f is approximated by a sequence $\{p_m(x)\}_m$ of rational polynomials with $e_f : \mathbf{N} \to \mathbf{N}$ a modulus of uniform convergence, that is $m \geq e_f(N)$ implies $|f(x) - p_m(x)| \leq 2^{-N}$ for $x \in [0, 1]$. Since rational polynomials can be coded by natural numbers, the sequence $\{p_m(x)\}_m$ may be considered as a number theoretic function.

In the terminology of 4.1, if $e_f : \mathbf{N} \to^{(k)} \mathbf{N}$ and $\{p_m(x)\}_m : \mathbf{N} \to^{(k)} \mathbf{N}$, then $\{a_n\}_n : \mathbf{N} \to^{(k+1)} \mathbf{N}$.

Bishop has pointed out the maximum value theorem, Hahn-Banach theorem, the ergodic theorem, etc. are provable with the help of "limited principles of omniscience" which is a special case of NOS (chapter 1, [5]). These theorems are provable in constructive mathematics plus NOS, thus the output, e.g. p in the maximum value theorem, is arithmetical in the input f. The maximum value theorem presented above is an LCM-refinement of this fact. We expect that the other theorems mentioned by Bishop are also proved in LCM in a similar way.

Many theorems in recursion theory seem to be difficult to prove in LCM, since often they maintain the existence of higher levels in the arithmetical hierarchy. However, Odifreddi pointed out to the first author that Kreisel basis theorem, which is a recursion theoretic refinement of König's lemma, would be provable in LCM, and so is Gödel's completeness theorem. This is actually the case. Kreisel basis theorem represents a typical case of level-conscious proof in LCM.

On the other hand, Ishihara [13] proved equivalence of some formulations of Hahn-Banach theorem, minimal value theorem, König's lemma, and a weaker version of limited principles of omniscience. This was done in Bishop style constructive mathematics and so it does not directly fit into LCM. However, this work strongly suggests possibility of *reverse mathematics of constructivity*. Simpson and his followers have developed a theory called "reverse mathematics," in which proof-theoretic strength of a theorem in mathematics is characterized modulo a weak system of second order arithmetic (see [18]). In the same vein, the objective of reverse mathematics of constructivity is to find semi-classical principles necessary to prove non-constructive theorems. For example, the minimal number theorem MNP and Σ_1^0-LEM are equivalent in **HA**.[2] Simpson has shown that Hilbert basis theorem is equivalent to well orderedness of ω^ω in a framework of reverse mathematics [19]. In the formal systems of reverse mathematics, LEM are free for use, but the comprehension and the induction are restricted. In our setting, the former is restricted and the latter are free for use. What is the implication of this difference? Is there any general framework in which we can consider these formal theories with weak principles of constructivity, comprehension and induction? Some framework of type theories and logic-coding technique such as Edinburgh LF might be good for our aim.

6   Computability Theories in Analysis and LCM

If $n = \lim_t f(t)$, then $f(0)$, $f(1)$, $\cdots$ is a recursive progression of *rational numbers* converging to n, that is, the convergence in Gold's sense already exists in computational analysis. Recently, many researchers in applied mathematics and engineering are becoming aware of importance of discrete systems such as cellular automata thanks to advancement of computers. However, this would not mean that discrete systems prevail over continuous systems. There is a possibility to understand discrete systems deeper by techniques and visions developed in the research of continuous systems like control theory, numerical analysis and applied mathematical analysis. In a sense, LCM uses continuous apparatus "limit" to understand a discrete system "logic". Relationship between LCM and computational analysis and numerical analysis must be an important research topic. Some definite relationships are already known.

6.1   Computability Theory of Discontinuous Functions

From the time of Brouwer, it is known that computational or constructive functions in analysis are continuous due to its finiteness. For example, values of discontinuous functions like the Gaussian function $[x]$ are not constructively evaluated. Common sense is that $[x]$ is discontinuous and so it is not computable.

However, Yasugi et al. [23] pointed out that there is a simple recursive functional F such that if x is a recursive real, that is, x is represented by a pair of Cauchy rational sequence $\{r_i\}$ and modulus of convergence m_1, both of which are recursive, then $F(x)$ is a code of recursive rational sequence $\{q_i\}$ converging to $[x]$. Even a recursive modulus of convergence m_2 for $\{q_i\}$ exists. Then,

---

[2] To show MNP by Σ_1^0-LEM, prove $\exists x. f(x) \leq k \Rightarrow$ MNP by induction on k, and instantiate k with $f(0)$. This proof is due to the referee of the paper.

what is non-computable? The mapping $x = \langle\{r_i\}, m_1\rangle \mapsto m_2$ is not recursive. Interestingly, it is limiting recursive. There is a pair of recursive functional G_1 and G_2 such that $G_1(x)$ or $G_2(x)$ is a modulus of convergence of m_2. We cannot decide which is a modulus among the two, but we can decide it by Σ_1^0-LEM. If we regard the Gaussian as a function from reals to integers, then it is limiting recursive. The value part $F(x)$ is recursive, since it is regarded as real-valued function and $[x]$ is defined as a limit of recursive sequence of rationals.

Other discontinuous functions are also analyzed more precisely in [23] by means of limiting recursive functionals and similar operators. It suggests that there might be some hierarchy of functions similar to the one we suggested in the previous section.

Yasugi and her colleagues also pointed out that discontinuous functions naturally exist in computational Fréchet spaces. Just as the Gaussian function is represented by Fourier series, discontinuous functions can be regarded as computational *points* in some function spaces. It is not known how this observation is related to LCM.

6.2   A Relation to BSS Theory

Yasugi's original motivation was so-called BSS-theory (Blum, Shub, Smale-theory) of computational complexity theory over reals [6], in which the Gaussian function is a "hello world" example. The theory is a counterpart of computational complexity on natural numbers. It concerns computational complexity modulo polynomial or algebraic computations. On the contrary, LCM concerns computability or limiting computability. Nonetheless, there is an interesting structural resemblance between them, if BSS is applied to traditional recursive analysis.

BSS theory starts with any given ordered ring or ordered field.[3] Let us take the recursive reals used in traditional recursive analysis as the base ordered field. Then, the discontinuous functions considered in [23] are all computable in BSS-theory. A computing machine in BSS-theory is a finite directed graph of nodes as the figure 2. (The example is "Derived Newton Machine" of [6].) It is considered as a flowchart with its own memory space, e.g. x, y, z in the example. The figure would be self-explanatory except the following conditions:

A1. the data in the memory space as x, y, z are members of the field;
A2. the functions g_1, g_2, h are rational maps.[4]

A characteristic of BSS machine is the branching box. Deciding $a > b$ or not is a typical instance of Δ_2^0-LEM, when a and b are recursive reals with recursive modulus of convergence. Thus, BSS-theory is beyond the traditional recursive analysis. However, it is LCM-analysis just as Yasugi's observation.

---

[3] There is a version the base ring (field) is not ordered, but we do not consider it here for simplicity.

[4] If we start with ring instead, polynomial maps are used.

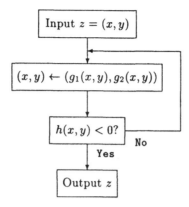

Fig. 2. A BSS Machine for Newton method

Let M be a BSS-machine. For simplicity, we assume the input is a single real x, and $f_M(x) = y$ represents the function defined by M. Furthermore, $f_M^*(x, n)$ denotes the trace function of f_M up to n-steps. Namely, if the value of $f_M(x)$ is computed by M with less than n-steps, then it returns $f_M(x)$ and otherwise it returns $\perp$, a special sign for undefinedness. Since a recursive real $x = \langle\{r_i\}, m_1\rangle$ is coded by a recursive function, we may regard the α of $\mathbf{HAL}(\alpha)$ representing it. Then it is not difficult to prove the following theorem.

Theorem 4. *For any BSS-machine M, we can effectively construct formulas F and G of $\mathbf{HAL}(\alpha)$ such that*

1. *F and G provably define functions respectively in $\mathbf{HA}$ and $\mathbf{HAL}$. Namely, $\mathbf{HA} \vdash \exists!e.F(n,e)$ and $\mathbf{HAL} \vdash \exists!d.G(n,d)$.*
2. *Let α be a recursive real. Assume that $f_M^*(\alpha, n)$ is not $\perp$. Let e and d be the natural numbers such that $F(n,e)$ and $G(n,d)$ hold. Then $\langle\{\{e\}^\alpha(i)\}_i, \lambda x.\{d\}^\alpha(x)\rangle$ represents a recursive real y and $f_M^*(\alpha) = f(\alpha) = y$.*

This theorem holds for *the extended BSS-machines*, that is, the condition A2 is relaxed to "g_1, g_2, h are computable real-valued functions in the sense of traditional recursive analysis, and their existence is provable in $\mathbf{HA}$". In a sense, this is an LCM-refinement of the path decomposition theorem of [6].[5] Roughly speaking, the theorem shows that BSS-functions definable without infinite loops are recursive in value-part, $\{r_i\}$ of a recursive real $\langle\{r_i\}, m_1\rangle$, but limiting recursive in modulus-part m_1 just as Yasugi's observation on the Gaussian function. If loops are present, even the value-part is limiting recursive in general.

---

[5] The path decomposition theorems hold for the extended BSS-machines by replacing "basic semi-algebraic sets" with finite intersections of sets defined by functions appearing in the nodes of machines."

7   Improvements of LCM Theory

The basic mathematical theory of LCM is still in its infancy. Especially, the realizability semantics by limiting BRFT and formal theory $\mathbf{HAL}(\alpha)$ are short for practice of, for example, animation of Hilbert invariant theory. Improvements of theories are necessary for such an aim. In this section, existing and necessary improvements are reviewed.

7.1   LCM Proofs as Nets of Guessing Streams

Our realizability semantics uses limiting BRFT. It is simple from the theoretical point of view, but it cannot be used in real practice. For example, assume we prove $A \wedge B$ from A and B by the conjunction introduction rule. Two realizers $\lim_s f(s)$ and $\lim_t g(t)$ will be assigned to the premises. Then a realizer of $A \wedge B$ is $\lim_u \langle f(u), g(u) \rangle$. However, it is not canonical in any sense. $\lim_u \langle f(u+7), g(u) \rangle$, $\lim_u \langle f(u+7), g(2^u) \rangle$, etc. are all correct realizers. There is no canonical way to merge (interleave) two "local times" s and t into a single global time u. Animation by realizers with such a merged time would be like speculating on behaviors of processes on a multi-task OS by observing the trace of single sequentialized time slices. Thus, we need some calculus of functions or processes in which limits are not merged into a single limit, but are executed concurrently. It would be a simple concurrent system with "change of mind" signal. With such a calculus, LCM proofs will be understood as nets of guessing streams. Each semi-classical principle produces a guessing stream on its local time. Then their guesses are communicated and transported through a net determined by constructive parts of proofs to produce an output stream. It is not difficult to imagine such a net of streams by reading proofs in Hilbert's invariant theory. However, formal calculi are yet to be known.

7.2   Berardi's Interpretation of LCM

Berardi has introduced a new version of his approximation theory based on our limit idea [4]. His theory uses limit over directed sets. This is already an improvements, but the most interesting features would be the fact that his interpretation gives some meaning even to atomic formulas. Ordinary realizability including ours do not give any good information to atomic formulas except its validity. Thus soundness of rules on atomic formulas is trivial. However, streams of guesses are passed through at such rules in his interpretation through atomic formulas. He has already given a notion of concurrent nets of guess streams. It is expected to show the way to the realistic LCM-proof animation discussed above. He is also extending his theory to cover NOS by repeated limits.

7.3   Akama's Construction and Limiting Computational Calculi

Akama has constructed limiting partial combinatory algebra (PCA) [1]. His construction is more mathematical than ours and seems mathematically interesting on its own. It may give new constructions of models of PCA, λ-calculus and even some fragments of lambda calculi for classical logic. In Akama's construction, even divergent recursive sequences are considered to have "non-standard" limits. This makes his theory more mathematically beautiful than ours. Since his model enjoys a good conservation result, realizers extracted via his limiting PCA are limiting recursive in the standard sense.

His work will be also important for practical proof animation, since it is expected that proof systems like PX system [9] could be built on such a calculus. Design of formal axiom systems for such proof systems is an interesting problem. Is it possible to axiomatize Akama's limiting PCA only by the vocabulary of PCA? The same question can be asked for BRFT. Is $\mathbf{Lim}(F)$ axiomatized only by the vocabulary of BRFT? If not, how are they axiomatized?

7.4   Higher Order LCM

An important open problem of LCM might be *limiting higher order typed calculi* and *higher order LCM*. Limiting operators on models of type-free calculi PCA and BRFT have been given. However, no type systems have been "limited". Starting with $\mathbf{Lim}(F)$, a PER model of second order polymorphic lambda calculus is given by a standard way. However, what we seek is a limiting construction of existing type systems. For example, is there a limit-operation like $\mathbf{Lim}(F)$ on BRFT on models of simply typed lambda calculus? If such a construction exists for models of Coq type system, this would have a practical implication. We would be able to interpret $\mathbf{Lim}(\mathrm{Coq})$ in Coq system itself. This would enable us to execute proof animation of LCM-extended Coq in the constructive Coq system itself. Recently, Akama gave a limit-construction of his style to any cartesian closed category with a weak natural number object. It is expected that his construction is applicable to a great number of computational calculi, possibly polymorphic calculi such as the one of Coq.

The problem of higher order LCM would be just as important as the problem of limiting higher order computational calculi. LCM is *predicatively classical*. Namely, semi-classical principles are applicable to predicative formulas even if NOS is used. Isn't there any possibility to extend it to higher order quantifications? We expect that a large amount of practical mathematics would not need such quantifications, but, if we can relax the restriction, it must be useful.

Acknowledgments

The first author is grateful to Prof. Z. Luo and the organizers of TYPES 2000 meeting for the chance of presenting this work. The first author expresses his sincere thanks to Akihiro Yamamoto. He pointed out the relation of learning theory to Hilbert's invariant theory and Berardi's work, and it initiated the whole research. We are very grateful to Yohji Akama, Stefano Berardi, Thierry Coquand, Hajime Ishihara, Shun-ichi Kimura, Nobuki Takayama, Piergiorgio Odifreddi, and Mariko Yasugi for helpful discussions on the subject. The first author is supported by No. 10480063, Monbusyo, Kaken-hi (the aid of Scientific

Research, The Ministry of Education). We are grateful to the referee for pointing out that MNP and Σ_1^0-LEM are equivalent giving us many other helpful comments. Finally, the first author wishes to dedicate this paper to the memory of late Prof. Masaya Yamaguchi. Without his philosophical influences, this work would not have been possible.

References

1. Akama, Y.: Limiting Partial Combinatory Algebras Towards Infinitary Lambda-calculi and Classical Logic, in Proc. of Computer Science Logic, Lecture Notes in Computer Science, Springer, 2001.
2. Beeson, M.: Foundations of Constructive Mathematics, Springer, 1985
3. Baratella, S. and Berardi, S.: Constructivization via Approximations and Examples, Theories of Types and Proofs, M. Takahashi, M. Okada and M. Dezani-Ciancaglini eds., MSJ Memories 2 (1998) 177–205
4. Berardi, S.: Classical logic as Limit Completion, -a constructive model for non-recursive maps-, submitted, 2001, available at http://www.di.unito.it/~stefano/Berardi-ClassicalLogicAsLimit-I.rtf.
5. Bishop, E.: Foundations of Constructive Analysis, McGraw-Hill, 1967
6. Blum, L. et al.: Complexity and Real Computation, Springer, 1997
7. Coquand, Th. and Palmgren, E.: Intuitionistic choice and classical logic, Archive for Mathematical Logic, 39 (2000) 53–74
8. Gold, E. M.: Limiting Recursion, The Journal of Symbolic Logic, 30 (1965) 28–48
9. Hayashi, S. and Nakano, H.: PX: A Computational Logic, 1988, The MIT Press
10. Hayashi, S., Sumitomo, R. and Shii, K.: Towards Animation of Proofs - Testing Proofs by Examples -, Theoretical Computer Science (to appear)
11. Hilbert, D.: Über die Theorie der algebraische Formen, Mathematische Annalen 36 473–531
12. Hilbert, D.: Theory of Algebraic Invariants, Cambridge Mathematical Library (1993)
13. Ishihara, H.: An omniscience principle, the König lemma and the Hahn-Banach theorem, Zeitschrift für Logik und Grundlagen der Mathematik 36 (1990) 237–240
14. Kohlenbach, U.: Intuitionistic choice and restricted classical logic, Math. Logic Quarterly (to appear)
15. Kohlenbach, U.: Two unpublished notes, November and December, 2000.
16. Nakata, M. and Hayashi, S.: Realizability Interpretation for Limit Computable Mathematics, 2001, Scientiae Mathematicae Japonicae (to appear)
17. Odifreddi, P. G.: Classical Recursion Theory North-Holland, 1989
18. Simpson, S. G.: Subsystems of Second Order Arithmetic, Springer, 1999
19. Simpson, S. G.: Ordinal numbers and the Hilbert basis theorem, Journal of Symbolic Logic, 53 (1988) 961–974
20. Strong, H. R.: Algebraically Generalized Recursive Function Theory, IBM journal of Research and Development, 12 (1968) 465–475
21. Takeuti, G.: Proof Theory, 2nd ed., North-Holland, 1987
22. Wagner, E. G.: Uniformly Reflexive Structures: On the Nature of Gödelizations and Relative Computability, Transactions of the American Mathematical Society, 144 (1969) 1–41
23. Yasugi, M., Brattka, V. and Washihara, M.: Computability aspects of some discontinuous functions, 2001, Scientiae Mathematicae Japonicae (to appear)

Formalizing the Halting Problem in a Constructive Type Theory

Kristofer Johannisson

Department of Computing Science, Chalmers University of Technology
SE-412 96 Göteborg, Sweden
krijo@cs.chalmers.se

Abstract. We present a formalization of the halting problem in Agda, a language based on Martin-Löf's intuitionistic type theory. The key features are:
- We give a constructive proof of the halting problem. The "constructive halting problem" is a natural reformulation of the classic variant.
- A new abstract model of computation is introduced, in type theory.
- The undecidability of the halting problem is proved via a theorem similar to Rice's theorem.

The central idea of the formalization is to abstract from the details of specific models of computation. This is accomplished by formulating a number of axioms which describe an abstract model of computation, and proving that the halting problem is undecidable in any model described by these axioms.

1 Introduction

The Classical Halting Problem. "Given a program and an input, does the program terminate for the given input?" This is one way of formulating the halting problem. It is a well known result that this problem is undecidable, i.e. that one can not write a program which solves the problem for an arbitrary pair of a program and an input. The undecidability of the halting problem is proved in most books on computability, e.g. [7] or [8].

To be a bit more formal, we could state the halting problem in the following way: there exists no decision program which decides termination. By "decision program which decides termination" we intuitively mean a program which, given a pair of a program code[1] and an input, returns either 0 or 1: 1 iff the program represented by the code terminates for the input, 0 iff the program does not terminate for the given input.

The standard proof for the halting problem then has the following general structure: Assume (for a contradiction) that we have a program *halts* which decides termination. Then construct (this step will depend on the choice of model of computation) a program liar which has the following contradictory property: liar terminates for input "code for liar" iff liar does not terminate for input "code for liar".

---

[1] Here we assume the existence of an encoding function from programs to values (e.g. natural numbers).

P. Callaghan et al. (Eds.): TYPES 2000, LNCS 2277, pp. 145–159, 2002.

Reformulating in Type Theory. This classical statement of the halting problem can be naturally reformulated into a constructive one, which can then in turn be formalized in Martin-Löf's type theory. The only modification we have to make is to change the definition of "decision program": we need to add to this definition a proposition which states that a decision program should always terminate (a very natural requirement). This proposition is actually a classical consequence of the original definition, i.e. the constructive reformulation is classically equivalent to the classical statement of the halting problem. We will discuss this in more formal detail in Sect. 3.

Abstraction. One should also note that the classical proof above depends on the choice of model of computation. In this formalization we will avoid that by taking an abstract approach. We formulate a number of axioms which state the requirements of any model of computation for which we want to prove the undecidability of the halting problem. These axioms will form an abstract model of computation. The proof will then depend only on these axioms, not on any specific model of computation. In order to get a proof for a specific model, one would have to prove that the axioms hold for that particular model.

Formalizing in Agda. Agda is an implementation of Martin-Löf's type theory.[2] Readers unfamiliar with Agda may want to check Sect. 1.3 below. The concepts of structures and signatures in Agda support the abstract approach described above. Structures are basically records, and signatures are record types. The abstract model will be formalized as a signature CompModel, the elements of which will be the axioms of the abstract model. The halting problem proof will then take an argument of the type CompModel (the proof will be contained in a package which takes a CompModel argument). In Agda, this would (schematically) have the following form:

$$\text{CompModel} \in \text{Type} = \mathbf{sig} \left\{ \begin{array}{l} \text{Some} \\ \text{axioms} \end{array} \right.$$

package HaltingProblem ($c \in$ CompModel) **where**
    proof $\in \ldots = \ldots$

If we implement a specific model of computation, we can prove the axioms of CompModel for this implementation – i.e. construct a structure of the type CompModel – and get a proof for the halting problem for the specific model:

$$\text{specificCM} \in \text{CompModel} = \mathbf{struct} \left\{ \begin{array}{l} \text{Show that the} \\ \text{specific model} \\ \text{satisfies the axioms} \end{array} \right.$$

---

[2] Note that Agda is not identical to Martin-Löf type theory, e.g. the Id-sets used by Martin-Löf cannot be implemented in Agda.

1.1 Related Works

Our approach is different from Constable and Smith [3], who also showed a way of formalizing basic recursive function theory in constructive theory. They prove a number of standard theorems for this theory, one of them about the undecidability of the halting problem. We will compare our proof in Agda to these theorems in Sect. 3.3. The main difference between our approach and [3] is that we work in a constructive type theory (Agda) which is consistent with classical logic. We cannot prove that there is no Agda-function (i.e. program) which solves the halting problem. Instead, we represent internally a model of computation – CompModel – and prove that there is no CompModel-program which solves the halting problem.

There are many other variants of abstract approaches to computability. A survey can be found in [6]. In this survey one can distinguish at least two directions: the works considered with generalizing recursion theory for other structures than the natural numbers, and the ones which give an axiomatic presentation of recursion theory. An example of the latter are Wagner's Uniformly Recursive Structures (URS) [9], which we will discuss in relation to CompModel in Sect. 2.

An early work by Boyer and Moore on a mechanical proof of the unsolvability of the halting problem can be found in [2].

1.2 Structure of this Work

The abstract model of computation, CompModel is discussed and defined in Sect. 2. We also compare it to URS in this section. Section 3 then presents the proof of the undecidability of the halting problem from the axioms in CompModel. Instead of a proof about the halting problem, it turns out to be natural to prove a theorem similar to that of Rice. We compare this theorem and proof to the Rice-theorem from [3].

1.3 Agda

Agda is a functional language with dependent types. It is also an implementation of a version of Martin-Löf's type theory (as we have already noted, the Id-sets used by Martin-Löf cannot be implemented in Agda). Information about Agda is to be found mostly on the internet, see [4], [5] and [1]

Notational Conventions. The Agda code in this report has been typeset in order to increase readability. For instance, we use the symbols "$\in$" and "Σ" instead of "::" and "Sigma". We also use different fonts: SanSerif for constants, *Italics* for variables and **bold** for reserved words. This typesetting is similar to the way Agda code appears in the proof editor Alfa, but not identical.

Signatures and Structures. Abstraction is supported in Agda by the concepts of signatures and structures. Structures are basically records – unordered collections of named elements. The type of a structure is a signature. There is also a package mechanism which facilitates modularization.

A signature is defined by giving the labels and types of its components, e.g. S = **sig** {fst $\in$ A; snd $\in$ B}. A structure is defined by giving the values of its components, e.g. s $\in$ S = **struct** {fst = a; snd = b}. We can select specific components of a structure, e.g. s.fst or s.snd. If e is an expression, then **open** s **use** p = fst, q = snd **in** e allows us to use p and q instead of s.fst or s.snd in e.

The MicroLib. There is a library of standard definitions for Agda, the "MicroLib", which will be used in this report. Examples from the MicroLib include first order logic (conjunction, disjunction, existential quantifier), reasoning about relations (transitivity, substitutivity) and setoids (sets with an equivalence relation). We will use the notation "$\times$" for conjunction (product) and "+" for disjunction (disjunct sum).

For existential quantification we use Σ, e.g. Σ P $(\lambda(x \in P) \to Q\ x)$ to express that there exists $x \in P$ such that $Q\ x$ holds. To express that $Q\ x$ holds for all $x \in P$ we simply use the type $(x \in P) \to Q\ x$.

1.4   Complete Proofs in Agda

The signature CompModel is given in Appendix A. However, the complete proofs in Agda are too large to be included here. They can be downloaded from the world wide web at `http://www.cs.chalmers.se/~krijo/agdaCode.html`. The relevant parts of the MicroLib can be found there as well.

2   The Abstract Model of Computation

In this section we will present CompModel, the abstract model of computation, as a signature in Agda. We will also compare CompModel to Wagner's concept of Uniformly Recursive Structures (URS) [9].

The intention is that CompModel should not be ad hoc for the halting problem, but that it should be a natural and abstract description of what constitutes a model of computation. Basically, the axioms of CompModel should meet the following requirements: They should (1) state the minimum requirements of any computation model, (2) enable the proof of the halting problem, and (3) still be natural.

It was not until CompModel was more or less finished that we found out about URS. We discuss URS and CompModel in Sect. 2.2 below.

2.1   The Signature

For the complete signature, see Appendix A. The basic components of CompMod-el are a set of programs, a set of values, a relation of computation, some programs (constant, looping, conditional) and some axioms which specify properties of the relation and the programs. Here we present and explain all elements.

Programs and Values.

$$\text{Prog} \in \text{Set}$$
$$\text{SetoidVal} \in \text{setoid}$$
$$\text{Val} = \text{set SetoidVal}$$

We have a set Prog of programs (machines, functions, ...) which perform com-putations, and a set of values which provides input and output for the programs. Val is the set component of the SetoidVal structure, which defines an equivalence relation on Val. (The signature setoid is defined in the MicroLib). We will denote this equivalence relation by $==$, and we will write $a == b$ instead of the Agda code `Equals SetoidVal a b`. This equivalence relation must be substitutive, i.e. if $a, b \in$ Val and $P \in$ Val $\to$ Set, then the implication $(a == b) \times P\, a \to P\, b$ should hold. We call this property substEqVal, which is defined using another definition from the MicroLib, isSubstitutive:

$$\text{substEqVal} \in \text{isSubstitutive Val (Equals SetoidVal)}$$

The expression Equals SetoidVal refers to the equivalence relation defined in SetoidVal.

Computation.

$$\text{computes} \in (p \in \text{Prog}) \to (inp, out \in \text{Val}) \to \text{Set}$$

The proposition computes p inp out states that the program p produces output out given the input inp. We cannot define this relation as a function, since it would not be total. However, we state that computes is a functional relation, i.e. that it is deterministic:

$$\text{pComputes} \in (p \in \text{Prog}) \to (inp, out_1, out_2 \in \text{Val}) \to$$
$$(\text{computes } p\ inp\ out_1) \times (\text{computes } p\ inp\ out_2) \to$$
$$out_1 == out_2$$

Encodings. Programs take only values (elements of Val) as input. However, when formalizing the halting problem, we will reason about programs that take other programs as input. Therefore, we will need an encoding of programs as values:

$$\text{encode} \in \text{Prog} \to \text{Val}$$

Interestingly enough, nothing is required of the encode function – the proof of the halting problem will work for any encode of the type Prog $\rightarrow$ Val. For instance, we do not have to state that encode is injective. This matter is further discussed in Sect. 3.3.

We will also need to encode pairs of values as single values (in order to compute functions of several arguments):

$$\text{pairCode} \in \text{Val} \rightarrow \text{Val} \rightarrow \text{Val}$$

As in the case of encode, we do not need to state that pairCode is an injective or otherwise "normal" encoding of pairs. However, while encode does not occur in any other definition in CompModel besides its own, pairCode is used when we state the properties of diagonalize:

diagonalize $\in$ Prog
pDiagonalize $\in$ ($inp \in$ Val) $\rightarrow$ computes diagonalize inp (pairCode inp inp)

diagonalize is a program which, given the input inp, should return the code for the pair (inp, inp).

Program Composition. Given two programs p_1 and p_2, one might want to construct a more complex program by putting them in a sequence:

$$\text{compose} \in \text{Prog} \rightarrow \text{Prog} \rightarrow \text{Prog}$$

The idea is that compose p_1 p_2 is a program which first gives its input to p_1. If this results in an output, that output is in turn given to p_2. The result of p_2 is the final result:

pCompose $\in$ ($p_1, p_2 \in$ Prog) $\rightarrow$ ($inp, out \in$ Val) $\rightarrow$ Σ Val
$\quad$ ($\lambda(interm \in$ Val) $\rightarrow$ (computes p_1 inp $interm$) $\times$
$\quad$ (computes p_2 $interm$ out)) $\leftrightarrow$
$\quad$ computes (compose p_1 p_2) inp out

Boolean Values. We will need at least two distinct elements of Val which will represent "true" and "false":

True $\in$ Val
False $\in$ Val
pTrueisnotFalse $\in$ Not (True == False)

Constant and Looping Programs. We will need constant and looping programs for the proof:

loop $\in$ Prog
pLoop $\in$ ($inp \in$ Val) $\rightarrow$ Not (Σ Val ($\lambda(out \in$ Val) $\rightarrow$ computes loop inp out))
const $\in$ Val $\rightarrow$ Prog
pConst $\in$ ($c \in$ Val) $\rightarrow$ ($inp \in$ Val) $\rightarrow$ computes (const c) inp c

The loop program is a program which never produces any output, regardless of its input. The program const c always terminates with the result c (for any input).

Conditional Programs. We introduce something which resembles an "if-statement" from ordinary programming languages:

$$\mathsf{IF} \in \mathsf{Prog} \to \mathsf{Prog} \to \mathsf{Prog} \to \mathsf{Prog}$$

IF is a function for constructing a "conditional" program given three other programs. The idea is that IF *cond then else* first gives its input to the program *cond*. Depending on the result (True or False), it gives the same input to either *then* or *else*. pIF and pIF2 states that this is indeed the case:

$$
\begin{aligned}
\mathsf{pIF} \in \; & (\mathit{cond}, \mathit{then}, \mathit{else} \in \mathsf{Prog}) \to (\mathit{inp}, \mathit{out} \in \mathsf{Val}) \to \\
& (\text{computes } \mathit{cond}\; \mathit{inp}\; \mathsf{True} \times \text{computes } \mathit{then}\; \mathit{inp}\; \mathit{out} \to \\
& \quad \text{computes } (\mathsf{IF}\; \mathit{cond}\; \mathit{then}\; \mathit{else})\; \mathit{inp}\; \mathit{out}) \\
& \times \\
& (\text{computes } \mathit{cond}\; \mathit{inp}\; \mathsf{False} \times \text{computes } \mathit{else}\; \mathit{inp}\; \mathit{out} \to \\
& \quad \text{computes } (\mathsf{IF}\; \mathit{cond}\; \mathit{then}\; \mathit{else})\; \mathit{inp}\; \mathit{out}) \\
\mathsf{pIF2} \in \; & (\mathit{cond}, \mathit{then}, \mathit{else} \in \mathsf{Prog}) \to (\mathit{inp}, \mathit{out} \in \mathsf{Val}) \to \\
& \text{computes } (\mathsf{IF}\; \mathit{cond}\; \mathit{then}\; \mathit{else})\; \mathit{inp}\; \mathit{out} \to \\
& \quad \text{computes } \mathit{cond}\; \mathit{inp}\; \mathsf{True} \times \text{computes } \mathit{then}\; \mathit{inp}\; \mathit{out} \\
& \quad + \\
& \quad \text{computes } \mathit{cond}\; \mathit{inp}\; \mathsf{False} \times \text{computes } \mathit{else}\; \mathit{inp}\; \mathit{out}
\end{aligned}
$$

There are two things to be noted: First, that if *cond* returns something else than True or False, then IF should not terminate. Second, that IF is lazy in the sense that, depending on the result of *cond*, we will run *either* then or else, not both.

2.2   CompModel and URS

URS. A Uniformly Reflexive Structure (Wagner's first version, [9]) consists of a basic set U and a gödelization $G \in U \to U^U$. Given $u, x \in U$, $(G(u))(x)$ means applying the function with index u to x. A shorter notation for $(G(u))(x)$ is $[u](x)$. Many-placed functions are defined by induction: for functions with two arguments we have $[u](x, y) = [[u](x)](y)$, and so on. Then there are three (and no more) axioms which the structure (U, G) must satisfy:

1. There exists $* \in U$ such that, for every $u \in U$, $[u](*) = * = [*](u)$. (In what follows, V denotes the set $U - *$)
2. There exists $\alpha \in U$ such that, for all $f, g, x \in V$, $[\alpha](f, g) \neq *$ and $[[\alpha](f, g)](x) = [f](x, [g](x)) = [[f](x)]([g](x))$ (c.f. the S combinator).
3. There exists $\psi \in U$ such that, for all $a, b, c, x \in V$,

$$[[\psi](c, b, a)](x) = \begin{cases} a \text{ if } x = c \\ b \text{ if } x \neq c \end{cases}.$$

It has been shown that the recursive functions constitute a URS.

There are many variants of URS (see the survey in [6]). For instance, there are URS where one separates functions and values as two different sets instead of just having one set U.

Comparison. How does CompModel compare to the URS? Clearly, there are some similarities: A URS contains a function α for composition (i.e. the S combinator), CompModel contains compose. Also, the ψ and $*$-element in a URS seem to correspond to IF and loop from CompModel. These similarities suggest that the concepts of composition and conditional / nonterminating programs (functions) are central when reasoning about computability. However, URS and CompModel are also quite different:

1. In a URS we have a set of indices / values and a function for applying function with a given index to a given value. In CompModel we have different sets for programs and values, and a function for encoding programs as values. There is no direct way of obtaining a program from a value. Computation is modeled as a relation.
2. One can prove quite a lot of standard computability theorems from the axioms of a URS. It is not clear exactly what will follow from the axioms of CompModel.
3. In any URS one can pick a subset of the set of values "... which are analogous to the integers and with respect to which we can perform the operations of primitive recursion and minimalization" [9, page 17]. CompModel, on the other hand, does not allow this internal representation of the recursive functions.

It is clear that CompModel is a first attempt, and that – though not inherently ad hoc for the halting problem – it is still quite limited compared to the well developed theory of the URS.

3   The Halting Problem in Agda

We will now give a proof of the undecidability of the halting problem using CompModel. The proof has the following structure: First, we introduce the notion of "computations" and decidable predicates for computations. Then, we prove a theorem for such predicates which is similar to Rice's theorem. The proof for the halting problem is then easily derived from this theorem. In Sect. 3.3 there are some notes on the encodings (the encode and pair components of CompModel) and on the connection to Rice's theorem.

As we have already noted in the introduction, the constructive statement of the halting problem is slightly different (but classically equivalent) from the classical one: we have to add something to the definition of the concept of "decision program" (Sect. 3.1 below).

The complete Agda code for the proofs in this section are not given, see Sect. 1.4 for more details.

3.1 Basic Concepts

Computations.

$$\text{computation} \in \text{Set} = \text{Prog} \times \text{Val}$$

A computation is a pair of a program and a value. This pair is supposed to represent the result of applying the program to the value. If the program produces an output when given the value as input, then the computation "results in" that value:

> results_in $(c \in \text{computation})(out \in \text{Val}) \in \text{Set} =$
> **open** c **use** $p = \text{fst}, inp = \text{snd}$ **in**
> computes $p\ inp\ out$

Using the relation results_in, we can define the predicate **terminates**:

$$\text{terminates } (c \in \text{computation}) \in \text{Set} =$$
$$\Sigma \text{ Val } (\lambda(out \in \text{Val}) \rightarrow \text{results\_in } c\ out)$$

A predicate for computations is just a function of the type computation $\rightarrow$ Set. This type can also be written as Pred computation, where Pred is defined in the MicroLib.

We will also need equality on computations:

> eqComp $(c_1, c_2 \in \text{computation}) \in \text{Set} =$
> terminates c_1 + terminates $c_2 \rightarrow$
> $\Sigma \text{ Val } (\lambda\ out \rightarrow \text{results\_in } c_1\ out \times \text{results\_in } c_2\ out)$

Two computations c_1 and c_2 are equal if the convergence of one of them implies that they both converge to the same value.

Decidability.

We say that a program *decProg* decides a predicate *pred* $\in$ Pred computation iff for all computations $c = (p, inp)$:

1. decProg gives output True for input (encode p, inp) iff *pred* holds for c
2. decProg gives output False for input (encode p, inp) iff *pred* does not hold for c
3. decProg is total in the sense that, given the input (encode p, inp), it will always terminate with output True or False.

Condition 3 is what distinguishes the constructive statement of the halting problem from the classical one. It can be classically derived from the first two conditions: By the law of the excluded middle, we have *pred* $c \lor \neg(pred\ c)$, therefore decProg will always terminate for an input (encode p, inp).

This definition of decidability is formalized in Agda:

decides (*pred* ∈ Pred computation)(*decProg* ∈ Prog) ∈ Set =
(c ∈ computation) →
open c **use** p = fst, inp = snd **in**
(computes decProg (pairCode (encode p) inp) True) +
computes decProg (pairCode (encode p) inp) False))
×
((computes *decProg* (pairCode (encode p) inp) True ↔ *pred c*) ×
(computes *decProg* (pairCode (encode p) inp) False ↔ Not (*pred c*)))

Given this definition, it is natural to define what it means for a predicate to be decidable:

decidable (*pred* ∈ Pred computation) ∈ Set =
Σ Prog (λ(*decProg* ∈ Prog) → decides *pred decProg*)

Substitutivity. It turns out that we need to define what it means for a predicate in Pred computation to be substitutive with respect to eqComp:

isSubst (*pred* ∈ Pred computation) ∈ Set =
(c_1, c_2 ∈ computation) → eqComp c_1 c_2 → *pred* c_1 → *pred* c_2

3.2   The Halting Problem Is Undecidable

We now want a proof of the proposition Not (decidable terminates). It is possible to prove this directly, but it turns out that by a slight modification of such a proof we obtain a proof of a more general proposition. As we have already mentioned, this general proposition is similar to Rice's theorem. We proceed by first stating and proving this general proposition, and then proving that the halting problem is an instance of this general proposition.

Nontrivial and Substitutive Predicates Are Undecidable. We will prove that any nontrivial and substitutive predicate for computations is undecidable. By nontrivial we mean that there is an example and a counterexample for the predicate, i.e., that there exist two different computations for which the predicate holds and does not hold, respectively. In Agda:

abstractHaltp (*pred* ∈ Pred computation)
(*pSubst* ∈ isSubst *pred*)
(*ex, counterex* ∈ computation)
(*predEx* ∈ pred *ex*)
(*predCounter* ∈ Not (pred *counterex*))
∈ Not (decidable *pred*) = ...

The arguments of abstractHaltp are: A predicate *pred*, a proof *pSubst* that *pred* is substitutive, an example *ex* and a proof *predEx* that *pred* holds for *ex*, a counterexample *counterex* and a proof that *pred* does not hold for *counterex*.

Then, in order to show Not (decidable terminates) one can just apply abstract-Haltp to the predicate terminates along with examples of /non/terminating computations (constructed using loop and const) and the corresponding proofs. This instantiation of abstractHaltp is described below in Sect. 3.2.

The Proof. The structure of the proof abstractHaltp above is as follows: We assume that the predicate *pred* is decidable, i.e. that there exists a program *decProg* which decides *pred*. Using *decProg*, we define a liar construction. Then, we prove some properties of liar, and use these properties in order to reach the desired contradiction. We give an outline of this proof below.

Assumptions. Since we want to prove Not (decidable *pred*), we assume decidable *pred* and derive a contradiction. The assumption means that we have *decProg* $\in$ Prog and *pDecProg* $\in$ decides *pred decProg*.

The Liar. The liar should be a program such that when applied to itself (i.e. when given the value encode liar as input), we can derive a contradiction.

In order to construct liar we need a somewhat technical construction which we will call constInp:

$$\text{constInp } (p \in \text{Prog})(c \in \text{Val}) \in \text{Prog} =$$
$$\text{compose (const } c) \ p$$

$$\text{pConstInp } (p \in \text{Prog})(c, inp, out \in \text{Val}) \in$$
$$\quad (\text{computes (constInp } p \ c) \ inp \ out) \leftrightarrow (\text{computes } p \ c \ out)$$
$$= \dots$$

The idea is that constInp $p \ c$ is a program which for all inputs acts as the program p when given c as input.

$$\text{liar} \in \text{Prog} = \text{IF (compose diagonalize decProg)}$$
$$\quad (\text{constInp } counterex.\text{fst } counterex.\text{snd})$$
$$\quad (\text{constInp } ex.\text{fst } ex.\text{snd})$$

Here we see the Agda notation for selecting components from a structure: *counterex*.fst refers to the fst component of the structure *counterex*. Note that (constInp *counterex*.fst *counterex*.snd) is a program which, for all inputs, behaves exactly like the computation *counterex*, and (constInp *ex*.fst *ex*.snd) behaves like *ex*. Informally, this means that when the conditional part of the IF-construction is True, liar is a counterexample for *pred*, and, correspondingly, that liar is an example when the condition is False.

Giving input *inp* to the program (compose diagonalize decProg) is the same thing as giving the input (*inp*, *inp*) to the program decProg (this is a lemma in the complete proof). Using this fact, we state the intuitive line of reasoning about liar above more formally in Agda:

$$
\begin{aligned}
&\text{pLiar } (inp \in \text{Val}) \in \\
&\text{computes } decProg \text{ (pairCode } inp\ inp) \text{ True} \rightarrow \\
&\quad \text{Not } (pred \text{ (struct \{fst} = liar; \text{ snd} = inp\})) \\
&\times \\
&\text{computes } decProg \text{ (pairCode } inp\ inp) \text{ False} \rightarrow \\
&\quad pred \text{ (struct \{fst} = liar; \text{ snd} = inp\}) \\
&= \ldots
\end{aligned}
$$

Contradiction. Let liarliar be the computation (liar, encode liar)[3]. We have assumed that *decProg* decides *pred*. According to the definition of decides, this means that decProg returns either True of False when given (liar, encode liar) as input.

Case a: decProg returns True for input (liar, encode liar) – Since *decProg* decides *pred*, we have that *pred* holds for liarliar. However, according to pLiar, we also have that *pred* does not hold for liarliar. Contradiction!

Case b: decProg returns False for input (liar, encode liar) – Since *decProg* decides *pred*, we have that *pred* does not hold for liarliar. However, according to pLiar, we also have that *pred* holds for liarliar. Contradiction!

This concludes the general proof.

The Halting Problem Instance. In order to prove that the halting problem is undecidable, we have to prove that terminates is a nontrivial and substitutive predicate. Then, we just apply abstractHaltp to these proofs.

It is trivial to give an example and a counterexample for terminates, since these are supplied in CompModel: the programs const and loop (see Sect. 2.1). We just apply them to any constant (i.e. True or False) in order to construct computations.

Proving the substitutivity of terminates is also easy, since two computations are equal when they both terminate with the same result or when they both diverge.

3.3 Notes

About the Encodings. Any natural or typical instantiation of the encoding functions encode $\in$ Prog $\rightarrow$ Val and pairCode $\in$ Val $\rightarrow$ Val $\rightarrow$ Val should be injective; an encoding which maps two (or more) different objects (programs or pairs) to the same value is obviously not what one would normally want. However, as noted above (Sect. 2.1), no such property of encode or pairCode is included in CompModel. In fact, there are no properties at all of the two encoding functions (except for the property pDiagonalize) in CompModel.

---

[3] Strictly speaking, this should be written as **struct** {fst = liar; snd = encode liar} in Agda.

Both encoding functions are used when defining decides (Sect. 3.1). In other words, if any of the encoding functions are pathological, then the meaning of decides (and decidable) will be pathological. The proof of Not (decidable terminates) will still be correct, but the proved proposition will not have the intended meaning.

An encoding which is not injective makes it harder for programs to distinguish encoded objects from each other. Consider the case when the function encode is a constant function: The proposition decides *pred decProg* states that decProg should give output True or False for the input pairCode (encode p) *inp*, depending on whether *pred* holds for the computation (p, inp) or not. Since encode is a constant function, decProg can not depend on the program component of computations. This means that a predicate *pred* $\in$ Pred computation which depends on the program component of computations will not be decidable. Since terminates is such a predicate, the undecidability of terminates is a simple consequence of the choice of encoding function.

Rice's Theorem. The general proof abstractHaltP seems to be rather similar to the result known as Rice's theorem. Constable and Smith [3] have formulated a version of Rice's theorem in their constructive type theory. Using an informal notion of classes, they say that for any type T, a class C_T is trivial iff $(\forall x : T.x \in C_T) \vee (\forall x : T.\neg(x \in C_T))$. Then, Rice's theorem states that for all types T, $C_{\overline{T}}$ is decidable iff $C_{\overline{T}}$ is trivial ($\overline{T}$ is the type of elements which, if they converge, converge to an element in T).

This version of Rice's theorem is, at least on an informal level, quite similar to abstractHaltP. Classes can be correlated to predicates, the type $\overline{T}$ to computation, and $C_{\overline{T}}$ to Pred computation. The "if" part of Rice's theorem is obvious, so the theorem essentially states that *"decidable $\rightarrow$ trivial"*. On the other hand, abstractHaltP says *"nontrivial $\rightarrow$ ¬decidable"*. Note that "nontrivial" here means that there exists an example and a counterexample. Classically, this is the negation of "trivial", but not constructively.

The proof of Rice's theorem in [3] is based on a proof that the halting problem is undecidable, i.e. that the class $\{x : \overline{T} \mid x \downarrow\}$ is not decidable for all types T which have members.

4   Conclusion

We have now seen an abstract model of computation in type theory and a proof of the undecidability of the halting problem for this model. Informally, CompModel seems to meet the requirements in Sect. 2 about being a natural and abstract description of what constitutes a model of computation. A natural way to be more formal about these requirements would be to instantiate CompModel using some standard model, e.g. Turing machines or the recursive functions (this is work in progress).

Another line of further research would be to try to prove some other theorems for CompModel. This, as well as the instantiation, might also suggest improvements or changes to the axioms in CompModel.

Acknowledgments

Many thanks to my supervisor Peter Dybjer and the members of the Programming Logic Group at Chalmers for all their help.

References

[1] Experiments in Machine-Checked Proofs, URL:
   http://www.cs.chalmers.se/Cs/Research/Logic/experiments.mhtml.
[2] R. S. Boyer and J. St. Moore. A mechanical proof of the unsolvability of the halting problem. *Journal of the Association for Computing Machinery*, 31(3):441–458, 1984.
[3] R. Constable and S. Smith. Computational foundations of basic recursive function theory. In *Third IEEE Symposium on Logic in Computer Science*, pages 360–371, Edinburgh, Scotland, July 1988.
[4] Catarina Coquand. Homepage for Agda. URL:
   http://www.cs.chalmers.se/~catarina/agda/.
[5] Thierry Coquand. Structured type theory. Postscript format, URL:
   http://www.cs.chalmers.se/~coquand/STT.ps.Z.
[6] A. P. Ershov. Abstract computability on algebraic structures. In *Algorithms in Modern Mathematics and Computer Science*, volume 122 of *Lecture Notes in Computer Science*, pages 397–420, Berlin, Heidelberg, New York, 1981. Springer-Verlag.
[7] Zohar Manna. *Mathematical Theory of Computation*. McGraw-Hill Book Company, 1974.
[8] George J. Tourlakis. *Computability*. Reston Publishing Company Inc., Reston, Virginia 22090, 1984.
[9] Eric G. Wagner. Uniformly Reflexive Structures: On the nature of gödelizations and relative computability. *Transactions of the American Mathematical Society*, 144:1–41, October 1969.

A CompModel

CompModel =

sig $\Bigg\{$

$\quad$ Prog $\in$ Set

$\quad$ SetoidVal $\in$ setoid

$\quad$ Val $=$ set SetoidVal

$\quad$ substEqVal $\in$ isSubstitutive Val (Equals SetoidVal)

$\quad$ computes $\in (p \in$ Prog$) \to (inp, out \in$ Val$) \to$ Set

$\quad$ pComputes $\in (p \in$ Prog$) \to (inp, out_1, out_2 \in$ Val$) \to$
$\qquad$ (computes $p\ inp\ out_1$) $\times$ (computes $p\ inp\ out_2$) $\to$
$\qquad$ $out_1 == out_2$

$\quad$ encode $\in$ Prog $\to$ Val

$\quad$ pairCode $\in$ Val $\to$ Val $\to$ Val

$\quad$ diagonalize $\in$ Prog

$\quad$ pDiagonalize $\in (inp \in$ Val$) \to$ computes diagonalize inp (pairCode inp inp)

$\quad$ compose $\in$ Prog $\to$ Prog $\to$ Prog

$\quad$ pCompose $\in (p_1, p_2 \in$ Prog$) \to (inp, out \in$ Val$) \to \Sigma$ Val
$\qquad$ $(\lambda(interm \in$ Val$) \to$ (computes $p_1 inp\ interm$)$\times$
$\qquad$ (computes $p_2\ interm\ out$)) $\leftrightarrow$
$\qquad$ computes (compose $p_1\ p_2$) $inp\ out$

$\quad$ True $\in$ Val

$\quad$ False $\in$ Val

$\quad$ pTrueisnotFalse $\in$ Not (True $==$ False)

$\quad$ loop $\in$ Prog

$\quad$ pLoop $\in (inp \in$ Val$) \to$ Not $(\Sigma$ Val $(\lambda(out \in$ Val$) \to$
$\qquad$ computes loop $inp\ out$))

$\quad$ const $\in$ Val $\to$ Prog

$\quad$ pConst $\in (c \in$ Val$) \to (inp \in$ Val$) \to$ computes (const c) $inp\ c$

$\quad$ IF $\in$ Prog $\to$ Prog $\to$ Prog $\to$ Prog

$\quad$ pIF $\in (cond, then, else \in$ Prog$) \to (inp, out \in$ Val$) \to$
$\qquad$ (computes $cond\ inp$ True $\times$ computes $then\ inp\ out \to$
$\qquad\quad$ computes (IF $cond\ then\ else$) $inp\ out$)
$\qquad$ $\times$
$\qquad$ (computes $cond\ inp$ False $\times$ computes $else\ inp\ out \to$
$\qquad\quad$ computes (IF $cond\ then\ else$) $inp\ out$)

$\quad$ pIF2 $\in (cond, then, else \in$ Prog$) \to (inp, out \in$ Val$) \to$
$\qquad$ computes (IF $cond\ then\ else$) $inp\ out \to$
$\qquad$ computes $cond\ inp$ True $\times$ computes $then\ inp\ out$
$\qquad$ $+$
$\qquad$ computes $cond\ inp$ False $\times$ computes $else\ inp\ out$

On the Proofs of Some Formally Unprovable Propositions and Prototype Proofs in Type Theory

Giuseppe Longo

Laboratoire d'Informatique, CNRS et École Normale Supérieure
45, Rue D'Ulm, 75005 Paris, France
http://www.di.ens.fr/users/longo

Introduction: Some History, Some Philosophy

At the age of 7 or 8, Gauss was asked to produce the result of the sum of the first n integers (or, perhaps, the question was slightly less general ...). He then proved a theorem, by the following method:

$$
\begin{array}{cccc}
1 & 2 & \dots & n \\
n & (n\text{-}1) & \dots & 1 \\
\hline
(n+1) & (n+1) & \dots & (n+1)
\end{array}
$$

which gives $\Sigma_1^n i = n(n+1)/2$.

Clearly, the proof is not by induction. Given n, a uniform argument is proposed, which works for any integer n. Following Herbrand, we will call **prototype** this kind of proof. Of course, once the formula is known, it is very easy to prove it by induction, as well. But, one must know the formula, or, more generally, the "induction load". A non-obvious issue in automatic theorem proving, as we all know.

Let's now speculate on the possible "cognitive" path which "brings to" (and gives certainty!) to this proof. The reader can surely see, in his mental spaces, the "number line", that is the well-ordered sequence of integer numbers. They are there, one after the other, in increasing order: you may see it on a straight line, it may oscillate, but it should be, for you, from left to right (isn't it? please check ... and give up doing mathematics, if you do not see the number line; see [Dehaene98] for some data about it). It seems as if little Gauss has had the mental courage to put it on paper and then ... reverse it. No induction, just the order and its inverse, and the proof works for any n, a perfectly rigorous proof.

Consider now a non-empty subset in your number line. You can surely see that this set has a least element (look and see: if there is an element, there is a least one among the finitely many preceding it, even if you may not know which one). The "observation" imposes itself, by the (well-)ordering of the line, as geometric evidence, a very robust one. Moreover, one does not need to know if and how the subset eventually goes to infinity: if it has one point somewhere (the set is not empty), this is at some finite point and, then, there is a smaller

P. Callaghan et al. (Eds.): TYPES 2000, LNCS 2277, pp. 160–180, 2002.

one which is the least of the "given" subset. In the conclusion, we will call this, a "geometric judgement".

In the few lines above, we hinted to an understanding of the ordering of numbers with reference to a mental construction, in space (or time). Frege would have called this approach "psychologism" (Herbart's style, according to his 1884 book). Poincaré instead could be a reference for this view on certainty and meaning of induction as grounded on intuition, possibly of space. In Brouwer's foundational proposal as well, the mathematician's intuition of the sequence of natural numbers, which founds Mathematics, relies on a phenomenal experience; however, this experience should be grounded on the "discrete falling apart of time", as "twoness" ("the falling apart of a life moment into two distinct things, one which gives way to the other, but is retained by memory", [Brouwer48]). Thus, "Brouwer's number line" originates from (a discrete form of) phenomenal time and induction derives meaning and certainty from it.

Intuition of ordering in space or time, actually of both, contributes to establish the number line, as an invariant of these active experiences: formal induction follows from, it doesn't found this intuition. This is my understanding of Poincaré's and Brouwer's philosophy. By recent scientific evidence (see [Dehaene98]), we seem to use extensively, in reasoning and computations, the "intuitive" number line; these neuropsychological investigations are remarkable facts, since they take us beyond the "introspection" that the founding fathers used as the only way to ground mathematics on intuition. We are probably along the lines of transforming the analysis of intuition from naive introspection to a scientific, objective, investigation of our cognitive performances.

Let's now go back to ... the sum of the first n integers. About eighty years later, Peano and Dedekind suggested that a proof, such as little Gauss', was certainly a fantastic achievement (in particular for such a young man), but that one had to prove theorems, in Number Theory, by some sort of "formal and uniform method", defined as a "potentially mechanisable" one, insisted Peano and Padoa. Then, they definitely specified "formal induction" as THE proof principle for Arithmetic (Peano Arithmetic, PA).

Frege set induction at the basis of his logical approach to mathematics; he considered it a key logical principle, and gave by this to PA the founding status that it still has. Of course, Frege thought that logical induction (or PA) was "categorical" (to put it in modern terms), that is that induction captured exactly the theory of numbers, or that everything was said within PA: this logical theory simply coincided, in his view, with the structure and properties of numbers (Frege didn't even make the distinction "theory vs. model" and never accepted it: the logic was exactly the mathematics).

We all know how the story continues. In his 1899 book (The Foundation of Geometry), Hilbert set geometry on formal grounds, as a solution of the incredible situation where many claimed that the rigid bodies could be not so rigid, that light rays could go along (possibly curved) geodesics .... Riemann's habilitation (under Gauss' supervision), in 1854 ([Riemann54]), had started this "delirium", as Frege had called the intuitive–spatial meaning of the axiom for geometry, [Frege84], p.20. Helmholtz, Clifford, Poincaré had insisted on this idea of Riemann's and on its possible relevance for the understanding physical action at distance (gravitation, in particular): "in the physical world nothing else takes place, but (continuous) variations of curvature of space" W.Clifford (1882 (!!)). For these mathematicians, meaning, as reference to phenomenal space, and its mathematical structuring preceded rigour and provided "foundation", see [Boi95], [Bottazzini95]: by mathematics, geometry in particular, they wanted to make the physical world intelligible, more then just deriving theorems by rigorous tools as formal/mechanical games of symbols. Hilbert had a very different foundational attitude: for the purposes of foundations (but only for these purposes), forget the meaning in physical spaces of the axioms of non-Euclidean geometries and interpret their purely formal presentation in PA. And his 1899 book contains one of the earliest and most remarkable achievements in "programming": he fully formalised a unified approach to geometry, by closely analysing several relative consistency issues, and "compiled" it in PA, by analytic tools. Formal rigour and effective-finitistic reduction are at the core of it.

Thus, on one hand, the geometrisation of physics, from Riemann to Einstein and Weyl (via Helmholtz, Clifford and Poincaré), brought to a revolution in that discipline, originating by breathtaking physico-mathematical theories (and theorems); on the other, the attention to formal, potentially mechanisable rigour, independent of meaning and intuition, gave us fantastic formal machines, from Peano and Hilbert to Turing and our digital computers.

The following year, at the 1900 Paris conference, Hilbert definitely contributed to give to PA (and to formal induction) their central status in foundation, by suggesting to prove (formally) the consistency of PA: then the consistency of the geometric axiomatisations would have followed from that of formal Number Theory (with no need of reference to meaning, in time, in space or whatever). Moreover, a few years later, he proposed a further conjecture, the "final" solution to all foundational problems, a jump into perfect rigour: prove the completeness of the formal axioms for Arithmetic. Independently of the heuristics of a proof, its certainty had to be ultimately given by formal induction.

However, there was more than this in the attitudes of many at the time. That is, besides foundation as "a-posteriori formalisation", the "potential mechanisation" of mathematics was truly dreamed, not only as a locus for certainty, but also as a "complete" method for proving theorems (as mentioned above, the Italian logic school firmly insisted on this, with their "pasigraphy", a universal formal language, a mechanisable algebra for all aspects of human reasoning). Or, the "sausage machine" for mathematics (and thought), as Poincaré ironically called it, could be put at work: provide pigs (or axioms) as input, produce sausages (or theorems) as output[1]. We know how the story of com-

---

[1] At that time, the very beginning of the century, Poincaré, in correspondence with Zermelo, hinted at the possible independence of the Continuum Hypothesis, CH, from Zermelo's formal axioms: he conjectured that there could be no sausage machine for Set Theory either. This was rigorously proved by Gödel and Cohen some 30 and 60 years later, respectively.

plete a-posteriori formalisation and, a fortiori, of potential mechanisation ended
... Hilbert's conjectures on the formally provable consistency, decidability and
completeness of PA turned out to be all wrong, and the 1931 proof of this fact
originated (incomplete, but) fantastic formal machines (by an early rigorous
definition of "computable function").

The consequence for Number Theory is that induction is incomplete and
that one cannot avoid infinitary machinery in proofs (in the rigorous sense of
Friedman, see [Friedman97], for example). In some cases, this can be described
in terms of the structure of "prototype proofs", as it will be proposed below.
Moreover, even the problem of the induction load, or of the prototype proof
in the inductive step, is a non-trivial issue in actual mechanisation. Clearly, a
posteriori, the induction load may be generally described within the formalism,
but its "choice", out of infinitely many possible ones, may require some external
heuristics (typically: analogies).

The aim of this paper is to focus on some specific limits of formalisation, by
a close analysis of some "concrete" but formally unprovable number-theoretic
statements. This is done in order to encourage a broadening of the tools for
proofs, and stress the role of "interaction" man/machine in proof-assistants and
proof-checking: singling out the fully un-formalisable fragments is a crucial component of the work in these areas. It may help to pass over to the machine
exactly the fully formalisable parts. Beyond the myth, and in full awareness
of the incompleteness of formalisms, we may further develop these remarkable
application of Proof Theory and Type Theory.

1   Herbrand's Prototype Proofs

"...when we say that a theorem is true for all x, we mean that for each x
individually it is possible to iterate its proof, which may just be considered a
prototype of each individual proof." Herbrand (1930), see [Goldfarb87], pp.288-
9. Little Gauss' theorem above is an example of such a proof. But any proof of a
universally quantified statement, over a structure that does not realize induction,
is a "prototype" (e.g., for any Euclidean triangle, the sum of the internal angles
is 180°: take a generic triangle, draw the parallel line to one side etc.). Similarly,
if you want to prove a property for any element of a (non-trivial sub-)set of
reals, of complex numbers .... But, in Number Theory, one has an extra and
very strong proof-principle: induction. Clearly, in 1930, Herbrand thought that,
in the special case of integer numbers, their universally quantified properties
could always be proved by induction: completeness of PA was the current belief
(with a few remarkable exceptions, such as H. Weyl, who – though hesitantly
– conjectured the incompleteness of PA in Das Kontinuum, 1918 (!), and also
stressed that the dream of potential mechanisation was a form of trivialisation
of mathematics).

But what is the difference between prototype proofs and induction?

In a prototype proof, you must provide a reasoning which uniformly holds for
all arguments, and this uniformity allows (and it is guaranteed by) the use of a

"generic" argument (see below). Induction provides an extra tool: the intended property doesn't need to hold for the same reasons for all arguments. Actually, it may hold for different reasons for each of them. One only has to give a proof for 0, and then provide a uniform argument to go from x to $x + 1$. That is, uniformity of the reasoning is required only in the inductive step: this is where a prototype proof steps in again, the argument from x to $x + 1$. Yet, the situation may be more complicated: in case of nested induction also this inductive step, a universally quantified formula, may be given by induction on x. However, after a finite number of nesting, one has to get to a prototype proof going from x to $x + 1$ (induction is logically well-founded).

Thus, induction provides a rigorous proof principle, which, over well-orderings, holds in addition to uniform (prototype) proofs, modulo the fact that, sooner or later, a prototype proof steps in. Note though that the prototype/-uniform argument in an inductive proof allows to derive, from the assumption of the thesis for x, its validity for $x + 1$, in any possible model. Moreover, as we shall see, by induction one may inherit properties from x to $x + 1$ (e.g., totality of a function of x, see below).

Yet, one more point should be mentioned. In an inductive proof, one must know in advance the formula (the statement) to be proved: little Gauss did not know it, for example. A non minor problem in automatic theorem proving .... Indeed, (straight) induction (i.e. induction with no problem in the choice of the inductive statement or load) is closer to proof-checking than to "mathematical theorem proving": proving a theorem, in mathematics, in general, is answering a question, not necessarily, not always, checking an already given formula.

Type Theory may help us to give a more rigorous description of what is a prototype proof. Propositions are types and proofs are terms, for us. Then a prototype proof, with a generic argument, is a term which may be uniformly instantiated by that argument: a schema for each individual proof. Let's see now a very informal definition, just a suggestion of what should be formalised:

Definition (Very Informal Type Theory, VITT). *Given a type A, a (closed) term P is **generic** and a (proof-)term N : [P/x]A is a **prototype**, if there exists a term M : A such that*

$$[P/x]M = N : [P/x]A.$$

This is surely VITT, as it is not mentioned: what are types and variables, exactly (1st order dependent types? Variable types? more?); what is equality? which kind of restriction is assumed in the substitution operation $[P/x]M$? A rigorous definition of prototype proofs as lambda-terms, in Girard's System F, and a few results (coherence, decidability), are given in [Longo00]. As for now, consider the informal definition above for what it is worth, with a first order understanding of variables and dependent types (if there is provably no way to make it rigorous, I will return the generous support for this invited lecture

...). The point is that many different answers are possible for each of the three questions above, and each would lead to different results.

Of course, in a prototype proof, the generic input P must be typed, possibly with reference to a semantics. It is clearly so in little Gauss' theorem above: n must be a standard integer (it must have the type of the integers). There is no way to carry on his proof for an arbitrary x, a variable of formal arithmetic, as the construction makes no sense over the non-standard part of a model (the required symmetry to obtain the constant sum $n+1$ is lost). A proof by induction, instead, in passing from x to $x + 1$, uses a formal variable, which is typed just as a first order entity and may be interpreted in any model. Thus, a prototype proof over the integers may need a strong information on types: the input is a standard integer, say. As we know, this is not formalisable in PA (by the Overspill Lemma: no predicate which holds for infinitely many values, may be valid only on the standard initial fragment of a model). This is why, in general, a prototype proof does not yield a formal inductive proof.

Intermezzo: Completing Incompleteness (Part I and II)

Inter-Part I: On the Incompletability of PA

Hilbert's concern, in proposing his famous wrong conjectures was twofold. First, in view of the semantic delirium of geometry, since the fall of the 2300 year old Euclidean empire, one had to retrieve certainty in logico-formal reasoning only, with no reference to meaning in space, time or whatever, as recalled above. Second, in his great mathematical rigour, he was aware of the general mess of the extraordinary mathematics of the XIX century: a fantastic but turbulent growth, where proofs (of valid theorems) were often grounded on hand-waving (Cauchy's work provides good examples for this ...) and theorems were not always true. Thus, Hilbert proposed to give a frame where one could decide "what is a proof". As a matter of fact, in Hilbertian systems one can give a decidable notion of proof: this is a key aspect of Hilbert's notion of formal systems. Then, as recalled above, he further conjectured that any proposition had to be decided, beginning with the describable propositions of the language of PA, of course.

As we all know too well, Gödel, by the I Incompleteness Theorem, proved that

If PA is (ω-)consistent, then it is incompletable

That is, no consistent formal extension of PA is complete, or that no consistent extension, with a decidable notion of proof, allows to decide all arithmetic propositions. Moreover, given such an extension, the formalised statement asserting its own consistency is one of the undecidable propositions (II Incompleteness Theorem).

It is easy though to give an example of **non** formal system, extending PA. Consider Arithmetic with

$$(\omega - rule) \quad \frac{A[n]}{\forall x A[x]} \; for \; all \; n \in \mathbb{N}$$

This system is complete, but it has a non decidable notion of proofs (yet, proofs are well-founded trees).

Note that, the ω-rule derives $\forall x A[x]$ from the assumption of the infinitely many instances of $A[n]$. This is different from any use of prototype proofs, where $\forall x A[x]$, over N, is obtained from a schematic (prototype) proof of $A[n]$, w.r.t. a generic (replaceable) $n \in N$.

Inter-Part II: Formal Proofs of Consistency

The firm formalists often insistingly remark that, after all, any unprovable statement can be proved in a "suitable" formal frame. Now, Gödel's theorem implies that the order of quantifiers cannot be reversed: for any statement, there exists an extension of PA which proves it ... (and not conversely.) As stated, this is trivially true, since, given a formal statement in PA, there exists for sure an extension of PA which proves it: add that very statement as a new axiom .... Yet, even in the case of the (trivial) extension (by the very statement, but by more interesting formal principles as well), there is a non minor problem: one has to prove the consistency of the intended extension! Often, in philosophical discussions, this fact is forgotten.

Consider, say, the formalised statement of the consistency of PA, $Cons_{PA} \equiv \neg Theor_{PA}(0 = 1)$ (i.e. "$0 = 1$" is not a theorem), as a typical unprovable proposition, given by Gödel's second theorem (yet the following argument applies a fortiori to the many formally unprovable propositions, that imply $Cons_{PA}$, such as the ones analysed below). One can surely formally derive $\neg Theor_{PA}(0 = 1)$ from a suitable, and consistent, formal frame: ZF, for example (Zermelo-Fraenkel formalised Theory of Sets). Even the axiom of infinity in ZF can be formally stated in a finitistic fashion; so, a Turing Machine, or our firm formalist, can mechanically derive $\neg Theor_{PA}(0 = 1)$ from the encoded version, in PA, of the axioms of ZF. Call the conjunction of these (encoded) axioms Ax_{ZF} and observe that:

$$PA \vdash (Ax_{ZF} \rightarrow \neg Theor_{PA}(0 = 1)) \tag{1}$$

since, by the various equivalence theorems and by Gödel's representation lemma, any Turing computable function can be fully represented in PA.

Can one then say that the consistency of PA has been formally proved, by finitistic-formal tools, as many claim? Well, PA is consistent if it generates no contradictions, that is if ... $Cons_{PA} \equiv \neg Theor_{PA}(0 = 1)$ **holds**. Does (1) prove this fact?

No, it only proves what is written, i.e. the formal implication $(Ax_{ZF} \rightarrow \neg Theor_{PA}(0 = 1))$. This statement implies the consistency of PA, as validity of $\neg Theor_{PA}(0 = 1)$, provided that ZF is ... consistent (otherwise, from Ax_{ZF}, one could derive everything, including false statements). Now, the consistency of ZF can be shown in either of the following ways:

A - Ax_{ZF} formally generates no contradiction

B - ZF, i.e. Ax_{ZF}, has a model.

A and B are well known to be equivalent, but the formalist who rejects to give meaning to formulae, in particular to the axiom of infinity, may insist about proving A formally. Easy, give a formal Set Theory with a stronger formal axiom of infinity .... and so on so forth towards a never ending regressing chain (in this approach the model construction is hidden or just implicit).

Consider then B. If one explicitly constructs or assumes to have a model of ZF, including of the axiom of infinity, then (1) does prove the consistency of PA. This is so, because this construction/assumption implies that Ax_{ZF} generates no false theorems, and because, if Ax_{ZF} has a model, then also $Cons_{PA} \equiv \neg Theor_{PA}(0 = 1)$ holds (and PA is consistent), by (1).

In summary, a purely formal derivation of $Cons_{PA}$, from whatever formal axiom system, *does not show* the consistency of PA, unless one involves the **meaning** of the required axiom of infinity, for example by giving a model of ZF (by the way, ZF axiom of infinity essentially says: "PA has a model").

I am here saying a triviality that everybody should be aware of. Yet, too often, in theorem proving in particular, people claim that they can prove formally the consistency of PA, or whatever formally unprovable property. Yes, of course, one can derive the implication in (1) or, more generally, given any formalised statement, one can propose some strong enough formal axioms, which mechanically imply it. But, this implication proves the statement, in a mathematical sense, if one assumes the consistency of ZF or of the intended stronger theory. In other words, just "writing the axioms" and computing is not sufficient: one also has to assume/prove that the derivation is sound, i.e. that axioms and rules are consistent (meaningful). This is one of the general reasons for the need of interaction man/machine in theorem proving and it shows up when one has to bootstrap the machine with a suitable formal frame (which may depend on the result one aims at), but also along the proof, as it will be argued below.

2   Concrete Incompleteness I: Normalisation

In 1958, Gödel gave a proof of the first combinatorial statement, unprovable in PA: normalisation for a typed extension of lambda-calculus, system T. Lambda-calculus and its effective extensions may be (easily) encoded in PA and, thus, the encoded Π_2^0 statement, "for each term, there is a normal form" (in system T), may be shown to hold, even though it is not formally provable, since it implies the consistency of PA (proving consistency of PA was Gödel's aim, in 1958).

Some claim that this theorem cannot be called an independent "concrete" or combinatorial statement of PA, as it is "too much" related to consistency. Well, in the Types' community, we believe that lambda-calculus has also a mathematical-combinatorial interest per se, not just for proving consistency of PA. Thus, we do not see the reasons for depriving our community of the first achievement in this sense: a provably true, but formally unprovable mathematical statement of PA. Paris-Harrington theorem, a remarkable result of 1977, is usually given this honour; however, Ramsey theorem, which underlies this finitary variation, is not less related to consistency, via Set Theory.

But how normalisation can ever be proved, if it is unreachable within PA? Easy (oh, no, very difficult), by a prototype proof, not by induction.

2.1 The Unprovability

I will briefly analyse now the "internal reasons" for the formal (PA) unprovability of normalisation. By this I mean an informal insight into the parts of the proof where non-encodable arguments are used. Clearly, the unprovability is rigorously shown by the formal implication: from normalisation derive consistency. Yet, one may try to spell out explicitly the places where the incompleteness phenomena shows up, along the proof of normalisation.

This exercise is analogue, by duality, and may serve as a guideline, to the everyday task in interactive theorem proving (proof-assistants, proof-checking). As a matter of fact, in order to feed a computer with parts of a proof (e.g. a very difficult combinatorial lemma, lots of very long computations ...) one must be able to isolate the fully formalisable parts in the intended logical frame, and have the computer develop or check them. That is, one must be able to point out or distinguish the non-computable from the computable, the essentially higher order from the first order, the use of axioms of infinity and their models from arguments in PA and so on so forth.

Of course, every theorem may allow many different proofs. In our case, this means that the non-encodable passages may be different, as their very nature may depend on the kind of assumptions and proof adopted. I will then focus on Tait-Girard's argument by "candidates of reducibility". This approach to normalisation applies both to Gödel's system T and to Girard's System F (see [Girard90]). Indeed, I will mostly refer to the presentation in [Girard90] for the discussion.

(Tait-)Girard's proof uses a very heavy induction load. That is, in order to prove that every term has a normal form, by induction, it adds to an inductive assumption on the type of a term (see below), two extra assumptions. Why is this done (and needed)?

The point is that in no way induction can be straightforwardly applied to terms (e.g. by an induction on the complexity or length of terms or whatever). This is due to several features of typed calculi. First, the arrow (in the implicative types) is "contravariant": in any formula such as "$\forall x(\rho \to \sigma)$", the properties expressed by ρ are "negated", and this increases the complexity of the type as formula and of the terms living in it. Second, in second order types (i.e. in the types of System F, where universal quantification is over type variables), the type variables may be instantiated by any type, including the one under consideration; this strong impredicative feature forbids any (inductive) stratification of types and terms living in them. In particular, terms may contain types and depend on them: this is one further reason which does not allow induction on "pure" terms. Yet, and surprisingly enough, the dependence of terms on types is very uniform and this is crucial to the proof (see [Girard90]; in a sense, the specific value of a variable type in a term may be disregarded - or all its values affect the computation in the same way: this fact requires some technicalities and it is fully spelled out by the Genericity Theorem in [Longo93]).

Then, the induction on types (not on terms) goes on by using a set of terms in the given types (the "candidates of reducibility"). The terms in such sets are supposed to be normalisable (first clause of the induction), but also to satisfy two further properties, not obviously related to normalisability (see [Girard90], p. 43, 116, 117); a "fine tuning" of these properties in extensions is a common and relevant practice, both in Logic and in Computing (see [Coquand88] for a classic; a non-obvious extension to "subtyping" may be found in [Castagna95]). The further key observation is that the properties in the induction load are not written at the theoretical level (System F or whatever second order system, see 2.2), but belong to the metalanguage. Then, along the proof, one collects the meta-linguistic sets of candidates of reducibility into a type and performs some computations, within system F (in a sense, one "brings down" the meta-theoretic notion to a theoretic one).

Finally, the very handling of second order collections may be understood, in set-theoretic terms, as the use of a proper Second Order Impredicative Comprehension Axiom. This axiom requires an essential blend of syntax and semantics (this is why the competent formalists reject it), in view of the semantic convention on variables (capital variables, say, must range on (sub-)sets of any model, small ones on elements of these sets). Thus, the proof, as given, uses a blend of meta-theory, theory and semantics and, by this, it lies outside PA or of any coding of System F into PA (and much beyond it, see below).

Remark: Unprovability and Hilbert's "Organisation of the Mathematical Discourse". The blend of meta-theory, theory and semantics is very common in mathematics and in every day language: the dream of an unique, definitive formal universe, where all of mathematics could be formalised, relies exactly on this three level distinction and on the conjecture that a well isolated theoretical level could completely describe mathematics. Now, Hilbert(-Tarski) organisation of the mathematical discourse into meta-theory, theory and semantics has been one of the remarkable ideas of the century, in Logic, but it does not describe an absolute objectivity. In [Longo01], it is compared to Euclid's organisation of space, by rigid figures and their homotheties, as for relevance. Clearly, the later is an extraordinary approach to physical space, a non-arbitrary, well motivated proposal, but a rather artificial one, as there is no such a thing as a rigid body and physical space is not closed under homotheties (according to Relativity Theory and current microphysics, since only the group of automorphisms of Euclidean geometry contains the homotheties).

Similarly, Hilbert's approach is so artificial that it has been instrumental to the invention of fantastic artefacts, Turing Machines and, then, our digital computers. These machines work just at one level, the formal-theoretical one. The incompleteness theorems proved for us that this organisation of the proof is not an absolute: first, Gödel Representation Lemma showed that the metatheory of PA could be fully encoded into PA itself and, thus, that it is "part of it"; later, the above proof of normalisation essentially used a blend of the various levels, as just pointed out. Again, it was extremely useful to invent such a rigorous way to analyse the proof (in a sense, I have been using it above), but it is just a technical tool and a temporary one: it lives nowhere, as, for us, humans, there is no such a thing as a "metalanguage" ("I may play chess according to certain rules. But I may also invent a game where I play with the rules themselves. The pieces of the game are then the rules of chess and the rules of the game are, say, the rules of logic. In this case, I have *yet another game*, not a *metagame*" [Wittgenstein68] p.319).

However, there may be a difference in method (even within the same "cage of language", as Wittgenstein would put it): proofs should be analysed *also* by non-mathematical *arguments*, e.g. by non-mathematised insights into the general structure and dynamics of thought, not only by logico-mathematical proofs. Mathematical Logic, yet the main tool for the foundational analysis, is still part of mathematics, by its method and its proofs, thus it cannot *completely* found it. It is somewhat surprising that many leading colleagues, who carefully avoid some consistent and expressive mathematical circularities (or vicious/virtuous circles, impredicativity or non well-foundeness, say), accept this conceptual and philosophical, severe circularity and only develop a metamathematical analysis of foundation. No proof can found the notion of proof. Or, as suggested by Wittgenstein ... "Hilbert's metamathematics will turn out to be a disguised Mathematics" (quoted in [Waismann31], see also [Heinzmann90], [Floyd98] for more on Wittgenstein and incompleteness).

2.2   Berardi-Altenkirch Normalisation of System F, in LEGO

As already mentioned, given any formally unprovable statement, one can surely invent a formal frame to prove it. As a matter of fact, this frame may even reconstruct the very path of the given proof. The unprovability will be then described by the "proof-theoretic strength" of the formalisation and the validity of the statement will rely on the consistency of the proposed formal theory.

This kind of analysis is one of the major contributions given by Mathematical Logic to the foundation of Mathematics. One of its main applications is the invention of systems to handle automatically as much mathematics as possible. In our case, the analysis amounts largely in displaying the exact "formal order" of the different constructions, understood above in terms of language, metalanguage, meta-metalanguage etc. It differs by this from the set-theoretic analysis, mentioned in 3.1 below.

S. Berardi started an analysis of Tait-Girard proof by higher order Arithmetic ([Berardi91]). T. Altenkirch completed it and fully encoded it into LEGO, a very interesting and effective type-theoretic proof-assistant (see [Altenkirch92]).

Let us first clarify what is meant by "higher order logical system". In the case of Arithmetic, sometimes people refer, say, to Second Order Arithmetic, as THE categorical theory of numbers. That is, to the non-formal theory where the interplay between induction and a full second order comprehension axiom for sets, allows to say formally: "any non-empty subset of N has a least element".

This is not a formal system, in the sense we attributed to Hilbert, since the notion of proof is not recursive: as a matter of fact, the system is categorical (and thus complete) and, by this, the set of "theorems" (which coincides with the valid propositions over N) cannot be recursively enumerable.

One may instead define a formal system handling higher order variables (PA_2 formalises set variables, PA_3 variables of sets of sets), and leave induction restricted at each level (number theoretic induction, induction over formulae containing set variables ... all treated differently, see below). In these systems, the notion of proof is decidable, or deductions are effective (even fully mechanisable by LEGO!), and, of course, Gödel's theorems apply to them: thus, they are incomplete. Now, PA_3 allows a universal quantification over (the sort of) *sets of sets of integers*, i.e. a universal quantification over $P(P(N))$ (the powerset of the powerset of N). By this, as we shall see, it provides the right (and minimal) formal frame for the specific theorem we are currently interested in, normalisation for System F (and a fortiori for Gödel's T).

Note finally that many prefer to call PA_2, PA_3 ..., first order *multisorted* systems (and "apparent" higher order ones, as extended first order systems with just different labels for variables). This is a matter of taste, as "proper" higher order systems are non-formal and yield non-effective deductions: the point is to be clear as for what one is talking about and to have clearly in mind that the core idea of Hilbert's formal systems relies on the decidability of the notion of proof. This is so in PA_2, PA_3 ..., while it is lost in "proper" (categorical or complete) higher order systems.

As already mentioned, Tait-Girard's proof uses an inductive definitions over "candidates of reducibility", as *subsets of types* (in the sense that types are "sets of terms" and they have the kind of the sets of integers). This induction may be described as a recursive definition of a function over sets, or a "third order induction". Now, first order induction works on integers, second order induction takes the *minimum fixed point* of monotone operators over sets of integers. The non-obvious concept of third order induction amounts to say that one has to take the minimum fixed point of monotone operators over sets: clearly, such an operator, a function, is a (single-valued) set of (pairs of) sets, similarly as a function over numbers is a (single-valued) set of (pairs of) integers. It is then a "third order" operator and it requires PA_3 to be formalised. Note that no less can be used, as normalisation for System F implies consistency of PA_2, not just PA ($= PA_1$).

Let's see more closely what happens along the proof, by hinting to the insightful formalisation by Berardi. As mentioned in 2.1, the key point is that the proof does not go by induction on terms (as first order entities), but by higher order induction. That is, the proof uses a combined induction on types (second order induction) and on operators on types (third order induction). In particular, the recursive definition of candidates of reducibility uses a map on types, $[.]_\rho$, from a candidate assignments ρ to candidate $[\sigma]_\rho$, for each type σ. This map is a (single-valued) set of (pairs of) sets. Or, more precisely, one considers the graph of the operator:

$$(\sigma, \rho)| \to [\sigma]_\rho$$

and this has the type of *sets of sets of integers*. The key step in the formalisation of Girard's proof is based on taking the minimum fixed point for this operator. Its construction (existence) implies the convergence of the normalisation algorithm on System F.

Call now $\forall x \exists y Norm(x, y)$ the (first order) formal statement of PA_1 that "for any (coded) term of F, there exists a (code for) its normal form" (more precisely, in $Norm(x, y)$, y is the code for the reduction sequence ending with a normal form.) Then $Norm(x, y)$ is a decidable predicate in x and y. Thus we have just pointed out that

$$PA_3 \vdash \forall x \exists y \text{Norm}(x, y) \tag{2}$$

Thus, *under the assumption that PA_3 is consistent* (more precisely: that it is 1-consistent), the proposition holds in the standard model (where number-codes of terms refer to actual terms), or

$$N \models \forall x \exists y \text{Norm}(x, y). \tag{3}$$

Now, fix $n \in N$ and consider $\exists y Norm(n, y)$. This is a Σ_1^0 predicate and any *valid Σ_1^0 predicate* over N is provable in PA_1 (easy: scan the integers till you find one satisfying the predicate). Then

$$given\ n \in N, \text{generic}, PA_1 \vdash \exists y \text{Norm}(n, y) \tag{4}$$

As already observed, no induction on n (as code of the term to be normalised) could be used in the proof: induction is entirely transferred at the level of types and functions on types. Of course, the reader must appreciate the difference between (2) and (4), a subtle but crucial difference: the PA_1 unprovability of $\forall x \exists y Norm(x, y)$ says that there is no way to prove it by first order induction on x. Thus, the proof in (4) is essentially a prototype proof, with generic input n, a standard integer or a true code for a term. And a detour must be taken, via PA_3 and its higher order forms of induction, over sets of types.

Observe also that (4) proves only that there is a computable function that, taken an integer n as code of a term, gives (the code) of its normal form, y. Yet, as stated, (4) does not prove that this "normalising function" is total (and this is where lies part of the logical complexity of the problem). An inspection of the proof of $\forall x \exists y Norm(x, y)$ within PA_3, a proof needed in order to assert its truth, shows that this function is indeed total[2]. Finally, PA_2 may suffice to "isolate"

---

[2] As already mentioned, induction, in contrast to prototype proofs, proves totality of Π_2^0 predicates "for free". Consider the following theorem: (D) $\forall x \exists y (2x < y)$. Of course, this has both a prototype proof ("For a generic n, take m = 2n+1") and an inductive one, in PA ("for 0 take 1; assume that, given x, you have y, then, for (x+1), set y' = y + 2"). Difference: the second proof inductively proves also that the map from x to y in (D) is total. The first one, instead, requires a further insight: one has to prove that "." and "+" are total (not too hard, in this case; very complex, as for the totality of the normalising function, since this is shown in PA_3).

the standard integers (PA_2 contains a predicate for them, exactly, i.e. it defines the type of integers), but this is not sufficient to formalise (4), as the proof that the normalising function is total entirely relies on the third order structure of the normalisation proof.

3 Concrete Incompleteness II: Kruskal-Friedman Theorem

Everybody knows what is a tree. Trees in Mathematics grow downwards. They are partial orders with a root on top (the largest element), and, for each node a in a tree T, $\{x/x > a\}$ is totally ordered. A tree-embedding h, form T to T' (notation: $T < T'$), is an injective map, which preserves upper bounds (i.e. h(sup$\{a,b\}$) = sup'$\{$h(a),h(b)$\}$).

Kruskal, in 1960, proved the following theorem (KT): *For any infinite sequence of finite trees $\{T_n/n < \omega\}$, there exist j and k such that $j < k < \omega$ and $T_j < T_k$.*

A non-obvious result. In 1981, Friedman proposed a finitary version of this fact, i.e. a variant that may be stated in PA. Here is a form of it:
For any n, there exists an m such that for any finite sequence $T_1, T_2, ..., T_m$, where T_i has at most $n(i+1)$ nodes, there exist j and k such that $j < k < m$ and $T_j < T_k$.
(FFF or Friedman's Finite Form, see [Harrington85], [Gallier91]).

This Π_2^0 statement of PA has a purely combinatorial nature, since in no apparent way is it related to consistency issues. Moreover, both KT and FFF have several interesting consequences in finitary combinatorics (e.g. in Term Rewriting Systems).

Let's try to sketch the "reasons for unprovability", as we did in the previous case. However, now, two radically different proofs of KT (and FFF) are available, each worth analysing, although briefly. Once more, the logical reasons for unprovability are grounded on Gödel's second theorem, since, surprisingly enough (and this is the remarkable insight of Friedman), FFF implies the consistency of PA (and much more). We will hint to the specific passages of the proofs where unprovability shows up. FFF easily follows from KT, by an application of Koenig's lemma ("any infinite finitely branching tree has a infinite path"). This lemma is conservative over PA, thus the problems are hidden along the proof of KT.

3.1 The Set-Theoretic Analysis

The usual, set-theoretic proof of KT goes by a strong non-effective argument. It is non-effective for several reasons.

First, one argues "ad absurdum", i.e. one shows that a certain set of possibly infinite sequences of trees is empty, by deriving an absurd if it were not so. More precisely, one assumes that the set of "bad sequences" (or sequences without ordered pairs of trees, as required in the statement of KT) is not empty and defines a minimal bad sequence from this assumption; then one shows that that minimal sequence cannot exist, as a "smaller" one can be easily defined from it.

Note that this minimal sequence is obtained by using a quantification on a set that is ... going to be proved to be empty, a rather non-effective procedure. Moreover, the empty-to-be set is defined by a Σ_1^1 predicate, well outside PA (a proper, impredicative second order quantification over sets, see the discussion on system F, 2.1).

For the non-intuitionist who accepts a definition of a mathematical object (a sequence in this case) ad absurdum, as well as an impredicatively defined set, the proof poses no problem. It is abstract, but very convincing (and relatively easy). The key non-arithmetisable steps are in the Σ_1^1-definition of a set and in the definition of a sequence by taking, iteratively, the least element of this set. Yet, the readers (and the graduate students to whom I lecture) have no problem in applying our shared mental experience of the "number line" to accept this *formally* non-constructive proof: from the assumption that the intended set is non-empty, one understands ("sees") that it has a least element, without caring of its formal (Σ_1^1-)definition. Or, if the set is assumed to have an element, then the way the rest of the set "goes to infinity" doesn't really matter, in order to understand that it must have a least element: the element supposed to exist (by the non-emptiness of the set) must be somewhere, in the finite, and the least one will be among the finitely many which precede it, even if there is no way to present it explicitly. Finally, the sequence defined ad absurdum, in this highly non-constructive way, will never be used: it will be absurd for it to exist. So its actual "construction" is irrelevant.

Of course, this is far away from PA, but it is convincing to anyone accepting the "geometric judgement" mentioned in the introduction: a non-empty subset of the number line has a least element (see the conclusion as well).

3.2   The Constructive Version

An intuitionistically acceptable proof of KT has been recently given in [Rathjen93]. This proof of KT is still not arithmetisable, of course, but it is "constructive", at least in the broad sense of infinitary inductive definitions, as widely used in the intuitionist community (see the seminal work by Martin-Löf; a classical introduction is in [Aczel78]).

The idea is to construct an effective "reverse embedding" of the partial order of finite trees (partially ordered by the tree-inclusion above) into a suitable ordinal representation system. This requires an insightful study of the combinatorial properties of tree-embeddings.

In short, let $FinBad_T$ be the set of *finite* bad sequences of trees (see 3.1). Once given a system of ordinals (ORS,$<$), Rathjen and Weiermann construct a function $f : FinBad_T \to ORS$ such that, if s and t are in $FinBad_T$, and t extends s strictly, then $f(t) < f(s)$ (in ORS). Clearly then, if there exists an infinite bad sequence, and thus an infinite ascending sequence in $FinBad_T$, then ORS would contain an infinite descending sequence.

The function f is actually primitive recursive and everything up till this point can be done in Intuitionist Arithmetic, HA. KT then follows from the proof that $(ORS,<)$ is well-founded and, in particular, that every primitive recursive sequence $p(0) > p(1) > p(2) > \ldots$ must terminate after finitely many steps. This proof is done in an intuitionistically acceptable formal frame, called ID_1, which extends induction along constructible ordinals and well beyond PA.

This is the non-formalisable part, in PA. As a matter of fact, PA cannot prove the Π_2^0 statement that $(ORS,<)$ is (primitive recursively) well-founded, since this well-ordering is sufficient to prove the consistency of PA (it actually implies induction well beyond ϵ_0, the ordinal of consistency for PA).

As in the previous case, the Π_2^0 statement of FFF cannot be proved by first order or ω-induction, i.e. within PA. In the approach by inductive definitions though, the difficulty is taken care by "pulling induction along the ordinals", well beyond ω, or even ϵ_0.

Conclusion

The proof-theoretic investigation of Mathematics has been one of the major achievements of XX century Science: it gave us mathematical rigour and modern computing. The latter, its major fall-out, is changing our live. Yet, we need to go beyond its techniques and philosophy. First because, in view also of its successful paradigms, Mathematics has now reached a remarkable level of rigour and we are no longer scared of the novel geometric intelligibility of physical space, which originated Frege's "royal way out" from the "delirium" in Geometry (see [Tappenden95] for more on Frege's view). Second, because we may take advantage, also in computing, by an enlargement of our foundational paradigms, beyond the traditional linguistic-finitistic certitudes. In a sense, we should try to bring together the two "foundational ways" that split at the end of XIX century, as I tried to summarise in the introduction. In short, the foundation of Mathematics lies in:

- Logic
- Formalisms
- Regularities of phenomenal space.

We all know what the first two points mean and their relevance. By the third, I mean the reference to a few regularities of phenomenal space, such as connectivity (Riemann) or symmetry (Weyl), but "ordering" as well. By the subsequent and constructed notion of well-ordering, it was meant here the (very strong) geometric judgement: "consider a *generic* non-empty subset of the integer number line, observe that it has a least element" (see [Dehaene98] for a neuropsychologist's experimental analysis of our mental number line, as a cognitive experience).

As a matter of fact, in Mathematics, we transform these regularities of space - that we happen to "see" - (as well as our cognitive approach to them by mental re-constructions), into explicit conceptual invariants (as "hypothesis", in Riemann's terminology). This process grounds mathematics, as a conceptual construction, in our "phenomenal lives" (as Weyl would put it): concepts and structures are the *result* of a cognitive/knowledge process. Then they are further extended, by language and logic: from connectivity we go to homotopy theory or to the topological analysis of dimensions, say. Symmetries lead us to duality and adjunctions in Categories. The ordering of numbers is extended into transfinite ordinals.

Of course, these notions may be formalised, each in some "ad hoc" way, as there is no Newtonian or ZF absolute universe. But evidence and foundation are not completely captured by the formalisations, since "The primary evidence should not be interchanged with the evidence of the 'axioms'; as the axioms are mostly the result already of an original formation of meaning and they already have this formation itself always behind them", [Husserl33]. Moreover, incompleteness tells us that the reference to this underlying and constitutive (not independent) meaning cannot be avoided in foundation, as the consistency issue is crucial in all formal derivations (see the Intermezzo, part II).

In this perspective, we need to ground mathematics also on a few "geometric judgements" which are not less solid than the logic ones: "symmetry" for example is at least as fundamental as "modus ponens", or it steps heavily into mathematical constructions (and in proofs, as pointed out by Girard - see [Girard01] for recent advances of his program in Logic and foundations). As already mentioned, physicists argue since long "by symmetry". More generally, modern Physics extended its analysis from the Newtonian "causal laws" (the analogue to the logico-formal and absolute "laws of thought", since Boole and Frege) to an understanding of phenomenal world by our active geometric structuring of it: from the conservation laws as symmetries (Noether's theorem), to the geodesics of Relativity Theory (see [Weyl27] for an early mathematical and philosophical insight into this). The normative nature of geometric structures is currently providing a further understanding even of recent advances in microphysics ([Connes94]). Our foundational analyses and their applications should also be enriched by this broadening of paradigm in scientific explanation: from laws to geometric intelligibility (grounded on accessibility of space, see [Longo02]). But in Logic as well, we have to move from viewing formal properties and logical laws as a linguistic description of an independent reality, to their appreciation as a result of a praxis: they are the constituted invariants of our practice of reasoning and language, as an open ended "game" between us and a world to be organised also by language.

In Number Theory, well-ordering, as a structuring of phenomenal space and time by integer numbers, is a founding "geometric judgement": its certainty is the consequence of a common conceptual construction, the number line, we all experience in our (mathematised) mental space (see above and the Introduction). When "formalised", as proper second order impredicative statement (induction plus second order full comprehension), it is highly ineffective. Instead, it is perfectly "robust" (and "effective" - as a mental construction) if seen as a "geometric judgement", related to the constructed order structure of integer numbers. Of course, considering - "seeing" - a *generic* non-empty subset is crucial, instead of formally taking *all* non-empty subsets (see also 3.1 and below). And, as a judgement, it provides a reliable geometric argument for consistency of PA: all other proofs use consistency of stronger theories, large ordinals or axioms of infinity. Clearly, it is not an alternative to the fine analysis of consistency provided by Proof-Theory (normalisation, relative consistency results ...) but is complements it, by grounding Mathematics on broader mental experiences.

It should be clear, though, that the mathematical constructions, such as the "number line", are "shared" cognitive performances as they are done in language and intersubjectivity, along history. They are "progressive conceptualisations", as suggested by Enriques ([Enriques01], [Faracovi82]), which originate from regularities of space, time and reasoning, and in no way the grounding of mathematics also on geometric judgements should let us forget the key role of language for our communicating human community. Only by language we can conceive and express never ending, discrete iteration, which we later place back into mental space (the number line). By action *and* language we organise (we "order") space and time. In general, intersubjective exchange, by language, is a core component of the *constructed* invariance and conceptual stability of Mathematics: that is, invariance is also the result of a conscious appreciation of "what we all share" (Poincaré), including a constructed mental image. Thus, in spite of this approach's debt to Brouwer's constructivism (but this debt has been filtered by the remarkable teaching by Dana Scott, J.-Y. Girard and Per Martin-Löf in Logic), we radically depart here from Brouwer's "languageless" mathematics, [vanDalen91]. As well as from the Platonism/formalism debate, the "new scholastic" of the XX century, as Enriques called it ([Enriques36], but Poincaré and Hermann Weyl should be quoted as well).

The reference to these leading geometers' critical (anti-formalist and anti-ontological) attitude w.r.to the main-stream foundational debate may be the occasion for a concluding remark concerning the use of "generic" elements in proofs (and in judgements). This is of course an essential notion in defining prototype proofs (and geometric judgements).

In Mathematical Logic, since Frege and Hilbert, by the prevailing algebraic approach and by the focus on Arithmetic, generic elements are dealt with as variables, formally handled by "for all introduction/elimination rules" and induction. Typically, in algebra, one proves $(x + y)^2 - 2xy = x^2 + y^2$, say, by formally manipulating two variables; then, the full generality of the result is obtained by the "for all introduction" rule. In this frame, "generic" means "variable ranging on all elements of the intended domain": in the proof of a universally quantified statement, just type and formally manipulate your variables and you are done. A crucial point, of course, is that "for all" is interpreted by ... "for all", along the proof. This proof-theoretic treatment of "for all" (as well as the naive set-theoretic interpretation[3]) is confirmed by the use of arithmetic induction: in order to prove a property for all numbers, prove it for 0, than extend *the proof* to *all* numbers by moving from x to $x + 1$.

---

[3] cf. the much more structured and informative interpretation of first and second order variables and quantification in categories, [Lawvere76], [Lambek86], [Asperti91].

However, when one has to prove a property of structures or objects, that do not form a well-ordered set, *all* right triangles, say, or *all* Riemann's manifolds, or *all* algebraically closed fields, or even *all* real numbers ... in no way the proof is done "for all". One considers a *generic* right triangle, or Riemann's manifold, or non-empty subset of bad-sequences (see 3.1) ..., gives the proof and, at the end, one observes, by scanning again the argument: note that I only used the very definition of the intended structure (or mathematical object), no more no less, and this shows that my proof has the right level of generality. The drawing on the sand of a Greek beach of a right triangle, the geometric proof by Pythagoras, the observation that the proof *depends only* on the right angle and *not* on the length or ratio of the sides, is the real birth of Mathematics. Note that, in order to give the geometric proof, the sides (their ratio) have to be given, as a specific right triangle (ratio of sides) *has* to been drawn. In no way *all* right triangles, with their different (ratio of) sides, are scanned, as, say, in inductive proofs; on the contrary, a (provably) generic one is used. The methodological difference is crucial and the key role given to Formal Arithmetic in foundation and in Proof Theory has been hiding it. Moreover, this contributed to the new scholastic in the philosophy of Mathematics: on one side, it confirmed the formalist lack of appreciation of the *construed* genericity of individual Mathematical objects and structures (formal variables are the "generic" signs); on the other it contributed to ontological commitments (the reference to *existing* sets of *all* right triangles, Riemann manifolds, real numbers ... instead of the analysis of their conceptual construction).

There is one more reason for going in this further direction, which stresses also the role of "geometric judgements" and generic structures (and should accompany the proof-theoretic analysis, when technically insightful). Mathematics is not only grounded on proofs, the main concern of late XIX and XX century Mathematical Logic, but it also goes by *construction of concepts and structures.* Indeed, new concepts and structures are required even along proofs, as shown by the permanent need of "new axioms" (or, also, when just trying to find the right induction load). The analysis of these constructions should not be left to the magic or the metaphysics of some ontological "intuition", but it should become part of a scientific investigation. This analysis I call "the cognitive component" of the foundation of mathematics and it is an ongoing project (see the research program "Géométrie et Cognition", Longo's web page, or [Longo02]).

Acknowledgements

I am greatly indebted to Stefano Berardi and Micheal Rathjen for several very stimulating and helpful discussions and e-mail messages. An anonymous referee and a disclosed one, Gilles Dowek, contributed to the revised version by their relevant and numerous comments. Of course, any mistake and the strong philosophical commitment remain of my own responsibility. This work has been partially supported by the "Action Cognitique" (MENRST), as part of the "Atelier Géométrie et Cognition".

References

[Aczel78] Aczel P., "An introduction to inductive definitions", Handbook of Mathematical Logic, Barwise ed., 1978.

[Altenkirch92] Altenkirch T. "A formalization of the Strong Normalization proof for System F in Lego", December '92.

[Asperti91] Asperti A., Longo G. Categories, Types and Structures, M.I.T. Press, 1991.

[Berardi91] Berardi S. "Girard's normalization in Lego", Univ. Torino, 1991.

[Boi95] Boi L. Le problème mathématique de l'espace, Springer, 1995.

[Bottazzini95] Bottazzini U., Tazzioli R., "Naturphilosophie and its role in Riemann's mathematics", Revue d'Histoire des Mathématiques n. 1, 3-38,1995.

[Brouwer48] Brouwer L. "Consciousness, Philosophy and Mathematics", 1948, in Collected Works vol. 1 (Heyting ed.), North-Holland, 1975

[Castagna95] Castagna G., Ghelli G. and Longo G. "A calculus for overloaded functions with subtyping", Information and Computation, 117(1):115–135, Feb. 1995.

[Connes94] Connes A. Non-commutative Geometry, Academic Press, 1994.

[Coquand88] Coquand T., Huet G. "The calculus of Constructions" Information and Computation, 76, 95 - 120, 1988.

[vanDalen91] van Dalen D. "Brouwer's dogma of languageless mathematics and its role in his writings" Significs, Mathematics and Semiotics (Heijerman ed.), N.H., 1991.

[Dehaene98] Dehaene S. The Number Sense, OxfordUP, 1998. (Review/article downloadable from http://www.di.ens.fr/users/longo, as well as all Longo's papers below)

[Enriques01] Enriques F. Problemi della Scienza, 1901.

[Enriques36] Enriques F. "Philosophie Scientifique", in Actes du Congrès International de Philosophie Scientifique, Paris, 1935, Hermann, Paris, vol.I, 1936.

[Faracovi82] Faracovi O. "Ragione e progresso nell'opera di Enriques", in Federigo Enriques. Approssimazione e verita', a cura di O.P. Faracovi, Belforte, Livorno, 1982.

[Floyd98] Floyd J. "Wittgenstein on Gödel and Mathematics", Conference on "Wittgenstein et les Fondements des Mathématiques", ENS, Paris, 1998 (to appear).

[Frege84] Frege G. The Foundations of Arithmetic, 1884 (Engl. transl. Evanston, 1980.)

[Friedman97] Friedman H. "Some Historical Perspectives on Certain Incompleteness Phenomena", May 21, 5 pages, draft, 1997.

[Gallier91] Gallier J., "What is so special about Kruskal's theorem and the ordinal Γ_0?" Ann. Pure. Appl.Logic, 53, 1991.

[Girard90] Girard J.Y., Lafont Y., Taylor P. Proofs and Types, Cambridge U. Press, 1989.

[Girard01] Girard J.Y., "Locus Solum", Mathematical Structures in Computer Science, vol.11, n.3, 2001.

[Goldfarb87] Goldfarb H., Jacques Herbrand: Logical Writings, 1987.

[Harrington85] Harrington L. et al. (eds) H. Friedman's Research on the Foundations of Mathematics, North-Holland, 1985.

[Heinzmann90] Heinzmann G. "Wittgenstein et le théorème de Gödel", Actes du Colloque sur "Wittgenstein et la philosophie aujourd'hui" (Klinschsiech), 1990.

[Husserl33] Husserl E., The origin of Geometry, part of Krysis, 1933.

[Lambek86] Lambek J., Scott P.J., Introduction to higher order Categorical Logic, Cambridge University Press, 1986.

[Lawvere76] Lawvere F.W. "Variable quantities and variable structures in topoi", in Algebra Topology and Category Theory: A collection of papers in honor of Samuel Eilenberg, A. Heller and M. Tierney (eds.), Academic Press, (101 121), 1976

[Longo00] Longo G. "Prototype proofs in Type Theory", in Mathematical Logic Quaterly (formely: Zeit. f. Math. Logik und Grund. der Math.),vol. 46, n. 3, 2000.

[Longo01] Longo G. "The reasonable effectiveness of Mathematics and its Cognitive roots", in New Interactions of Mathematics with Natural Sciences (L. Boi ed.), Springer, 2001.

[Longo02] Longo G. "Space and Time in the foundation of Mathematics, or some challenges in the interaction with other sciences", invited lecture at the First AMS/SMF meeting, Lyon, July 2001 (to appear).

[Longo93] Longo G., Milsted K. and Soloviev S., "The genericity theorem and parametricity in the polymorphic Lambda-calculus" Theor. Comp. Sci. vol. 121, 1993.

[Paris78] Paris J., Harrington L., "A mathematical incompleteness in Peano Arithmetic", Handbook of Mathematical Logic, Barwised ed., 1978.

[Rathjen93] Rathjen M., Weiermann A. "Proof-theoretic investigations on Kruskal's theorem." Annals of Pure and Applied Logic,60, 49–88, 1993.

[Riemann54] Riemann B. "On the hypothesis which lie at the basis of geometry", 1854 (English transl. by W. Clifford, Nature, 1873).

[Tappenden95] Tappenden J. "Geometry and generality in Frege's philosophy of Arithmetic" Synthese, n. 3, vol. 102, March 1995.

[Waismann31] Waismann F., Wittgenstein und der Wiener Kreis, Frankfurt a. M., Suhrkamp, 1967 (Waismann's notes: 1929-31). (English translation: Wittgenstein and the Vienna circle : conversations recorded by Friedrich Waismann, Barnes & Noble Books, New York, 1979)

[Weyl27] Weyl H. Philosophy of Mathematics and of Natural Sciences, 1927 (Engl. transl., Princeton University Press, Princeton, New Jersey, 1949).

[Wittgenstein68] Wittgenstein L. Philosophical Remarks. Engl. transl. by G. E. M. Anscombe, Barnes & Noble, New York, 1968.

Changing Data Structures in Type Theory: A Study of Natural Numbers

Nicolas Magaud and Yves Bertot

INRIA Sophia Antipolis

Abstract. In type-theory based proof systems that provide inductive structures, computation tools are automatically associated to inductive definitions. Choosing a particular representation for a given concept has a strong influence on proof structure. We propose a method to make the change from one representation to another easier, by systematically translating proofs from one context to another. We show how this method works by using it on natural numbers, for which a unary representation (based on Peano axioms) and a binary representation are available. This method leads to an automatic translation tool that we have implemented in *Coq* and successfully applied to several arithmetical theorems.

1 Introduction

Mechanical theorem provers can be used to develop provably correct software and formalize large bodies of mathematics. Proofs for mathematical objectives are eventually very useful for proofs of software components, when the correctness of algorithms rely on arbitrarily complex mathematical notions, but very often the proof styles are different. For plain mathematics, simplicity of concepts is of paramount importance. For software proofs, efficiency plays a more important role.

When considering natural numbers, the difference of style between mathematical proofs and software proofs is embodied in two different representations of numbers. With the *unary* representation, natural numbers are described as an inductive set with two constructors: 0 and the successor function. The simplicity is perfect, inductive proofs only have two cases, and this representation is preferred in most proof developments about the mathematical properties of natural numbers. But the representation of a number has a size proportional to the number itself. Computers use a more compact representation, called the binary representation. A natural number is either 0, or a sequence of *bits*, ones and zeros, that starts with a *one*. Inductive proofs on this structure naturally have four cases: two base cases corresponding to 0 and 1 and two other cases corresponding to numbers of the form $2 \times x$ and $2 \times x + 1$ when x is already a number different from 0.

Up until now, a large number of results have been established about natural numbers, where the proofs make an intensive use of the unary representation of natural numbers. However, it may sometimes be desirable to switch to the

P. Callaghan et al. (Eds.): TYPES 2000, LNCS 2277, pp. 181–196, 2002.
© Springer-Verlag Berlin Heidelberg 2002

more efficient binary representation. Should all results be proven again, or is there a way to re-use all the proofs that have already been done, either by translating them from one data-structure to another, or by establishing these results for an abstract data type that can be shown to be "made concrete" in both representations?

This study concentrates on natural numbers, but it should apply to many other contexts where initial choices of concrete data types might need to be re-considered as software evolves.

2   Proof Representation

In type-theory based logical systems, the Curry-Howard isomorphism establishes a correspondence between logic and typed λ-calculus, where types are logical formulas and λ-terms are proofs. This isomorphism makes it possible to identify the implication with a function type. Moreover, universal quantification is identified with a general notion of *dependent* function types, where the type for the returned value depends on the input value.

For instance, let us consider a proof of "transitivity of implication" for propositions P, Q, and R. The statement has the following shape:

$$(R \Rightarrow P) \Rightarrow (P \Rightarrow Q) \Rightarrow R \Rightarrow Q. \tag{1}$$

The proof for this statement is represented by the following term:

$$\lambda th_1 : (R \Rightarrow P); th_2 : (P \Rightarrow Q); th_3 : R.(th_2 \ (th_1 \ th_3)). \tag{2}$$

According to Heyting-Kolmogorov's semantics, we can also associate a computational interpretation to proofs. We can consider the term (2) as a function, which constructs a proof of Q when given proofs th_1, th_2, th_3, respectively for $R \Rightarrow P$, $P \Rightarrow Q$, and R. Just respecting the type discipline, we know that applying th_1 to th_3 is well-formed and returns a term of type P (i.e., a proof of P), then applying th_2 finishes the proof.

Proof Translation: If P', Q', and R' are the same properties as P, Q, and R, but now expressed with respect to a new concrete representation, it is easy to reuse the proof (2) to obtain the following statement:

$$(R' \Rightarrow P') \Rightarrow (P' \Rightarrow Q') \Rightarrow R' \Rightarrow Q' \tag{3}$$

A proof of this statement then becomes:

$$\lambda th_1' : (R' \Rightarrow P'); th_2' : (P' \Rightarrow Q'); th_3' : R'.(th_2' \ (th_1' \ th_3')). \tag{4}$$

Application to Induction Principles: In practice, when one chooses a data type, proof tools automatically generate induction principles. As a result, the proof structures are influenced by the chosen data types and some theorems available in one formalization will be missing in the other one. One important part of the translation work is to recover theorems corresponding to induction principles from the old setting in the new setting.

Convertibility: Reasoning about inductive types also involves computing recursively over these types. In type-theory based proof systems, the tradition is to include computation in the conversion mechanisms that are automatically used by the type-checker to compare two terms. Because the recursion mechanism is intimately linked with the structure of the data types, this has a tremendous impact on the translation of proofs: terms that are convertible in one setting may not be convertible in the new setting. If P_A and P_B are considered equivalent in the initial setting, the following term represents a correct proof:

$$\lambda th_1 : (P_A \Rightarrow Q); th_2 : (R \Rightarrow P_B); th_3 : R.(th_1 \ (th_2 \ th_3)). \tag{5}$$

The statement being proven is the following one:

$$(P_A \Rightarrow Q) \Rightarrow (R \Rightarrow P_B) \Rightarrow R \Rightarrow Q. \tag{6}$$

When changing over to a new data-structure, P'_A, and P'_B, obtained by translating P_A and P_B respectively may not be convertible anymore. It then becomes necessary to modify the structure of the proof term (5) and replace the implicit equivalence between P_A and P_B by the explicit application of a theorem $th'_4 : P'_B \Rightarrow P'_A$ that we will have to prove on the side. In this case, the translation of (6) becomes the following term:

$$\lambda th'_1 : (P'_A \Rightarrow Q'); th'_2 : (R' \Rightarrow P'_B); th'_3 : R'.(th'_1 \ (th'_4 \ (th'_2 \ th'_3))).$$

In this paper, we describe the practical problems we have encountered in an attempt to transform theorems from the unary representation to the binary representation. We first recall the basics of inductive data type descriptions in *Coq* (section 3), we then show how to make the conversion steps in the proof explicit (section 4). In a third stage, we study how to represent functions and their computational behavior (section 5). We finally present the tools we need to change from one inductive data type to another, focusing on natural numbers and their unary and binary representations (section 6). We then present an overview of related work and examine the perspectives of this work.

3   Inductive Types in *Coq*

3.1   General Characteristics of *Coq*

The *Coq* system provides an implementation of the *Calculus of Inductive Constructions* [3, Chap.4]. It is a typed λ-calculus with dependent types [16] with capabilities for inductive definitions [14]. *Coq* provides an interactive mode for developing proofs by backward chaining, with a collection of proof commands called tactics, that decompose goal logical formulas into simpler sub-goals. Each tactic makes it possible to perform elementary reasoning steps. Eventually, all proof steps are combined in a λ-term where the reasoning steps are represented using λ-calculus constructions.

Inductive types are defined by giving a collection of functions whose co-domain is the type being defined and whose arguments may be in the inductive type being defined (in fact, there are precise constraints on how the inductive type may be used in the arguments but we will use only a simple form of inductive types here). These functions are called constructors. For instance, unary natural numbers are defined in the following manner:

```
Inductive nat : Set := O : nat | S : nat -> nat.
```

The inductive type is called nat, the functions given to define it are O (a function with no argument, i.e., an element in the type) and S (a unary function). The automatic tools in *Coq* use this definition to automatically construct a *structural induction principle* and recursion operator that makes it possible to define recursive functions. The induction principle has the following shape:

```
nat_ind
 : (P:(nat->Prop))(P O)->((n:nat)(P n)->(P (S n)))->(n:nat)(P n)
```

In the calculus of inductive constructions, the recursion operator associated to an inductive type is integrated in the conversion mechanisms that make it possible to compare terms. It can be assimilated with a term nat_rec with the following type:

```
nat_rec
 : (P: (nat -> Set))(P O)->((n:nat)(P n)->(P (S n)))->(n:nat)(P n)
```

The reduction rules associated with this recursion operator are as follows:

$$(\text{nat\_rec } P \, v_0 \, f \, O) \rightarrow v_0 \quad (\text{nat\_rec } P \, v_0 \, f \, (S \, n)) \rightarrow (f \, n \, (\text{nat\_rec } P \, v_0 \, f \, n))$$

The definition of recursive functions is not always translated in terms of the recursion operators. Instead, the definition of a recursive function can be viewed as giving rise directly to reduction rules. For instance the factorial function is defined as follows:

```
    Fixpoint fact [n:nat] : nat :=
        Cases n of O => (S O)
                | (S p) => (mult (fact p) (S p))
    end.
```

Corresponding reduction rules are:

$$(\text{fact } 0) \rightarrow (S \, 0) \quad (\text{fact } (S \, n)) \rightarrow (\text{mult } (\text{fact } n) \, (S \, n))$$

The logical system does not distinguish between convertible terms. Put in other terms, (fact 0) and (S 0) are identified. As a result, convertibility may hide some parts of the reasoning steps, a particularity that has been thoroughly exploited in *reflexion* or *two-level* approaches [8, 4].

4   Making Conversions Explicit

Some reasoning steps may be missing in proof terms because of the convertibility rules which may identify syntactically different terms. We start by describing through an example what implicit computational steps we would like to make explicit in a proof term. We then present the method developed to achieve this goal and describe an algorithm. We conclude by showing what happens when the algorithm is applied to an example.

4.1   An Example

Throughout this section, we consider the theorem plus_n_0.

$$\forall n \in \mathsf{nat} \quad n = (\mathsf{plus}\ n\ \mathsf{O}) \tag{7}$$

A proof (as a λ-term) of this property is :

$$\lambda n : \mathsf{nat}.(\mathsf{nat\_ind}\ (\lambda n0 : \mathsf{nat}.n0 = (\mathsf{plus}\ n0\ \mathsf{O}))$$
$$(\mathsf{refl\_equal}\ \mathsf{nat}\ \mathsf{O})$$
$$\lambda n0 : \mathsf{nat}; H : (n0 = (\mathsf{plus}\ n0\ \mathsf{O})).$$
$$(\mathsf{f\_equal}\ \mathsf{nat}\ \mathsf{nat}\ \mathsf{S}\ n0\ (\mathsf{plus}\ n0\ \mathsf{O})\ H)\ n)$$

It proceeds by induction on n, through the principle called nat_ind. The term (refl_equal nat O) is a proof that $\mathsf{O} = \mathsf{O}$. However, thanks to convertibility rules, it is also a proof of $\mathsf{O} = (\mathsf{plus}\ \mathsf{O}\ \mathsf{O})$. In the same way,

$$\lambda n0 : \mathsf{nat}; H : (n0 = (\mathsf{plus}\ n0\ \mathsf{O})).\ (\mathsf{f\_equal}\ \mathsf{nat}\ \mathsf{nat}\ \mathsf{S}\ n0\ (\mathsf{plus}\ n0\ \mathsf{O})\ H)$$

is a proof of

$$\forall n0 : \mathsf{nat} \quad n0 = (\mathsf{plus}\ n0\ \mathsf{O}) \Rightarrow (\mathsf{S}\ n0) = (\mathsf{S}\ (\mathsf{plus}\ n0\ \mathsf{O}))$$

It is also a proof of $\forall n0 : \mathsf{nat} \quad n0 = (\mathsf{plus}\ n0\ \mathsf{O}) \Rightarrow (\mathsf{S}\ n0) = (\mathsf{plus}\ (\mathsf{S}\ n0)\ \mathsf{O})$ which is the expected type for the third argument of the induction principle nat_ind.

4.2   Computing Expected and Proposed Types

The first step in our work is to extract the implicit computational steps from the proof terms. To this end, we use a technique derived from Yann Coscoy's work [5] on proof explanation. This technique is also very close to the methods for type-checking dependent types in the original ALF system [11]. We need to determine positions within the proof term where expected and proposed types differ. We restrict our work to concrete datatypes and concrete equalities and will not be able to work with an abstract algebraic "book" equality. Here we focus on Leibnitz's equality which is the basic equality of Coq.

4.3   Expected and Proposed Types

The algorithm we present in the next section uses intensively the notions of
expected and proposed type for a term. Given a term t, its expected type T_E is
the type required for the whole term (whose t belongs to) to be well-typed. Its
proposed type T_P is the one inferred by the type system.

Let us consider the example of an application $(t_1\ t_2)$. If the types inferred by
the system are the following ones:

$$t_1\ :\ T\ \rightarrow\ U\qquad\&\qquad t_2\ :\ T'$$

Then the expected type for t_2 is the argument T of the functional type $T\ \rightarrow\ U$,
whereas its proposed type is T'. The application $(t_1\ t_2)$ is well-typed provided
the expected and proposed types for t_2 are convertible. When using the formulas-
as-types (and proofs-as-terms) analogy, we see that the expected type is a *proof
obligation*, i.e. it states what we are required to prove. The proposed type is a
proof witness, i.e. it can be viewed as the formula whose term (here t_2) is actually
proven.

4.4   The Algorithm

The algorithm is designed to work on proof terms which are well-typed. Its aim
is to locate positions where implicit ι-reduction steps occur in the proof term. It
eventually builds a new proof term, with no more implicit (ι-reduction) steps.
All these steps are recorded as logic variables that need to be proven on the side.

The terms we consider in this study are constants, variables, inductive types
and their constructors, λ-abstractions, product types, and applications. The calculus of constructions also contains a case analysis operator (**Cases**) but it involves important complications and we treat this operator in a separate section.

The algorithm takes a statement T, its proof t, and a context C as input. It
proceeds by recursive case analysis on t, and returns a new term and a list of
conjectures.

- **Variables, Constants, Inductives Data Types, Constructors**
  All these constructs remain unchanged and are simply returned.
- **Abstraction** $t \equiv \lambda x : A.b$
  By hypothesis, the proof term we consider is well-typed. This ensures that
  T can be reduced (via weak-head normal form computation) to a product
  $\forall x : A'.B$. Moreover, we know that A and A' are convertible modulo $\beta\delta\iota$-
  reduction. We choose to add $x : A'$ in the context C.
  We then call the algorithm recursively with new input: b instead of t, B
  instead of T, and $x : A'$; C instead of C.
- **Application** $t \equiv (h\ u_1 \ldots u_n)$
  Here, we only know that the expected type for the whole application is T.
  There is no expected type for the function h. However, we can compute its
  proposed type T_P. It is necessarily a product since the initial term is well-
  typed. We can infer the expected type TE_{u_1} for the first argument. We then

compare that type with TP_{u1}, that can be inferred by the system. If the types TE_{u_1} and TP_{u_1} are convertible by $\beta\delta$ conversion the algorithm simply stores an intermediary value $v_1 = u_1$. If these two types are not convertible by $\beta\delta$ conversion, then the algorithm constructs a new conjecture c_1 whose type is $\forall x_1 : t_1, \ldots x_k : t_k.TP_{u_1} \to TE_{u_1}$ where the variables x_i are the variables that appear free in TP_{u_1} and TE_{u_1} and whose type is given in C; it stores an intermediary value $v_1 = (c_1\ x_1 \cdots\ x_k\ u_1)$. This process is repeated for all variables u_i, each time creating an intermediary value v_i and, possibly, a conjecture c_i. In the end, the algorithm constructs a term $v = (h\ v_1 \ldots v_n)$. If this term v has type T, then the algorithm returns v and the new list of conjectures. If this term has a type T' different from T, then the algorithm constructs yet another conjecture c of type $\forall x_1 : t_1, \ldots x_k : t_k.T' \to T$ and returns the term $(c\ x_1 \cdots\ x_k\ (h\ v_1 \cdots v_n))$ and a list of conjectures containing c and all the c_i's.

In the initial setting, all the conjectures are simple consequences of the convertibility rule. In the target setting, these conjectures will need to be proven.

4.5   What Happens to Our Example ?

In our example, the proof term of plus_n_0 is transformed into the following one:

λn : nat.(nat_ind

$\qquad (\lambda n0$: nat.$n0 = ($plus $n0$ $O))$

$\qquad (Ha\ ($refl_equal nat $O))$

$\qquad \lambda n0$: nat; H : $(n0 = ($plus $n0$ $O))$.

$\qquad\qquad (Hb\ n0\ ($f_equal nat nat S $n0\ ($plus $n0$ $O)\ H))\ n)$

Two conjectures Ha and Hb are generated in order to relate expected and proposed types in the branches of the application of the induction principle.

$$Ha : O\ =\ O\ \Rightarrow\ O\ =\ (\text{plus } O\ O)$$
$$Hb : \forall n : \text{nat}.(S\ n)\ =\ (S\ (\text{plus } n\ O))\ \Rightarrow\ (S\ n)\ =\ (\text{plus } (S\ n)\ O)$$

We see they can be proven easily, by first introducing the premises and then using the reflexivity of Leibniz's equality. This works because the terms on both sides of the equality are convertible modulo $\beta\delta\iota$-reduction.

5   Representing Functions

When translating proofs from one setting to the other, it is also necessary to find the corresponding representations for the functions that are used in the proofs. The corresponding representation for a given function may use a completely different algorithm, for instance to benefit from the characteristics of the new data structure. When the algorithm changes, we still need to express that the initial function and its representation in the new setting do represent the same function. This will be done by showing that the new function satisfies equalities that characterize the behavior of the initial function. We still need to show how to find these equalities.

We describe this on the example of addition. In the unary setting, addition works by moving as many S's in front of the second argument as there are S's in the first argument. This is expressed with a function that has this form:

```
Fixpoint plus[n:nat] : nat -> nat :=
   Cases n of 0 => [m:nat] m
    | (S p) => [m:nat](S (plus p m))
   end.
```

This notation is specific to *Coq*, and is very close to a let rec or letrec definition in ML.

As we already mentioned, such a definition adds to the convertibility rules in the proof system, so that (plus (S n) m) and (S (plus n m)) actually are convertible. In some theorems, the first term may be provided when the second term is requested. If we work in the binary setting, it is most likely that this convertibility will not hold anymore. On the other hand, it is possible to prove the equality that expresses that both terms are equal.

Generating Equalities: From a given function definition, it is possible to derive a collection of equalities that express the actual convertibility rules added in the system. These equalities are constructed by an analysis on the text of the definition. A first step to producing these equalities is to isolate the fixpoint equality as already studied in [2]. For the plus example, this equality has the following form.

$$\forall n : \text{nat. (plus } n) = \text{Cases } n \text{ of}$$
$$\begin{array}{ll} 0 & => [m\text{:nat}]m \\ |(S\ p) & => [m\text{:nat}](S \text{ (plus } p\ m)) \\ \text{end} \end{array}$$

However, this equality is still very linked to the unary data structure, because it makes use of the Cases construct. For this Cases construct to be valid, it is necessary to work in a setting where O and S are data type constructors. To abstract away from the data structure, we exhibit equalities corresponding to each conversion rule in the Cases construct. In our example the function performs two different tasks, depending on whether n is 0 or (S p) for some p.

If n is 0, then the left-hand side of the equality becomes (plus 0). The right-hand side is simplified according to corresponding rule in the Cases construct. The equality becomes as follows:

$$(\text{plus } 0) = [m : \text{nat}]m$$

If n is (S p), the equality becomes:

$$\forall p : \text{nat. (plus (S } p)) = [m : \text{nat}](S \text{ (plus } p\ m))$$

In fact, since type-theory based proof systems usually do not compare functions extensionally, it is better to modify these equalities to avoid comparing functions, but only values of these functions when these values do not have a function type. The equalities thus become as follows:

$$\forall m : \mathsf{nat}. \ (\mathsf{plus} \ 0 \ m) = m$$
$$\forall p, m : \mathsf{nat}. \ (\mathsf{plus} \ (\mathsf{S} \ p) \ m) = (\mathsf{S} \ (\mathsf{plus} \ p \ m))$$

Now, if we want to translate a proof about addition in the unary setting to another setting, we need to prove these equations for the new addition. These equations can then be used to simulate convertibility.

6   Mapping Binary Representation to the Unary Setting

The binary representation of natural numbers is inspired from the representation of integer numbers proposed by P. Crégut as an implementation basis for the Omega tactic, one of the most useful decision procedures for numerical problems provided in *Coq*.

6.1   Constructing the Bijections

The set of strictly positive numbers is the set generated by 1 and the two functions $x \mapsto 2 \times x$ and $x \mapsto 2 \times x + 1$. This set is constructed using the following inductive definition:

```
Inductive positive : Set :=
  one: positive              (* 1 *)
| pI: positive ->positive    (* 2x+1, x>0 *)
| p0: positive ->positive.   (* 2x, x>0 *)
```

The whole set of inductive numbers, including 0 is then described as the disjoint sum of {0} and strictly positive numbers.

```
Inductive bin : Set := zero : bin | pos : positive -> bin.
```

From these definitions, the structural induction principles that are generated basically express the following statement:

$$P \ (0) \wedge P \ (1) \wedge (\forall p \in \mathsf{positive} . \ P \ (p) \Rightarrow P \ (2p+1)) \wedge (\forall p \in \mathsf{positive} . \ P \ (p) \Rightarrow P \ (2p))$$

$$\Rightarrow \ \forall n \in \mathsf{bin} . \ P \ (n)$$

Seen as sets, types bin and nat are isomorphic, but their induction principles establish a strong distinction between them. To reduce this distinction, we need to exhibit a function S', defined in the binary setting, representing the successor function. Our definition of S' relies on an auxiliary definition for strictly positive numbers, aux_S':

```
Fixpoint aux_S' [n:positive] : positive :=
 Cases n of
  one (* 1 *)              => (p0 one)       (* 2*1 *)
 | (p0 t) (* 2*t t>0 *)    => (pI t)         (* 2*t+1 *)
 | (pI t) (* 2*t+1 t>0 *) => (p0 (aux_S' t)) (* 2*(t+1) *)
 end.
```

```
Definition S' : bin -> bin :=
 [n:bin] Cases n of zero => (pos one) |
                    (pos u) => (pos (aux_S' u)) end.
```

We can now construct two functions from elements of one data type to the other:

$$\text{BtoN} : \text{bin} \to \text{nat} \qquad \text{NtoB} : \text{nat} \to \text{bin}.$$

The function BtoN also relies on an auxiliary definition that takes care of strictly positive numbers.

```
Fixpoint aux_BtoN [x:positive] : nat :=
Cases x of one    => (S 0)
         | (p0 t) => (mult (S (S 0)) (aux_BtoN t))
         | (pI t) => (S (mult (S (S 0)) (aux_BtoN t)))
end.
```

```
Definition BtoN :=
 [x:bin] Cases x of zero => 0 | (pos p) => (aux_BtoN p) end.
```

On the other hand, defining NtoB is a simple matter of repeating S' the right number of times:

```
Fixpoint NtoB [n:nat] : bin :=
 Cases n of 0 => zero | (S x) => (S' (NtoB x)) end.
```

Then, we need to express that these two functions are the inverse of each other, or at least that NtoB is the left inverse of BtoN.

$$\text{NtoB\_inverse} : \forall a \in \text{bin} \quad \text{NtoB}(\text{BtoN}(a)) = a$$

This proof relies on an auxiliary theorem NtoB_mult2. This theorem re-phrases the interpretation of p0 as multiplication by 2 in the unary setting.

6.2   Mapping the Induction Principle

With the equality NtoB_inverse, proving an induction principle for the binary setting that has the same shape as the unary setting induction principle is easy. This theorem we name new_bin_ind has the following statement:

$$\forall P : \text{bin} \to Prop.(P \text{ zero}) \wedge (\forall x : \text{bin}.P(x) \Rightarrow (P \ (S' \ x))) \Rightarrow \forall x : \text{bin}.P(x)$$

To prove this, we take an arbitrary x in bin and hypotheses $P(zero)$ and

$$\forall b : \mathsf{bin} \quad P(b) \Rightarrow P(S'(b))$$

(let us call this hypothesis h) and we replace x with $\mathsf{NtoB}(\mathsf{BtoN}(x))$. We then prove the result by induction over $\mathsf{BtoN}(x)$. The base case is $P(\mathsf{NtoB}(0))$, which is equivalent to $P(\mathsf{zero})$, while the step case requires $P(\mathsf{NtoB}(S(n)))$ knowing that $P(\mathsf{NtoB}(n))$ holds. Here $\mathsf{NtoB}(S(n))$ is the same as $S'(\mathsf{NtoB}(n))$ by the definition of NtoB and we can use the hypothesis h.

All the inductive reasoning steps in the unary representation can then easily be translated to the binary setting by relying on this new induction principle. In this sense, we have made it possible to abstract away from the actual structure of the data type: proofs done in the binary setting can have a single base case and a single recursive case, exactly like the proofs done in the unary setting.

6.3   Mapping Recursive Functions

In this section, we work on the example of addition. The purpose of changing to a binary representation of numbers is to use efficient algorithms for the basic operations like addition. For this reason, addition is described by a very different algorithm - the well-known carry adder - that happens to be structural recursive in the binary data structure.

To show that this addition algorithm represents faithfully the addition function as implemented in the unary setting we only need to show that it satisfies the characteristic equations of unary addition, as translated in the binary setting. Naming Bplus this binary addition, the equations are:

$$\mathsf{Bplus\_S:} \quad \forall p, q : \mathsf{bin} \; \mathsf{Bplus}(S'(p), q) = S'(\mathsf{Bplus}(p, q))$$
$$\mathsf{Bplus\_0:} \quad \forall p : \mathsf{bin} \quad \mathsf{Bplus}(\mathsf{zero}, p) = p$$

The first equation is proven by induction on the binary structure, actually following the recursive structure of the function Bplus. Proving the second equation is very easy, since there is no recursion involved in this base case. However, since the addition function in the binary setting has been provided manually, this proof may require user-guidance.

The equations Bplus_S and Bplus_0 can be added to the database of an automatic rewriting procedure that will automatically repeat rewriting with these equations (oriented left to right). Termination is ensured by the fact that termination of ι-reduction was already ensured for addition in the unary setting.

6.4   Proving Conjectures

In an earlier stage of our work (section 4), we make ι-conversions occurring in proofs explicit. The result of this work is a new presentation of proofs accompanied with a collection of conjectures that correspond to the reasoning steps that are performed by ι-conversion in the initial proof. The next stage of our work is to prove translations of these conjectures in the new setting.

In our example on the proof of plus_n_O, this yields two conjectures:

Ha' : zero = zero $\rightarrow$ zero = Bplus(zero, zero)
Hb' : $\forall n$: bin. $n = $ Bplus$(n,$ zero$) \Rightarrow$ S'$(n) = $ Bplus(S'$(n),$ zero)

These conjectures are always easily proven automatically, by simply applying rewriting operations repeatedly with the characteristic equations proven in section 6.3. In *Coq* version 6.3.1, the proof is done using the proof command:

```
Repeat (Intro; AutoRewrite [simplification]);Auto.
```

This proof command assumes that users have added all the characteristic equations in the theorem database named simplification as they are proven.

6.5   Bookkeeping

As correspondences are established between functions and theorems in one setting and functions and theorems in the other setting, we need to maintain a lookup table that is used to perform the translation of statements and proofs from one setting to the other. From what we have seen so far, this lookup table has the following shape:

unary setting	binary setting	unary setting	binary setting
nat	bin	nat_ind	new_bin_ind
O	zero	S	S'
plus	Bplus	Ha	Ha'
Hb	Hb'		

This table is augmented every time a conjecture or a theorem is proven. The user may also explicitly add new elements that are provided manually, for instance when implementing multiplication and the characteristic equations for this operation. Here again, efficient multiplication on the binary structure is very different from multiplication on the unary structure.

If all the conjectures have been proven, all the theorems used by a given theorem have already been translated and proven, and the proof term for this theorem does not contain Case constructs, then the proof automatically obtained by translation according to the lookup table is always well-typed. Of course there remains the problem of handling the presence of Cases constructs. This is the object of the next section.

7   Case Constructs

Case analysis operators can appear in a proof term for various reasons. They are introduced if the proof of a logical property is performed by case analysis on an element of an inductive data type; they appear if proofs involve the one-to-one and disjointness properties of the constructors of an inductive data type; they are also used to describe computation.

We study under which conditions these case analysis steps can be treated and how they are handled in a very pragmatic way. In most cases, structural analysis can be replaced with the application of an induction theorem.

7.1 Translating the "Case" Tactic on a Goal of Sort Prop

In this case, we have a goal, say $P(n)$ to prove and we decide to proceed by case analysis on the element n which belongs to an inductively defined data type. In the example of unary numbers, we replace the Cases construct, for instance:

$$\text{Cases } n \text{ of } O => h0 \mid (S\ p) => (hr\ p) \text{ end}$$

by the application of the induction principle

$$(\text{nat\_ind } [n : \text{nat}](P\ n)\ h0\ [n : \text{nat}][\_ : (P\ n)](hr\ p))$$

There is no step hypothesis since we perform case analysis rather than induction on the data. The translation of this expression in the binary integers setting is straightforward, replacing nat_ind with new_bin_ind.

7.2 Translating Injectivity and Disjointness Properties of Constructors

The problem with injectivity and disjointness of constructors is that usual proofs for these properties rely on the ι-reduction behavior of the recursors. For instance, the recursor for natural numbers has the following behavior:

$$(\text{nat\_rec } P\ h0\ hr\ O) \rightarrow h0$$
$$(\text{nat\_rec } P\ h0\ hr\ (S\ p)) \rightarrow (hr\ p\ (\text{nat\_rec } P\ h0\ hr\ p))$$

We have not been able to construct an object simulating this behavior in the binary setting. On the other hand, we have been able to recognize injectivity and disjointness proofs by pattern-matching. The patterns for these reasoning steps are respectively:

$$(\text{LET ? ? ? (f\_equal } \ldots) \ldots) \quad (\text{LET ? ? ? (eq\_ind } \ldots) \ldots)$$

Then we build a new proof term by repeatedly applying the two following statements, that are easily proven using their counterparts in the unary setting and the bijections NtoB and BtoN:

$$\text{bin\_inj} : \forall n, m : \text{bin.}\quad (S'\ n) = (S'\ m) \rightarrow n = m$$
$$\text{bin\_discr} : \forall n : \text{bin.}\quad \neg \text{zero} = (S'\ n)$$

7.3 The "Case" Tactic on a Object of Sort Set

This use of the "Case" tactic to build a computational object is an extension of the previous case for injectivity and disjointness. It can be transformed into a nat_rec theorem application. However, as we can not provide the user with a suitable translation of such reduction steps, we can not translate such constructs for the time being.

8   Conclusion

We have implemented an ML module that performs the technique presented in this paper automatically. It has been tested on a small collection of theorems (e.g. commutativity and associativity of addition) that have all been translated automatically.

Researchers in the field of programming languages have already considered the issue of concrete data type representation and correctness of abstraction. Among them, Wadler proposed a new concept called *Views* [17] in order to reconcile the conflict between data abstraction and pattern matching in programming languages.

Other researchers have studied the topic of proof transformation. In particular, Madden [10] and Anderson [1] show how automatic transformation of proofs can lead to optimized programs extracted from these proofs. Richardson [15] even uses this technique to change data structures in a proof and the program that is extracted from this proof. His tool automatically obtains an algorithm for addition in the binary setting, but this algorithm is less efficient than the hand-crafted algorithm we study.

Another related topic revolves around proofs by analogy. Informal mathematics contain frequently formulas of the form *One would show P in the same manner that Q*. Melis and Whittle [13] propose a technique to construct a proof of a theorem from a model. Their approach is distinguished by the level of abstraction that they take. They rely on *proof-plans* that basically are very high-level proof procedures.

All this work around proof transformation is obviously related to the problem of transferring proofs from one theorem prover to another. Felty and Howe [7] show that proof transformations make it possible to use results established in different provers at the same time. Denney [6] proposes a mechanism to generate proof scripts for *Coq* from proof descriptions given in HOL.

As our plan was to provide a practical tool usable in the *Coq* system, our study mainly focuses on *Coq* Type Theory features. We did not take into account some other Type Theory features which are not available in the *Coq* system. In particular, we did not try to use the idea of recursion-induction, or tools described in McBride's PhD thesis [12] to derive "induction principles" which are direct elimination rules for function instances.

Our solution that goes through a translation of proof terms is not the only solution available. An alternative is to rely more on the bijection between the two representations, thus generalizing the way we have proven the induction principle new_bin_ind in this paper. If this approach is taken, then one needs to show that the bijections between the two representations establish a morphism with respect to all the operations that take part in the translation. Thus, it is necessary to prove a theorem that has the following statement:

morphism_plus : $\forall x, y$: bin. BtoN(Bplus(x, y)) = plus(BtoN(x), BtoN(y))

Fortunately, this theorem is easy to obtain from the proofs of the characteristic equations as done in section 6.3 and the induction principle new_bin_ind. Once the isomorphisms are established, proving the translated theorems is relatively easy: one simply needs to replace every variable x of type bin in the statement with the term $\mathsf{NtoB}(\mathsf{BtoN}(x))$ and then use the initial theorem by specializing it on $\mathsf{BtoN}(x)$. Complications occur when the statement of the theorem contains elaborate inductive types, conjunctions, disjunctions, existential quantifications. We still have to study this method and compare its applicability with the method described here.

One area where the two methods may have different advantages is the case when the initial setting and the final setting are not exactly isomorphic. For instance, we still need to compare this with an implementation of natural numbers where they are really represented with lists of boolean values. In this case, two different lists may represent the same natural numbers. What happens here is that the syntactic equality, often called Leibnitz's equality, is in correspondence with an arbitrary equivalence relation: terms that can be distinguished in the final setting correspond to identical terms in the initial setting.

References

1. Penny Anderson. Representing proof transformations for program optimization. In *International Conference on Automated Deduction*, LNAI 814, Springer-Verlag, 1994.
2. Antonia Balaa and Yves Bertot. Fix-point equations for well-founded recursion in type theory. In Harrison and Aagaard [9], pages 1–16.
3. B. Barras, S. Boutin, C. Cornes, J. Courant, J.C. Filliâtre, E. Giménez, H. Herbelin, G. Huet, C. Muñoz, C. Murthy, C. Parent, C. Paulin, A. Saïbi, and B. Werner. The Coq Proof Assistant Reference Manual – Version V6.1. Technical Report 0203, INRIA, August 1997. revised version distributed with Coq.
4. S. Boutin. Using reflection to build efficient and certified decision procedures. In Martin Abadi and Takahashi Ito, editors, *TACS'97*, LNCS 1281, Springer-Verlag, 1997.
5. Yann Coscoy. A natural language explanation for formal proofs. In Christian Rétoré, editor, *Logical Aspects of Computational Linguistics*, LNAI 1328, Springer-Verlag, 1996.
6. Ewen Denney. A prototype proof translator from HOL to Coq. In Harrison and Aagaard [9], pages 108–125.
7. Amy P. Felty and Douglas J. Howe. Hybrid interactive theorem proving using nuprl and hol. In *International Conference on Automated Deduction*, number 1249 in LNAI. Springer-Verlag, 1997.
8. H. Geuvers, F. Wiedijk, and J. Zwanenburg. Equational reasoning via partial reflection. In Harrison and Aagaard [9], pages 163–179.
9. J. Harrison and M. Aagaard, editors. *Theorem Proving in Higher Order Logics: 13th International Conference, TPHOLs 2000*, LNCS 1869, Springer-Verlag, 2000.
10. Peter Madden. The specialization and transformation of constructive existence proofs. In N.S. Sridharan, editor, *Proceedings of the Eleventh International Joint Conference on Artificial Intelligence*, pages 131–148. Morgan Kaufmann, 1989.
11. Lena Magnusson. *The Implementation of ALF - a Proof Editor based on Martin-Löf's Monomorphic Type Theory with Explicit Substitutions*. PhD thesis, Chalmers University of Technology / Göteborg University, 1995.

12. Conor McBride. *Dependently Typed Functional Programs and their Proofs*. PhD thesis, University of Edinburgh, 1999.
13. Erica Melis and Jon Whittle. Analogy in inductive theorem proving. *Journal of Automated Reasoning*, 22:117–147, 1999.
14. Christine Paulin-Mohring. Inductive Definitions in the System Coq - Rules and Properties. In M. Bezem and J.-F. Groote, editors, *Proceedings of the conference Typed Lambda Calculi and Applications*, LNCS 664, Springer-Verlag, 1993. LIP research report 92-49.
15. Julian Richardson. Automating changes of data type in functional programs. In *Proceedings of KBSE-95*. IEEE Computer Society, 1995.
16. Simon Thompson. *Type Theory and functional Programming*. Addison-Wesley, 1991.
17. Philip Wadler. Views: A way for pattern matching to cohabit with data abstraction. In *Proceedings, 14th Symposium on Principles of Programming Languages POPL'87*. ACM, 1987.

Elimination with a Motive

Conor McBride

Department of Computer Science
University of Durham

Abstract. Elimination rules tell us how we may exploit hypotheses in the course of a proof. Many common elimination rules, such as ∨-**elim** and the induction principles for inductively defined datatypes and relations, are parametric in their conclusion. We typically instantiate this parameter with the goal we are trying to prove, and acquire subproblems specialising this goal to particular circumstances in which the eliminated hypothesis holds. This paper describes a generic tactic, Elim, which supports this ubiquitous idiom in interactive proof and subsumes the functionality of the more specific 'induction' and 'inversion' tactics found in systems like COQ and LEGO[6, 7, 15]. Elim also supports user-derived rules which follow the same style.

1 Introduction

Elimination rules tell us how we may exploit hypotheses in the course of a proof. Some elimination rules are 'direct', in that the hypotheses they eliminate determine the conclusion which follows, for example, modus ponens or the rules which project the left and right conjuncts from a conjunction. Others are parametric in their conclusions, telling us how to make use of a hypothesis, whatever our goal. Here are two well known examples:

$$C \ : \ \mathsf{Prop} \qquad\qquad\qquad P \ : \ \mathbb{N} \to \mathsf{Prop}$$

$$\text{∨-elim} \ \frac{A \vee B \quad \begin{array}{c} A \\ \cdots \\ C \end{array} \quad \begin{array}{c} B \\ \cdots \\ C \end{array}}{C} \qquad \text{ℕ-induction} \ \frac{n \ : \ \mathbb{N} \quad P\,0 \quad \begin{array}{c} k \ : \ \mathbb{N} \quad P\,k \\ \cdots\cdots\cdots \\ P\,(\mathsf{s}\,k) \end{array}}{P\,n}$$

The first of these explains how to exploit a disjunctive hypothesis $A \vee B$ in the course of a proof of some proposition C. The latter follows from the disjunction if it holds in each of the circumstances where the disjunction holds. The second shows how to build a proof that property P holds for all natural numbers n by providing two components of proofs of P which correspond to the components—0 and s—from which all such n's are built. These rules are well adapted to a goal-directed style of proof, precisely because they are parametric in their conclusions. We can instantiate the parameter—C on the left, P on the right— to match the goal at hand. The 'elimination rules' discussed in this paper are just

P. Callaghan et al. (Eds.): TYPES 2000, LNCS 2277, pp. 197–216, 2002.

those in this parametric style. As many have observed, the freedom to choose the conclusion gives these rules great flexibility, but it complicates their usage. This paper describes a generic proof tactic which resolves that complexity.

In the setting of type theory, we can refer to an elimination rule by the term, *ruleProof*, which is its proof [13]. The type of such a term is the statement of the rule. Dependent function spaces express both implication and universal quantification, and are used to indicate parameters and premises alike: I write $\forall$ as the type-former, and allow (but do not demand) the usual $\rightarrow$ abbreviation in the case where the bound variable is not used. The above rules might appear thus:

$$\vee\text{-elim} : \forall A, B, C : \text{Prop.}\ A \vee B \rightarrow (A \rightarrow C) \rightarrow (B \rightarrow C) \rightarrow C$$
$$\mathbb{N}\text{-induction} : \forall P : \mathbb{N} \rightarrow \text{Prop.}\ P\,0 \rightarrow (\forall k : \mathbb{N}.\ P\,k \rightarrow P\,(\mathsf{s}\,k)) \rightarrow \forall n : \mathbb{N}.\ P\,n$$

$\vee$-**elim**'s parameter C requires no arguments and can thus be inferred at each usage by standard refinement tactics based on first-order unification. For example, we might use a tactic like LEGO's Refine to perform the elimination step in the proof that $\vee$ is commutative [11]. The first three ?'s below act as placeholders for the rule's parameters—unification infers A and B from the type of H, and C from the goal. The other ?'s stand for the proofs of the branches—these become subgoals.

$$H\ :\ P \vee Q \vdash ?\ :\ Q \vee P$$
$$\textbf{Refine}\ \vee\text{-\textbf{elim}}\ ???\ H\ ??$$
$$H\ :\ P \vee Q \vdash ?\ :\ P \rightarrow Q \vee P$$
$$H\ :\ P \vee Q \vdash ?\ :\ Q \rightarrow Q \vee P$$

$\mathbb{N}$-**induction**, however, is parametric in a predicate P. One might infer it by higher-order unification—Miller's 'pattern' unification would suffice [19, 18]. Today's refinement tactics force us to supply P by hand. For simple examples, this is just a cut-and-paste operation, with the $\forall$ in the goal becoming a λ in P:

$$\vdash ?\ :\ \forall x : \mathbb{N}.\ x + 0 = x$$
$$\textbf{Refine}\ \mathbb{N}\text{-\textbf{induction}}\ (\lambda x : \mathbb{N}.\ x + 0 = x)??$$
$$\vdash ?\ :\ 0 + 0 = 0$$
$$\vdash ?\ :\ \forall k : \mathbb{N}.\ k + 0 = k \rightarrow \mathsf{s}\,k + 0 = \mathsf{s}\,k$$

The rule, partially applied, returns a proof of type $\forall n : \mathbb{N}.\ P\,n$ for the given P, and the goal has exactly that shape. Cut-and-paste elimination is simple to implement, but its behaviour depends on the left-to-right order of hypotheses, and it only proves goals which are universal in the variables abstracted by the rule. It fails to support more complex eliminations involving, for example, inversion principles for relations and recursors for dependent datatypes. Consider this inversion principle for $\leq$:

$$\Phi\ :\ \mathbb{N} \rightarrow \mathbb{N} \rightarrow \text{Prop}$$

$$\leq\text{-\textbf{inversion}}\quad \cfrac{m \leq n \qquad \Phi\,k\,k \qquad \cfrac{j \leq k}{\cdots\cdots\cdots}\ \Phi\,j\,(\mathsf{s}\,k)}{\Phi\,m\,n}$$

Parameter Φ abstracts two arbitrary numbers. How do we invert a hypothesis which relates more specific numbers, such as h below?

$$\leq\text{-inversion} : \forall \Phi : \mathsf{N} \to \mathsf{N} \to \mathsf{Prop}.$$
$$(\forall k : \mathsf{N}.\ \Phi\, k\, k) \to (\forall j, k : \mathsf{N}.\ j \leq k \to \Phi\, j\, (\mathsf{s}\, k)) \to$$
$$\forall m, n : \mathsf{N}.\ m \leq n \to \Phi\, m\, n$$

$$\vdash ?\ :\ \forall x : \mathsf{N}.\ \forall h : x \leq 0.\ x = 0$$

Here, one argument of $\leq$ is specifically 0, not just any number—the goal has the wrong shape for cut-and-paste. We can overcome this problem by a traditional technique, which I learned from James McKinna whilst working for my MSc [15]. Transform the goal to the right shape! We must generalise the 0 to an arbitrary y, but we can 'undo' this effect by constraining y to equal 0 *propositionally*. The transformed goal, clearly equivalent, allows a cut-and-paste elimination, with easy subgoals.

$$\vdash ?\ :\ \forall x, y : \mathsf{N}.\ \forall h : x \leq y.\ y = 0 \to x = 0$$
$$\textbf{Refine } \leq\text{-inversion } (\lambda x, y : \mathsf{N}.\ y = 0 \to x = 0)\ ??$$
$$\vdash ?\ :\ \forall k : \mathsf{N}.\ k = 0 \to k = 0$$
$$\vdash ?\ :\ \forall j, k : \mathsf{N}.\ j \leq k \to (\mathsf{s}\, k) = 0 \to j = 0$$

Even this technique becomes problematic in the presence of type dependency. The family $\mathsf{Fin}\, n$ of n-element datatypes provides a good example.

$$\frac{n\ :\ \mathsf{N}}{\mathsf{Fin}\, n\ :\ \mathsf{Type}}$$

$$\frac{}{\mathsf{f0}_n\ :\ \mathsf{Fin}\, (\mathsf{s}\, n)}$$

$$\frac{x\ :\ \mathsf{Fin}\, n}{\mathsf{fs}_n\, x\ :\ \mathsf{Fin}\, (\mathsf{s}\, n)}$$

$$\Phi\ :\ \forall n : \mathsf{N}.\ \mathsf{Fin}\, n \to \mathsf{Type}$$

$$\mathsf{Fin\text{-}elim}\quad \frac{x\ :\ \mathsf{Fin}\, n \qquad \dfrac{k\ :\ \mathsf{N}}{\Phi\, (\mathsf{s}\, k)\, \mathsf{f0}_k} \qquad \dfrac{y\ :\ \mathsf{Fin}\, k \quad \Phi\, k\, y}{\Phi\, (\mathsf{s}\, k)\, (\mathsf{fs}_k\, y)}}{\Phi\, n\, xs}$$

Suppose we wish to prove some theorem about a specific finite set:

$$\vdash ?\ :\ \forall x : \mathsf{Fin}\, (\mathsf{s}\, 0).\ P[x]$$

We might transform the goal as before, changing the type of x to $\mathsf{Fin}\, n$ for arbitrary n, with a constraint $n = \mathsf{s}\, 0$, but the expression $P[x]$ will cease to be well-typed for many choices of P. We might also try to avoid changing the type of x by adding another constraint:

$$\vdash ?\ :\ \forall n : \mathsf{N}.\ \forall y : \mathsf{Fin}\, n.\ \forall x : \mathsf{Fin}\, (\mathsf{s}\, 0).\ n = \mathsf{s}\, 0 \to y = x \to P[x]$$

However, the usual definition of equality in intensional type theory does not permit the statement $y = x$, as the two have different types. If we wish to follow the method of generalisation with equational constraints, and to cope with type dependency, we must be careful about how we express those constraints. I address this problem by adopting a non-standard definition of equality with a more liberal formation rule, permitting the above transformation. A cut-and-paste elimination then gives:

$$\vdash ? \; : \; \forall k : \mathbb{N}. \, . s \, k = s \, 0 \; \rightarrow \; f0_k = x \; \rightarrow \; P[x]$$
$$\vdash ? \; : \; \forall k : \mathbb{N}. \, \forall y : \text{Fin } n.$$
$$(\forall x : \text{Fin } (s \, 0). \, k = s \, 0 \; \rightarrow \; y = x \; \rightarrow \; P[x]) \; \rightarrow$$
$$\forall x : \text{Fin } (s \, 0). \, s \, k = s \, 0 \; \rightarrow \; fs_k \, y = x \; \rightarrow \; P[x]$$

By solving the constraints, these can be simplified to:

$$\vdash ? \; : \; P[f0_0]$$
$$\vdash ? \; : \; \forall y : \text{Fin } 0. \, (\forall x : \text{Fin } (s \, 0). \, 0 = s \, 0 \; \rightarrow \; y = x \; \rightarrow \; P[x]) \; \rightarrow \; P[fs_0 \, y]$$

The first requires P to hold for the only value in Fin $(s \, 0)$. A further elimination and simplification would prove the second vacuously.

This paper gives a generic analysis of elimination rules and describes a generic tactic, Elim, for applying them. Its contribution is thus more practical than theoretical. It is addressed to providers of theorem-proving technology, in that it describes a general treatment of a common idiom, and to users of such technology, in that it illustrates the benefits afforded by that treatment. It works with the standard presentation of datatypes and relations [5,9], but also offers support for any theorems which characterise information by the leverage it exerts on an arbitrary goal.

Elim's job is to automate the application of elimination rules, mechanising not just cut-and-paste, but the goal transformations beforehand. It implements the conceptual proof step 'use this rule to eliminate that hypothesis' for any rule which is parametric in its conclusion, whether it is a standard induction or inversion principle, manufactured by machine, or a user's own. To develop Elim, it is necessary to consider the shape and usage of these rules uniformly, not just for particular standard patterns. For any elimination with any rule, we shall need to discuss

- *what* we eliminate: we must select something for the rule to analyse
- *why* we eliminate it: we must fit the parametric conclusion to the goal at hand.

The thing to be analysed, I call the **target** of the elimination. The target corresponds to the 'major premise' of an elimination rule for a logical connective, the expression analysed in a 'case' construct, the hypothesis to invert, the variable on which to do induction and so on. The value instantiating the parameter which governs the rule's conclusion, I call the **motive** of the elimination. The motive corresponds to the 'induction predicate', the return type of a case expression or a fold operator, the context of a rewrite and other similar notions.

For a given elimination, the Elim tactic automatically constructs a motive which captures the necessary permutation and generalisation of the goal's hypotheses, adding appropriate constraints. It has taken many years and much analysis to arrive at a design which delivers sensible behaviour with a simple interface. The next section addresses the issues which arise and seeks to justify the choices I have made in the implementation of Elim, described in section 3.

Section 4 sketches the companion tactic, Unify, which automates the simplification of equational constraints in the subgoals arising from Elim. The combination of Elim and Unify delivers precisely the effects achieved in each of the examples above. Section 5 gives more complex examples of these tactics in action. The paper closes by contemplating how we might change the theorems we choose to prove, given a tactic to support elimination with a motive.

2 Designing Elim

This section describes the analysis which underpins Elim. We seek to allow users to say 'use this to eliminate that' and for the machine to deliver a sensible step of reasoning. We first need to understand how to specify elimination rules so that Elim can apply them.

2.1 Anatomy of an Elimination Rule

The statement of an elimination rule is given by a functional type. We can analyse the various purposes served by its premises into a number of groups, subject to certain dependency relationships. Every elimination rule resembles the following, up to a dependency-respecting permutation of the arguments:

Group	Rule	Dependencies
parameters	$\forall \vec{a} : \vec{A}.$	
pattern variables	$\forall \vec{x} : \vec{X}.$	$\vec{X}$ may only depend on $\vec{a}$
preconditions	$\forall \vec{c} : \vec{C}.$	$\vec{C}$ may only depend on $\vec{a}, \vec{x}$
motive variable	$\forall \Phi : \forall \vec{i} : \vec{I}.\, U.$	$\vec{I}$ may only depend on $\vec{a}$
branches	$\forall \vec{b} : \vec{B}.$	$\vec{B}$ may only depend on $\vec{a}, \Phi$
	$\Phi \vec{p}$	$\vec{p}$ may only mention $\vec{a}, \vec{x}$
		must mention all the $\vec{x}$

Φ abstracts the **indices** $\vec{i} : \vec{I}$ over some universe of types U. The names 'parameters' and 'indices' come from Dybjer's analysis of inductive families [9]. I use de Bruijn's telescope notation $\forall \vec{a} : \vec{A}$ to abbreviate the sequence $\forall a_1 : A_1 \ldots \forall a_n : A_n$ [8].

Elim will require such a rule, together with enough information to instantiate the parameters, pattern variables and preconditions. These three groups play different rôles in an elimination: the parameters $\vec{a}$ select a particular instance of a polymorphic rule; the pattern variables $\vec{x}$ and preconditions $\vec{c}$ together tell us what is being eliminated. Elim constructs a motive appropriate to the current goal, its treatment of hypotheses depending on their usage in the values corresponding to these groups, then solves the goal, leaving the branches as subgoals. These subgoals refine our information about the patterns $\vec{p}$—and hence the pattern variables $\vec{x}$—wherever they apply the motive, but we cannot see inside the proofs of the preconditions.

Let us quickly review our examples. In ∨-**elim**, A and B are parameters, the proof of $A \vee B$ is the precondition, C is the motive variable and the remaining premises are branches. For ℕ-**induction**, n is the pattern variable, P is the motive variable and the two branches explain n in terms of 0 and s. ≤-**inversion** tells us what we can learn about pattern variables m and n under the precondition that $m \leq n$—motive variable Φ abstracts over m and n, but not the proof of the precondition. Fin-**elim** has pattern variables n and x, and Φ once again stands for the motive.

2.2   Types for Targets

The raw type of an elimination rule does not quite determine the analysis of its premises into the above groups. We are free to consider all the pattern variables to be parametric, allowing all the preconditions to become branches. Furthermore, although Elim needs to know the values of the $\vec{a}$, $\vec{x}$ and $\vec{c}$, we have not specified the intended format for presenting this information. These types do not say what constitutes a legitimate target.

Experienced humans can tell the intended usage of a rule if it has a familiar structure and a meaningful name. However, it is often counterproductive to require intelligence of machines when obedience will do. Elimination rules come in many shapes and sizes. Some, like 'double induction' principles, take more than one target. We should not expect the machine to guess what we have in mind. We should rather develop the language to say clearly what we mean, adding some form of 'user manual' to the logical statement of an elimination rule.

The modus vivendi of type theory is to specify the meaningful usage of a term via its type. We need to be able to specify what kind of expression a rule targets, and to instantiate that specification. Fortunately, dependent types can describe expressions—the following new constants will suit our purpose:

$$\frac{t \,:\, T}{\mathsf{Target}_T\, t \,:\, \mathsf{Type}} \qquad \frac{t \,:\, T}{\mathsf{on}_T\, t \,:\, \mathsf{Target}_T\, t}$$

These resemble the formation rule and constructor for a family of singleton types. I shall typically suppress the subscripted premises.

We may insert a sequence of redundant premises, $\mathsf{Target}\, t_1 \rightarrow \ldots \mathsf{Target}\, t_n \rightarrow$, into each elimination rule, to characterise the sequence of expressions which the rule takes as targets. The t's are the rule's **formal targets**. They must determine all the information required—parameters, pattern variables and preconditions—when they are unified with the **actual targets**, $e_1 \ldots e_n$, supplied when `Elim` is called. The arguments generated correspondingly are, of course, $(\mathsf{one}_1) \ldots (\mathsf{one}_n)$, and these determine the $\vec{a}, \vec{x}, \vec{c}$ by a mechanism similar to that for implicit syntax [21]. By adding formal targets to our elimination rules, we explain exactly how they should be invoked. The code packages which generate standard rules for datatypes and relations can easily be adapted to indicate their standard usage.

$$\vee\text{-elim} : \forall A, B, C : \mathsf{Prop}. \, \forall h : A \vee B. \, \mathsf{Target} \, h \, \rightarrow$$
$$(A \rightarrow C) \rightarrow (B \rightarrow C) \rightarrow C$$

$$\mathbb{N}\text{-induction} : \forall P : \mathbb{N} \rightarrow \mathsf{Prop}. \, P \, 0 \, \rightarrow \, (\forall k : \mathbb{N}. \, P \, k \rightarrow P \, (\mathsf{s} \, k)) \, \rightarrow$$
$$\forall n : \mathbb{N}. \, \mathsf{Target} \, n \, \rightarrow \, P \, n$$

$$\leq\text{-inversion} : \forall \varPhi : \mathbb{N} \rightarrow \mathbb{N} \rightarrow \mathsf{Prop}.$$
$$(\forall k : \mathbb{N}. \, \varPhi \, k \, k) \, \rightarrow \, (\forall j, k : \mathbb{N}. \, j \leq k \rightarrow \varPhi \, j \, (\mathsf{s} \, k)) \, \rightarrow$$
$$\forall m, n : \mathbb{N}. \, \forall l : m \leq n. \, \mathsf{Target} \, l \, \rightarrow \, \varPhi \, m \, n$$

$$\mathsf{Fin}\text{-elim} : \forall \varPhi : \forall n : \mathbb{N}. \, \mathsf{Fin} \, n \, \rightarrow \, \mathsf{Type}.$$
$$(\forall k : \mathbb{N}. \, \varPhi \, (\mathsf{s} \, k) \, \mathsf{f0}_k) \, \rightarrow$$
$$(\forall k : \mathbb{N}. \, \forall y : \mathsf{Fin} \, k. \, \varPhi \, k \, y \rightarrow \varPhi \, (\mathsf{s} \, k) \, (\mathsf{fs}_k \, y)) \, \rightarrow$$
$$\forall n : \mathbb{N}. \, \forall x : \mathsf{Fin} \, n. \, \mathsf{Target} \, x \, \rightarrow \, \varPhi \, n \, x$$

We now have a very flexible language in which to write the 'user manual' for an elimination rule. The above examples simply pick out a single premise, but formal targets can specify more complex shapes to be matched. The target premises also tell us how to mechanise our analysis of an elimination rule precisely, because they make a cut in the dependency graph between $\vec{a}, \vec{x}, \vec{c}$—exactly the premises the Target t's must determine and hence depend on—and the remainder $\varPhi, \vec{b}$. Below the cut, we may now take the parameters $\vec{a}$ to be exactly those on which the types of $\varPhi$ and the branches depend, and the pattern variables $\vec{x}$ to be the remainder of those on which the patterns $\vec{p}$ in the conclusion depend.

2.3   Targets, Hypotheses, and the Context

Each individual application of `Elim` needs to know which rule to apply—we can nominate this by supplying its proof *ruleProof*—and which actual targets to plug into its formal targets, explaining which hypotheses are to be eliminated. Hypotheses can be found in two places: typically, a goal sits within a context $\varGamma$ of 'external' hypotheses, whilst the goal itself is prefixed by a telescope of universally quantified 'internal' hypotheses $\vec{h} : \vec{H}$,

$$\varGamma \vdash ? : \forall \vec{h} : \vec{H}. \, G$$

The hypotheses in $\varGamma$ have established names, and persist not only for this goal, but for all its subgoals. The internal hypotheses are transient from proof step to proof step and might not have explicit names if there is no type dependency. Current proof systems like LEGO and COQ take a low-level view of this separation. The context collects the λ-bindings under which we work; tactic Intros creates a new λ, moving the leftmost h into the Γ, and giving it a permanent name. This view fits the pure λ-calculus, where hypotheses are used by *application*. However, with inductive structures, we often want to *analyse* a hypothesis into its components.

We can use the context/goal split to reflect this conceptual distinction. The permanence of the context makes it a good place to store hypotheses which have been fully analysed. The transience of the goal suits hypotheses awaiting further analysis, and those—e.g., accumulation parameters—whose usage may change within an induction. Elim will thus often seek its target within the goal. We had best define a new tactic which performs 'naming' without 'making permanent'.

TACTIC Names $x_1 \ldots x_n$

Rename the leftmost n hypotheses in the goal to $x_1 \ldots x_n$ (which must be distinct and fresh with respect to the context). Hypotheses previously unnamed should bear their new names explicitly.

Once our hypotheses have names, we can eliminate them. In the general situation, $\Gamma \vdash ? : \forall \vec{h} : \vec{H}. G$, an **actual target** is any expression e which is well-typed under $\Gamma; \vec{h} : \vec{H}$, and which unifies with the corresponding formal target. The Elim tactic should thus be given *ruleProof*, together with a sequence of actual targets $e_1 \ldots e_n$, unifying with the formal targets $t_1 \ldots t_n$ to determine *ruleProof*'s $\vec{a}, \vec{x}, \vec{c}$ arguments.

We have arrived at a style of proof which analyses the goal's hypotheses with Elim, then moves the permanent components into the context. Permanence may not come in strict left-to-right order. We can see the context/goal split as a cut in the dependency graph which declares the permanence/transience of hypotheses. The 'making permanent' tactic should reflect this view:

TACTIC FixHyp $h_1 \ldots h_n$

Move goal hypotheses $h_1 \ldots h_n$ into the context along with just enough others to maintain the dependency order—the downward closure of the h's. This is easily implemented in any system which allows us to claim new lemmas for later proof: claim a permutation of the goal which has the y's shuffled as far to the left as dependency permits; solve the goal via the claim; apply left-to-right introduction on the claim.

Now we can use the context/goal split to indicate what should be fixed outside an induction and what should be allowed to vary. For example, consider using **foldr** (where $[X]$ is the type of lists of X's)

$$\textbf{foldr} \ : \ \forall X, T : \mathsf{Type}. \ \forall t_{\mathsf{nil}} : T. \ \forall t_{\mathsf{cons}} : X \to T \to T. \ \forall xs : [X]. \ \mathsf{Target} \ xs \to T$$

to develop 'append'. We may choose to fix the second list and vary the first.

$$\vdash ? \; : \; \forall A : \mathsf{Type}. \, [A] \; \rightarrow \; [A] \; \rightarrow \; [A]$$

Names A as bs

$$\vdash ? \; : \; \forall A : \mathsf{Type}. \, \forall as, bs : [A]. \, [A]$$

FixHyp bs

$$A \; : \; \mathsf{Type}; bs \; : \; [A] \vdash ? \; : \; \forall as : [A]. \, [A]$$

Elim foldr as

$$A \; : \; \mathsf{Type}; bs \; : \; [A] \vdash ? \; : \; [A]$$
$$A \; : \; \mathsf{Type}; bs \; : \; [A] \vdash ? \; : \; A \; \rightarrow \; [A] \; \rightarrow \; [A]$$

2.4   Constructing the Motive

Let us now turn our attention to the details of constructing the motive for an elimination, with equational constraints where necessary, for a general goal and for a running example:

$$\Gamma \vdash ? \; : \; \forall \vec{h} : \vec{H}. \, G \qquad\qquad \vdash ? \; : \; \forall x : \mathbb{N}. \, \forall h : x \leq 0. \, x = 0$$
$$\text{Elim } ruleProof \; e_1 \ldots e_n \qquad\qquad \text{Elim } \leq\text{-inversion } h$$

$$\leq\text{-inversion} : \forall \Phi : \mathbb{N} \rightarrow \mathbb{N} \rightarrow \mathsf{Prop}.$$
$$(\forall k : \mathbb{N}. \, \Phi \, k \, k) \; \rightarrow \; (\forall j, k : \mathbb{N}. \, j \leq k \rightarrow \Phi \, j \, (\mathsf{s} \, k)) \; \rightarrow$$
$$\forall m, n : \mathbb{N}. \, \forall l : m \leq n. \, \mathsf{Target} \, l \; \rightarrow \; \Phi \, m \, n$$

Unifying the formal targets of our rule with the actual targets generates a substitution σ which maps the $\vec{a}, \vec{x}, \vec{c}$ arguments of the rule to expressions over the $\vec{h}$. We will be able to make a behavioural analysis of the h's corresponding to their usage within $\sigma\vec{a}, \sigma\vec{x}, \sigma\vec{c}$. Let us, for the moment, presume that no h's are used within the rule's parameters, the $\sigma\vec{a}$. In the above elimination with **foldr**, the element type A is, fortunately, in the context. I shall deal with 'parametric' hypotheses later.

The motive Φ must abstract over the indices $\vec{i} : \vec{I}$, and the rule gives a proof of $\Phi \vec{p}$. When Φ is applied in the rule's branches, we hope to learn something new about the instantiated patterns, $\sigma\vec{p}$. We must choose a motive which connects the $\vec{i}$ with the $\sigma\vec{p}$. A good first guess is the following (writing $\mapsto$ to indicate a definition, reserving $=$ for propositions):

$$\Phi \; \mapsto \; \lambda \vec{i} : \vec{I}. \, \forall \vec{h}' : \vec{H}'. \, \vec{i} = \sigma'\vec{p} \; \rightarrow \; G'$$

$$\Phi \; \mapsto \; \lambda m, n : \mathbb{N}. \, \forall x' : \mathbb{N}. \, \forall h' : x' \leq 0. \, m = x' \; \rightarrow \; n = 0 \; \rightarrow \; x' = 0$$

where the $\vec{h}'$ are local copies of the $\vec{h}$, and σ' is the substitution $[\vec{h}'/\vec{h}] \circ \sigma$, which uses the local copies where σ uses the originals. Similarly, G' is $[\vec{h}'/\vec{h}]G$. I write an equation between sequences $\vec{i} = \sigma'\vec{p}$ to indicate a sequence of equations

$i_1 = \sigma'p_1 \ldots i_n = \sigma'p_n$, each a separate premise. I shall delay the definition of $=$, but presume a proof for all x, $\mathsf{refl}\ x : x = x$. This gives

$$\Gamma; \vec{h}:\vec{H} \ \vdash \ \begin{aligned} &ruleProof\ \sigma\vec{a}\ \sigma\vec{x}\ \sigma\vec{c}\ (\lambda\vec{i}:\vec{I}.\ \forall\vec{h}':\vec{H}'.\ \vec{i} = \sigma'\vec{p} \to \Psi')\ \vec{b} \\ &\quad : \forall\vec{h}':\vec{H}'.\ \sigma\vec{p} = \sigma'\vec{p} \to \Psi' \end{aligned}$$

$$x:\mathbb{N}; h:x \leq 0 \ \vdash \ \begin{aligned} &\leq\text{-}\mathbf{inversion}\ (\lambda m, n:\mathbb{N}\ldots x' = 0)\ b_1\ b_2\ x\ 0\ l\ (\text{on } l) \\ &\quad : \forall x':\mathbb{N}.\ \forall h':x' \leq 0.\ x = x' \to 0 = 0 \to x' = 0 \end{aligned}$$

with the b's as yet unknown, but not affecting the overall type. The motive has generalised over the hypotheses and inserted equational constraints. In order to get our proof of Ψ, we must apply this term. Instantiating the $\vec{h}'$ with the original $\vec{h}$ turns $\sigma'\vec{p}$ back into $\sigma\vec{p}$, making the equations reflexive, and turns G' back into G.

$$\Gamma; \vec{h}:\vec{H} \ \vdash \ \left(\begin{aligned} &ruleProof\ \sigma\vec{a}\ \sigma\vec{x}\ \sigma\vec{c}\ (\lambda\vec{i}:\vec{I}.\ \forall\vec{h}':\vec{H}'.\ \vec{i} = \sigma'\vec{p} \to G')\ \vec{b} \\ &\qquad\qquad \vec{h}\ (\mathsf{refl}\ \sigma p_1)\ \ldots (\mathsf{refl}\ \sigma p_n) \end{aligned} \right) : G$$

$$x:\mathbb{N}; l:x \leq 0 \ \vdash \ \left(\begin{aligned} &\leq\text{-}\mathbf{inversion}\ (\lambda m, n:\mathbb{N}\ldots x' = 0)\ b_1\ b_2\ x\ 0\ h\ (\text{on } h) \\ &\qquad\qquad x\ h\ (\mathsf{refl}\ x)\ (\mathsf{refl}\ 0) \end{aligned} \right) : x = 0$$

We can manufacture appropriate candidates for the $\vec{b}$ by claiming lemmas which are given by plugging our motive into the $\sigma\vec{B}$. The goal can then be solved:

$$\Gamma \vdash ?_1 : [\lambda \ldots G'/\Phi]\sigma B_1$$
$$\ldots$$
$$\Gamma \vdash ?_n : [?_1/b_1, \ldots, ?_{n-1}/b_{n-1}, \lambda \ldots G'/\Phi]\sigma B_n$$

$$\Gamma \vdash \left(\begin{aligned} &\lambda\vec{h}:\vec{H}.\ ruleProof\ \sigma\vec{a}\ \sigma\vec{x}\ \sigma\vec{c}\ (\lambda\vec{i}:\vec{I}.\ \forall\vec{h}':\vec{H}'.\ \vec{i} = \sigma'\vec{p} \to G') \\ &\qquad ?_1\ \ldots ?_n\ \vec{h}_v\ (\mathsf{refl}\ \sigma p_1)\ \ldots (\mathsf{refl}\ \sigma p_n) \end{aligned} \right) : \forall\vec{h}:\vec{H}.\ G$$

$$\vdash ?_1 : \forall k:\mathbb{N}.\ \forall x':\mathbb{N}.\ \forall h':x' \leq 0.\ k = x' \to k = 0 \to x' = 0$$
$$\vdash ?_2 : \forall j, k:\mathbb{N}.\ j \leq k \to \forall x':\mathbb{N}.\ \forall h':x' \leq 0.\ j = x' \to (\mathsf{s}\ k) = 0 \to x' = 0$$

$$\vdash \lambda x:\mathbb{N}.\ \lambda h:x \leq 0.\ \leq\text{-}\mathbf{inversion}\ (\lambda m, n:\mathbb{N}\ldots x' = 0)\ ?_1\ ?_2\ \ldots\ (\mathsf{refl}\ 0)$$
$$: \forall x:\mathbb{N}.\ \forall h:x \leq 0,\ x = 0$$

Now that we have seen the basic idea, let us do better. Recall the motive we chose:

$$\Phi \mapsto \lambda m, n:\mathbb{N}.\ \forall x':\mathbb{N}.\ \forall h':x' \leq 0.\ m = x' \to n = 0 \to x' = 0$$

When eliminating l in $\forall x : \mathbb{N}.\ \forall l : x \leq 0.\ldots$, we do need to generalise the 0, but the x is already abstracted. Our current 'brute force' construction *equates* the local x' with the index m when we could *replace* x' by m. We can remove this unnecessary duplication by partially solving the matching problem inherent in the constraints $\vec{i} = \sigma'\vec{p}$. Wherever we find an equation $i = h'$, indicating that local hypothesis h' duplicates index i, we may replace h' by i, provided the two have the same type. We then discard the $\forall h'$ and remove the equation—the corresponding h and $(\mathsf{refl}\ h)$ must also disappear from the rule application. We should search left to right, as earlier matches may unify later types. In our example, x' is indeed replaced by m:

$$\Phi \mapsto \lambda m, n : \mathbb{N}.\ \forall h' : m \leq 0.\ n = 0 \ \to\ m = 0$$

$$\vdash ?_1 \ :\ \forall k : \mathbb{N}.\ \forall h' : k \leq 0.\ k = 0 \ \to\ k = 0$$
$$\vdash ?_2 \ :\ \forall j, k : \mathbb{N}.\ j \leq k \ \to\ \forall h' : j \leq 0.\ (\mathsf{s}\ k) = 0 \ \to\ j = 0$$

One annoying detail remains: we eliminated h, so why is h' still there? The local copy of h remains in the motive, and hence reappears in the subgoals, but we have learned nothing new about it—no equations constrain it. This is because $\leq$-**inversion** eliminates an inequality l, but l does not appear in the patterns—it is a precondition. Fortunately, we may identify those h's used only in preconditions $\sigma\vec{c}$, and remove as many as possible from the motive. Scanning right to left, if an h' occurs in $\sigma'\vec{c}$, but not $\sigma'\vec{a}, \sigma'\vec{x}$, and nothing depends on it, then remove the $\forall h'$ from the motive and the corresponding h from the rule application. The right-to-left scan ensures that later removable hypotheses do not 'protect' earlier ones. In our example, h' occurs only in precondition $\sigma' l$, so it can go:

$$\Phi \mapsto \lambda m, n : \mathbb{N}.\ n = 0 \ \to\ m = 0$$

$$\vdash ?_1 \ :\ \forall k : \mathbb{N}.\ k = 0 \ \to\ k = 0$$
$$\vdash ?_2 \ :\ \forall j, k : \mathbb{N}.\ j \leq k \ \to\ (\mathsf{s}\ k) = 0 \ \to\ j = 0$$

This technique is easily adapted to ensure that **parametric hypotheses**—those h's occurring in the instantiated parameters $\sigma\vec{a}$—are also correctly handled. For example, A is parametric in the following:

$$\vdash ? \ :\ \forall A : \mathsf{Type}.\ \forall as, bs : [A].\ [A]$$
$$\mathtt{Elim}\ \mathbf{foldr}\ as$$

Parametric hypotheses must be kept outside the motive, together with any others on which they depend: abstracting parameters within the motive leaves subgoals which demand more polymorphism than the branches of the elimination provide. We can make a cut in the dependency graph of the $\vec{h}$—the lower part, $\vec{h}_a : \vec{H}_a$, is the downward closure of the parametric hypotheses; the upper part, $\vec{h}_v : \vec{H}_v$ contains the remaining, **varying** hypotheses. In our example, the behavioural grouping and consequent analysis of the goal are as follows:

group	v	σv		goal part	our example
parameter	X	A	$\cdots$	$\vec{h}_a : \vec{H}_a$	$A : \mathsf{Type}$
precondition	xs	as	$\cdots$	$\vec{h}_v : \vec{H}_v$	$as, bs : [A]$
				G	$[A]$

We may apply the same basic technique except that we must start with local copies $\vec{h}'_v$ only of those hypotheses above the cut. The basic motive is thus

$$\Phi \mapsto \lambda \vec{i}\colon \vec{I}.\ \forall \vec{h}'_v\colon \vec{H}_v.\ \vec{i} = \sigma' \vec{p} \to G'$$

$$T \mapsto \forall as', bs'\colon [A].\ [A]$$

We may then improve the motive as before. T has no equations to remove, but the hypothesis as is used only in a precondition. We acquire

$$T \mapsto \forall bs'\colon [A].\ [A]$$

The proof term is built in the same way, except that the instantiated branches may depend on the $\vec{h}_a$. Each subgoal $?_i$ should thus be $\forall$-abstracted over the $\vec{h}_a$, with corresponding argument in the rule application becoming $(?_i\ \vec{h}_a)$.

$\vdash ?_1\ :\ \forall A\colon \mathsf{Type}.\ \forall bs'\colon [A].\ [A]$
$\vdash ?_2\ :\ \forall A\colon \mathsf{Type}.\ A \to (\forall bs'\colon [A].\ [A]) \to \forall bs'\colon [A].\ [A]$

$\vdash\ \lambda A\colon \mathsf{Type}.\ \lambda as, bs\colon [A].\ \mathbf{foldr}\ A\ (\forall bs'\colon [A].\ [A])\ (?_1\ A)\ (?_2\ A)\ as\ (\mathbf{on}\ as)\ bs$
      $:\ \forall A\colon \mathsf{Type}.\ \forall as, bs\colon [A].\ [A]$

Now η-reduce the proof term for a professional finish:

$\vdash\ \lambda A\colon \mathsf{Type}.\ \lambda as\colon [A].\ \mathbf{foldr}\ A\ (\forall bs'\colon [A].\ [A])\ (?_1\ A)\ (?_2\ A)\ as\ (\mathbf{on}\ as)$
      $:\ \forall A\colon \mathsf{Type}.\ \forall as, bs\colon [A].\ [A]$

These last seven pages contain the analysis of elimination which underpin the Elim tactic. The next page gives its implementation. It has taken seven years to write.

3   Implementing Elim

TACTIC Elim *ruleProof* $e_1 \ldots e_n$

For a given goal

$$\Gamma \vdash \forall \vec{h}\colon \vec{H}.\ G$$

ruleProof and the e's must be well-typed in $\Gamma; \vec{h}\ :\ \vec{H}$. Execute the phases—**targetting, motive construction, refinement**—detailed below.

Targetting. Up to permutation, the type of *ruleProof* must be of form

$$\forall \vec{v}\colon \vec{V}.\ \mathsf{Target}_{T_1}\ t_1\ \to\ \ldots \mathsf{Target}_{T_n}\ t_n\ \to\ R$$

where every v_i occurs either in some T_j or some t_j. These $\vec{v}$ are the $\vec{a}, \vec{x}, \vec{c}$ of our rule, but rather than making this separation now, we can make the corresponding analysis of the h's directly, once the $\vec{v}$ are known and R, the rest of the rule, is instantiated. For $i = 1 \ldots n$, unify each formal target t_i with the corresponding actual target e_i, accumulating a substitution σ which maps all the $\vec{v}$ to terms over $\Gamma; \vec{h} : \vec{H}$. Now, check that σR is, up to permutation, of the form

$$\forall \Phi : \forall \vec{i} : \vec{I}. \; U. \; \forall \vec{b} : \vec{B}. \; \Phi \, \vec{p}$$

where U is a universe (e.g. Prop, Type), and the $\vec{p}$ are independent of $\Phi, \vec{b}$. Without loss of generality, I presume that $ruleProof$'s arguments are so ordered. Of course, we actually permute the application generated, rather than the type.

Motive Construction. Find the parametric hypotheses: cut the dependency graph of $\vec{h} : \vec{H}$—below, the fewest $\vec{h}_a : \vec{H}_a$ including the h's used in $\sigma \vec{I}$ or $\sigma \vec{B}$; above, $\vec{h}_v : \vec{H}_v$, the rest. Find the removable hypotheses: cut $\vec{h}_v : \vec{H}_v$—below, the fewest $\vec{h}_m : \vec{H}_m$; above, $\vec{h}_r : \vec{H}_r$ such that each h_r occurs in $\sigma \vec{v}$ but not $\sigma \vec{p}$ or G.

Construct the fully generalised motive with hypotheses $\vec{h}_m : \vec{H}_m$, and constraints $\vec{E}$, the sequence of equations $\vec{i} = \sigma \vec{p}$:

$$\lambda \vec{i} : \vec{I}. \; \forall \vec{h}_m : \vec{H}_m. \; \vec{E} \; \to \; G$$

Remove unnecessary duplication: for each equation from left to right, if E_j reads $i_j = h$ for some $\forall h : I_j$ in the motive, update the motive by substituting i_j for h, then deleting E_j and the $\forall h : I_j$. Let ρ be the accumulated substitution, $\vec{h}' : \vec{H}'$ the remaining hypotheses and $\vec{E}'$ the remaining equations. Let $\vec{r}$ be the sequence of terms (refl p) corresponding to each equation $i = p$ in $\vec{E}'$.

Let ϕ, the actual motive for the elimination, be the shortest η-reduced form of

$$\lambda \vec{i} : \vec{I}. \; \forall \vec{h}' : \vec{H}'. \; \vec{E}' \; \to \; \rho G$$

Refinement. For each branch $b_i : B_i$, claim a lemma

$$\Gamma \vdash ?_i \; : \; \forall \vec{h}_a : \vec{H}_a. \; [\vec{?}/\vec{b}, \phi/\Phi] B_i$$

Let $\vec{g}$ be the corresponding sequence of terms $(?_i \, \vec{h}_a)$. Solve the goal with the shortest η-reduced form of

$$\Gamma \vdash \lambda \vec{h} : \vec{H}. \; ruleProof \; \sigma \vec{v} \; (\text{on } e_1) \; \ldots \; (\text{on } e_n) \; \phi \, \vec{g} \, \vec{h}' \, \vec{r} \; : \; \forall \vec{h} : \vec{H}. \; G$$

4   Stating and Solving Equations

The Elim tactic needs a notion of equality for its constraints. Over the years, there have been many accounts of equality in type theory: some are inductive in character, others more computational [13, 14, 23, 10, 20, 1]. We need not be prescriptive about which equality to use. However, that choice determines Elim's efficacy. When working with dependent types, and particularly with subsets of inductive families, we often need to express constraints where later equations compare elements of types which are only *propositionally* equal given the earlier equations. Recall our Fin (s 0) example

$$\vdash ? \; : \; \forall x : \mathsf{Fin}\,(\mathsf{s}\,0).\,P[x]$$
$$\text{Elim Fin-elim } x$$

The Elim tactic computes this motive, in which y and x have different types:

$$\Phi \mapsto \lambda n : \mathbb{N}.\; \lambda y : \mathsf{Fin}\,n.\; \forall x : \mathsf{Fin}\,(\mathsf{s}\,0).\,n = \mathsf{s}\,0 \;\to\; y = x \;\to\; P[x]$$

With the usual definitions of equality, $y = x$ is not well-typed. One solution is to use the following **heterogeneous** notion of equality:

$$\frac{a \,:\, A \quad b \,:\, B}{a = b \;:\; \mathsf{Prop}} \qquad \frac{\begin{array}{l} a : A \\ \Phi \;:\; \forall a' : A.\, a = a' \;\to\; \mathsf{Type} \end{array}}{}$$

$$\frac{a \,:\, A}{\mathsf{refl}\,a \;:\; a = a} \qquad \text{=-elim} \;\; \frac{q \,:\, a = a' \quad \Phi\,a\,(\mathsf{refl}\,a)}{\Phi\,a'\,q}$$

$$\begin{aligned} \text{=-elim} \;:\; & \forall A : \mathsf{Type}.\, \forall a : A.\, \forall \Phi : \forall a' : A.\, a = a' \to \mathsf{Type}. \\ & \Phi\,a\,(\mathsf{refl}\,a) \to \\ & \forall a' : A.\, \forall q : a = a'.\, \mathsf{Target}\, q \to \Phi\,a'\,q \\ \text{=-elim} \; & A\,a\,\Phi\,\phi\,a\,(\mathsf{refl}\,a)\,\_ \mapsto \phi \end{aligned}$$

This definition permits the *statement* of any equation, regardless of type. On the other hand, =-**elim** only permits the *use* of those equations where the type is shared. This is not the elimination rule one would expect with an inductive definition. It targets and indexes over only the homogeneous part of the relation, which is, of course, where all the inhabitants can be found. This gives exactly the same strength as Martin-Löf's equality extended with Altenkirch and Streicher's 'uniqueness of identity proofs' axiom [23, 16].

Once we can formulate the equations, Elim can construct constrained motives, even for examples like the above, and carry out its elimination. The resulting subgoals contain more instantiated versions of those constraints as hypotheses. These are unification problems [22], and we can implement a tactic Unify which simplifies them. This tactic was first described (under the name Qnify) in [15], so I shall not give details here. It solves first-order constraints involving variables and datatype constructors. It does so by repeatedly eliminating equational hypotheses, using Elim with rules which instantiate the schemes shown in figure

1. These schemes can be instantiated for every inductive family of datatypes—I omit the proofs for reasons of space, but the interested reader may refer to [16]. Unify executes explicitly the unification which is implicit in Coquand's system of pattern matching for inductive families, as partially implemented by Lena Magnusson in the ALF system [4, 12].

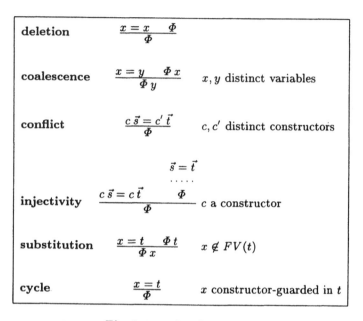

Fig. 1. Unify's rule schemes

TACTIC Unify

Unify operates on goals of form, up to permutation,
$$\Gamma \vdash \forall \vec{x}{:}\vec{X}.\ \vec{s} = \vec{t} \rightarrow G$$

such that $\vec{s}, \vec{t}$ inhabit the same telescope and are in **constructor form** over the $\vec{x}$—composed solely of x's and constructors. Unify repeatedly eliminates the leftmost equation with the appropriate rule from figure 1. This process either proves the goal (if there is no unifier), or computes a most general unifier σ over a reduced set of hypotheses $\vec{x}'$ and leaves a subgoal,
$$\Gamma \vdash \forall \vec{x}'{:}\vec{X}'.\ \sigma G$$

For example,

$$\vdash ? \ : \ \forall k{:}\mathbb{N}.\ s\,k = s\,0 \rightarrow f0_k = x \rightarrow P[x]$$
Unify
$$\vdash ? \ : \ P[f0_0]$$

It is usually sensible to follow `Elim` with `Unify`.

TACTIC `ElimUnify` *ruleProof* e_1 ... e_n

Execute `Elim` *ruleProof* e_1 ... e_n, then `Unify` on each subgoal.

For example,

$$\vdash ? \; : \; \forall x : \mathsf{Fin}\,(\mathsf{s}\,0).\,P[x]$$
`ElimUnify` **Fin-elim** x
$$\vdash ?_1 \; : \; P[\mathsf{f0_0}]$$
$$\vdash ?_2 \; : \; \forall y : \mathsf{Fin}\,0.(\forall x : \mathsf{Fin}\,(\mathsf{s}\,0).\,0 = \mathsf{s}\,0 \;\rightarrow\; y = x \;\rightarrow\; P[x]) \;\rightarrow\; P[\mathsf{fs_0}\,y]$$
`ElimUnify` **Fin-elim** x $\qquad$ (proves $?_2$ outright)

5   Example: An Induction Principle for $+$

In effect, `Elim` turns a theorem into a tactic. The standard induction and inversion principles for datatypes and relations give us basic proof methods for reasoning about those notions, but we can use `Elim` to give tactical support for more complex proof methods by manufacturing the elimination rules which describe them.

A standard proof idiom is to prove properties of functions by induction on their arguments. The pattern we routinely follow is to apply exactly the inductions and case analyses which replicate the recursive structure and pattern analyses performed by the function itself. In each resulting case, the arguments of the function are sufficiently instantiated for it to reduce, and the inductive hypotheses provide a (hopefully useful) characterisation of the results of recursive calls. For example, consider the following definition of $+$,

$$0 + y \mapsto y$$
$$(\mathsf{s}\,x) + y \mapsto \mathsf{s}\,(x + y)$$

It is quite common to prove general properties of '$x + y$' by structural induction on x. The constructor forms instantiating x in the subgoals are exactly those which allow $+$ to rewrite. This is a coincidence which we must engineer by selecting the right induction. We can express this pattern of reasoning about $+$ in a single theorem which explicitly combines the analysis, the rewriting and the characterisation of recursive calls. The key is to see $x + y = z$ as if it were a three-place inductively defined relation, with recursive calls becoming recursive premises. We can reason about this relation by means of its induction principle

$$\Phi \; : \; \forall x, y, s : \mathbb{N}. \; \mathsf{Type}$$

$$\textbf{+-induction} \;\; \dfrac{x + y = s \quad \dfrac{b \; : \; \mathbb{N}}{\cdots\cdots} \quad \dfrac{a + b = c \quad \Phi\, a\, b\, c}{\cdots\cdots\cdots\cdots}}{\Phi\, x\, y\, s}$$

$$\textbf{+-induction} \;\; : \;\; \forall \Phi : \forall x, y, s : \mathbb{N}. \; \mathsf{Type}.$$
$$(\forall b : \mathbb{N}. \; \Phi\, 0\, b\, b) \;\to$$
$$(\forall a, b, c : \mathbb{N}. \; a + b = c \;\to\; \Phi\, a\, b\, c \;\to\; \Phi\, (\mathsf{s}\, a)\, b\, (\mathsf{s}\, c)) \;\to$$
$$\forall x, y, s : \mathbb{N}. \; \forall q : x + y = s. \; \mathsf{Target}\; q \;\to\; \Phi\, x\, y\, s$$

Of course, we must prove this rule by induction on x, because that is the induction it effectively packages. Indeed, it seems plausible that such rules might be proven automatically, each time we define a function. Similar techniques in proving properties of inductively defined functions appear in James McKinna's thesis [17].

+-induction targets an equation, but applications of + are seldom so usefully abstracted in their natural habitat. We can prepare an application for targetting with the following rather useful variation on the theme of COQ's **Pattern** tactic [3].

TACTIC $\mathsf{AbstEq}\, x\, q\, e$

Given a goal,

$$\Gamma \;\vdash ? \; : \; \forall \vec{h} : \vec{H}. \, G$$

with $\vec{h}$ the shortest prefix of internal hypotheses such that $\Gamma ; \vec{h} : \vec{H} \;\vdash e \; : \; E$, and x, q fresh names, claim

$$\Gamma \;\vdash ?' \; : \; \forall \vec{h} : \vec{H}. \, \forall x : E. \, \forall q : e = x. \, [x/e]G$$

where every occurrence of e in G has been replaced by x. The tactic fails if this abstraction is not well-typed. The original goal is solved with

$$\Gamma \;\vdash \lambda \vec{h} : \vec{H}. \, ?' \; \vec{h}\, e\, (\mathsf{refl}\, e)$$

We may now use our higher-level characterisation of + to 'choose the right induction' in examples like the proof of associativity, which begins

$$\vdash ? \; : \; \forall x, y, z : \mathbb{N}. \; (x + y) + z = x + (y + z)$$
$$\mathsf{AbstEq}\; s\; q\; (x + y)$$
$$\vdash ? \; : \; \forall x, y, s : \mathbb{N}. \; \forall q : x + y = s. \; \forall z : \mathbb{N}. \; s + z = x + (y + z)$$
$$\mathsf{ElimUnify}\; \textbf{+-induction}\; q$$
$$\vdash ? \; : \; \forall b, z : \mathbb{N}. \; b + z = 0 + (b + z)$$
$$\vdash ? \; : \; \forall a, b, z : \mathbb{N}. \; (a + b) + z = a + (b + z) \;\to$$
$$(\mathsf{s}\, (a + b)) + z = (\mathsf{s}\, a) + (b + z)$$

The rest is standard, and indeed, the underlying induction is the usual one [2]. The point is that we have chosen to do 'the induction which unfolds $+$', rather than *an* induction which happens to unfold $+$.

With the advent of dependent types in programming, it is not unusual to find functions occurring in the type signatures of other functions. The 'new' functions must respect the computational behaviour of the 'old' ones, and this can place quite a burden of awareness on the programmer who eschews machine support. Consider developing, for example, the 'vector prefix' function:

$$\vdash ? \;:\; \forall A : \mathsf{Type}.\; \forall m, n : \mathbb{N}.\; \forall xs : \mathsf{Vect}\, A\, (m+n).\; \mathsf{Vect}\, A\, m$$

where $\dfrac{A : \mathsf{Type} \quad n : \mathbb{N}}{\mathsf{Vect}\, A\, n \;:\; \mathsf{Type}} \qquad \dfrac{A : \mathsf{Type}}{\mathsf{vnil} \;:\; \mathsf{Vect}\, A\, 0} \quad \dfrac{x : A \quad xs : \mathsf{Vect}\, A\, n}{\mathsf{vcons}\, x\, xs \;:\; \mathsf{Vect}\, A\, (\mathsf{s}\, n)}$

Our aim is to write the function which extracts the first m elements of a vector of length at least m. Rather than viewing the behaviour of the $+$ in the type as a burden, we can take it as a useful clue, and proceed as follows:

FixHyp A; AbstEq $s\, q\, (m+n)$; ElimUnify $+$-**induction** q
$\qquad A : \mathsf{Type} \vdash ?_1 \;:\; \forall b : \mathbb{N}.\; \forall xs : \mathsf{Vect}\, A\, b.\; \mathsf{Vect}\, A\, 0$
FixHyp xs; Refine vnil
$\qquad \ldots \vdash ?_2 \;:\; \forall a, b : \mathbb{N}.\; (\forall xs : \mathsf{Vect}\, A\, (a+b).\; \mathsf{Vect}\, A\, a) \;\to$
$\qquad\qquad\qquad\qquad\qquad \forall xs : \mathsf{Vect}\, A\, (\mathsf{s}\, (a+b)).\; \mathsf{Vect}\, A\, (\mathsf{s}\, a)$
Names $a\, b\, h$; FixHyp h
$\qquad \ldots; a, b : \mathbb{N}; h : \forall xs : \mathsf{Vect}\, A\, (a+b).\; \mathsf{Vect}\, A\, a$
$\qquad \vdash ? \;:\; \forall xs : \mathsf{Vect}\, A\, (\mathsf{s}\, (a+b)).\; \mathsf{Vect}\, A\, (\mathsf{s}\, a)$
ElimUnify Vect-**cases** xs
$\qquad \ldots \vdash ? \;:\; \forall x : A.\; \forall xs : \mathsf{Vect}\, A\, (a+b).\; \mathsf{Vect}\, A\, (\mathsf{s}\, a)$
FixHyp $x\, xs$; Refine $(\mathsf{vcons}\, x\, (h\, xs))$

6   Conclusion and Further Work

The Elim tactic is a generic tool which applies elimination rules with parametric conclusions to solve the goal at hand, based on the folklore technique of generalisation with equational constraints. By providing the means to describe the intended usage of such a rule, through redundant Target t premises, many specific tactics can be replaced by theorems with an appropriate 'user manual'. Elim removes the need for the separate 'induction' and 'inversion' tactics currently found in LEGO and COQ, and can be used as a component in more complex tactics which, for example, simplify equational hypotheses (like Unify) or carry out rewriting.

Of potentially greater long term significance is the support given by Elim to user-derived rules which capture higher-level patterns of reasoning, such as induction or inversion on function applications. As we build functions compositionally, so can we also build high-level tools for reasoning about their use. This approach has proved effective in a number of experiments, such as the development of first-order unification in my PhD thesis, [16], but there is much yet to do before it can be called a *method*. The key technological component underpinning this future investigation is now in place.

To my mind, though, the most important potential benefit from good elimination technology is the declarative power it gives to programming, especially with dependent types. The existing notations for programming are not designed to accommodate a scenario where the analysis of intermediate values affects the types of the resulting branches as well as the choice between them. Nor are they well placed to exploit the power of dependent types to express derived notions of analysis which are not based on datatype constructors, extending Wadler's notion of **view** [24]. If the Elim tactic allows us to make higher-level proof steps, what are the corresponding program constructs?

Acknowledgements

This work would not have been possible without a considerable inheritance of technology from the Coq project, in particular from Cristina Cornes. Much of the detail was worked out under the supervision of Healfdene Goguen, Martin Hofmann and Rod Burstall, to whom I also owe a debt of gratitude. However, it was James McKinna who planted the seeds which grew into this work, and it is his grant (UK EPSRC GR/N 24988/01) which continues to support it: this paper is for him.

References

1. Thorsten Altenkirch. Extensional equality in intensional type theory. In *LICS'99*, 1999.
2. Alan Bundy. The Automation Of Proof By Mathematical Induction. In Alan Robinson and Andrei Voronkov, editors, *Handbook of Automated Reasoning*. Elsevier Science, 1999.
3. L'Équipe Coq. The Coq Proof Assistant Reference Manual. http://pauillac.inria.fr/coq/doc/main.html, Apr 2001.
4. Thierry Coquand. Pattern Matching with Dependent Types. In *Proceedings of the Logical Framework workshop at Båstad*, June 1992.
5. Thierry Coquand and Christine Paulin. Inductively defined types. In P. Martin-Löf and G. Mints, editors, *COLOG-88. International Conference on Computer Logic*, LNCS 417, pages 50–66. Springer-Verlag, 1990.
6. Cristina Cornes. *Conception d'un langage de haut niveau de répresenatation de preuves*. PhD thesis, Université Paris VII, 1997.
7. Cristina Cornes and Delphine Terrasse. Inverting Inductive Predicates in Coq. In *Types for Proofs and Programs'95*, LNCS 1158, Springer-Verlag, 1995.
8. N.G. de Bruijn. Telescopic Mappings in Typed Lambda-Calculus. *Information and Computation*, 91:189–204, 1991.
9. Peter Dybjer. Inductive Sets and Families in Martin-Löf's Type Theory. In G. Huet and G. Plotkin, editors, *Logical Frameworks*. CUP, 1991.

10. Martin Hofmann. *Extensional concepts in intensional type theory*. PhD thesis, Laboratory for Foundations of Computer Science, University of Edinburgh, 1995.
11. Zhaohui Luo and Robert Pollack. LEGO Proof Development System: User's Manual. Technical Report ECS-LFCS-92-211, Laboratory for Foundations of Computer Science, University of Edinburgh, May 1992.
12. Lena Magnusson. *The implementation of ALF—A Proof Editor based on Martin-Löf's Monomorphic Type Theory with Explicit Substitutiton*. PhD thesis, Chalmers University of Technology, Göteborg, 1994.
13. Per Martin-Löf. A theory of types. manuscript, 1971.
14. Per Martin-Löf. *Intuitionistic Type Theory*. Bibliopolis·Napoli, 1984.
15. Conor McBride. Inverting inductively defined relations in LEGO. In E. Giménez and C. Paulin-Mohring, editors, *Types for Proofs and Programs'96*, LNCS 1512, pages 236–253, Springer-Verlag, 1998.
16. Conor McBride. *Dependently Typed Functional Programs and their Proofs*. PhD thesis, University of Edinburgh, 1999.
17. J. McKinna. *Deliverables: A Categorical Approach to Program Development in Type Theory*. PhD thesis, Laboratory for Foundations of Computer Science, University of Edinburgh, 1992.
18. Dale Miller. A Logic Programming Language with Lambda-Abstraction, Function Variables and Simple Unification. *Journal of Logic and Computation*, 2(4):497–537, 1991.
19. Dale Miller. Unification under a mixed prefix. *Journal of Symbolic Computation*, 14(4):321–358, 1992.
20. Christine Paulin-Mohring. Définitions Inductives en Théorie des Types d'Ordre Supérieur. Habilitation Thesis. Université Claude Bernard (Lyon I), 1996.
21. Robert Pollack. Implicit syntax.
22. Alan Robinson. A Machine-oriented Logic Based on the Resolution Principle. *Journal of the ACM*, 12:23–41, 1965.
23. Thomas Streicher. Investigations into intensional type theory. Habilitation Thesis, Ludwig Maximilian Universität, 1993.
24. P. Wadler. Views: A way for pattern matching to cohabit with data abstraction. In *POPL'87*. ACM, 1987.

Generalization in Type Theory Based Proof Assistants

Olivier Pons*

CEDRIC-IIE (CNAM)
18 Allée Jean Rostand 91025 Evry, France
pons@cnam.fr

Abstract. This paper describes a mechanism to generalize mathematical results in type theory based proof assistants. The proposed mechanism starts from a proved theorem or a proved set of theorems (a theory) and makes it possible to get less specific results that can be instantiated and reused in other contexts.

1 Introduction

In mathematics, it is common to try to generalize previously obtained results in order to reuse them in other contexts and to get more structured representations. To do so, we have to find which parts of a proof can be generalized, we have to understand the analogies between different mathematical structures and the analogies between proofs on these structures, and finally we have to study how these similarities can be useful in the development of new theories. For example, the basic results of arithmetic can be generalized to get algebraic theories like groups or rings.

Today, mechanical theorem provers are more and more used for large-scale formalizations of mathematics and programming language semantics and for algorithms certification and it seems interesting to investigate how the generalization process used in mathematical development can be used in this context.

The most difficult task is to detect analogies between objects. The formalization of this problem relies on the notion of anti-unification introduced by Plotkin [Plo69]. The problem is, given two terms t_1 and t_2 to find the tuple (g, θ, σ) where g is a term and θ and σ are two substitutions such as $\theta(g) = t_1$ and $\sigma(g) = t_2$. The term g is a called a generalization of t_1 and t_2. Finding the most specific generalization of two terms is now a well-understood problem for first order theories but not for higher order systems. Contributions to the understanding of the higher order case are for example Pfenning's studies of anti-unification for a subset of terms of the Calculus of Constructions [Pfe91] and the categorical approach of Hasker and Reddy [HR92] who have shown that in the case of second order theory, the complete set of most specific generalizations can be computed.

---

* This work was done during a postdoctoral appointment at Universidade do Minho (Portugal). It was supported by the Portuguese Science Foundation (Fundação para a Ciencia e a Technologia) under the Fellowship PRAXIS-XXI/BPD/22108/99.

P. Callaghan et al. (Eds.): TYPES 2000, LNCS 2277, pp. 217–232, 2002.

In this paper we follow a more pragmatic way. Our aim is not to find the least general generalization of two terms but simply to help the user to define generalizations of already proved theorems without assuming future instantiations of these theorems.

Our experiments have been done using the Coq system [Bar99] but the idea is generic and could be adapted to any system based on type theory that represent theorems as types and proofs as terms such as Nuprl [Con86] or Lego [LEG].

Formally our problem can be summed up as follows : We start with a triple (s, t, l) where s is a type, t is a term of type s, and l is a set of free identifiers occurring in s. We want to find a triple (s', t', l') where l' is the smallest set containing l such that :

- abstracting all the identifiers of l' in s and t we get respectively s' and t'.
- t' has type s'

The practical meaning of l', s' and t' will be highlighted in the first example.

The next section exemplifies our basic generalization mechanism in the case of the abstraction of an operator in a theorem. The third section formally describes the mechanism. The fourth section exemplifies type abstraction. The fifth shows how to structure our generalization. The sixth and seventh sections examine the correctness and the limitations of the basic mechanism and proposes a first extension. The eighth examines the abstraction mechanism for inductive objects. In the ninth section we examine related works and the tenth section gives a conclusion and enumerates the points that deserve a more thorough treatment in further work.

2   A Quick Overview

2.1   Abstraction

As an example, we will start with a basic lemma about multiplication of natural numbers. Suppose that we have proved the following property :

$$mult\_permute :$$
$$\forall\ n, m, p : nat.(\text{mult}\ n\ (\text{mult}\ m\ p)) = (\text{mult}\ m\ (\text{mult}\ n\ p))$$

We wish to generalize the statement of the theorem $mult\_permute$ to get a new statement in which the operator will not be $mult$ any more but an unspecified binary operator on natural numbers. For that purpose we will abstract the identifier $mult$ in the statement $mult\_permute$. The generalized statement[1] we obtain has the form:

$$\forall f : nat \rightarrow nat \rightarrow nat$$
$$\forall n, m, p : nat\ \Rightarrow$$
$$(f\ n\ (f\ m\ p)) = (f\ m\ (f\ n\ p)).$$

---

[1] In which we denote the abstracted operator with f to avoid the confusion with $mult$.

But this statement is not true any more. The problem is to find the properties that must be satisfied by the operator f in order to ensure that the new statement is still a theorem. Here, intuitively the statement will remain true for any associative and commutative function on integers but how to find these properties in the general case ? In order to justify our answer let us quickly recall how to work with the Coq proof system. When building a mathematical theory in the Coq proof assistant we start by defining the objects of this theory (terms of the Calculation of Inductive Constructions (CIC)). Then we state (in the CIC) some properties of these objects. To prove these properties we have a set of commands (tactics). The application of a tactic to an initial goal generates a list of subgoals. The user then chooses a subgoal in this list and applies a new tactic on this subgoal. The produced subgoals are then inserted in the list of remaining subgoals. The proof finishes when the resulting list is empty. The system uses the list of tactics to build the proof object (a term of the CIC) and check the complete proof by type-checking this term. Thus, we have several representations of the same proof. The list of tactics (called a proof script) reflects the strategy of the user and is generally the only thing he is interested in. Unfortunately, in the general case the script does not give enough information[2] to support automatic proof analysis and manipulation. Therefore, to find which properties must be satisfied by the abstracted operator, it is necessary to examine the (generated) proof term of the initial theorem. We will also use this term to build a proof of the generalized statement. Thus let us observe the proof term of *mult_permute*:

$$
\begin{aligned}
&\lambda n, m, p : nat \\
&(\boxed{\texttt{eq\_ind\_r}}\ nat\ (mult\ (mult\ m\ n)\ p) \\
&\quad \lambda n_0 : nat\ (mult\ m\ p) = n_0 \\
&\quad (\boxed{\texttt{eq\_ind\_r}}\ nat\ (mult\ n\ m) \\
&\qquad \lambda n_1 : nat\ (mult\ n\ (mult\ m\ p)) = (mult\ n_1\ p) \\
&\qquad (\boxed{\texttt{mult\_assoc\_l}}\ n\ m\ p) \\
&\quad (mult\ m\ n) \\
&\qquad (\boxed{\texttt{mult\_sym}}\ m\ n)) \\
&\quad (mult\ m\ (mult\ n\ p)) \\
&\quad (\boxed{\texttt{mult\_assoc\_l}}\ m\ n\ p))
\end{aligned}
$$

In addition to *mult*, *nat*, and $=$, which already appeared in the initial statement, the free identifiers which appear in the proof are *eq_ind_r*, *mult_assoc_l* and *mult_sym*. Therefore, in order to build the abstracted statement, it is necessary to determine which of the free identifiers are related to the identifier *mult*, which has been abstracted in the statement.

Thus, we seek their statements (*i.e.* their types) in the current environment. We find respectively:

---

[2] Mainly because it may contain automatic decision procedures.

eq_ind_r : $\forall A : Set.\forall x : A.\forall P : (A{\rightarrow}Prop)(P\ x){\rightarrow}\forall y : A.y = x{\rightarrow}(P\ y)$

$mult\_assoc\_r$: $\forall n, m, p : nat.(\boxed{\text{mult}}\ (\boxed{\text{mult}}\ n\ m)\ p) = (\boxed{\text{mult}}\ n\ (\boxed{\text{mult}}\ m\ p))$

$mult\_sym$: $\forall n, m : nat.(\boxed{\text{mult}}\ n\ m) = (\boxed{\text{mult}}\ m\ n)$

It appears that only the last two statements refer to the identifier *mult*. Thus, we deduce that the only properties of the multiplication which are used in the proof are symmetry and left associativity, which correspond to *mult_sym* and *mult_assoc_l* [3].

Thus, it is also necessary to abstract these two identifiers in the generalized statement in order to express the constraints on the operator f. We get the following statement:

$Lemma\ f\_permute$:
$\forall f : nat{\rightarrow}nat{\rightarrow}nat.$
$\quad \forall f\_assoc : \forall n, m, p : nat.(f\ n\ (f\ m\ p)) = (f\ (f\ n\ m)\ p).$
$\quad \forall f\_sym : \forall n, m : nat.(f\ n\ m) = (f\ m\ n)$
$\quad \forall n, m, p : nat.(f\ n\ (f\ m\ p)) = (f\ m\ (f\ n\ p))$

The proof of this theorem is obtained by abstracting *mult*, *mult_sym* and *mult_assoc_l* in the initial proof term. This leads to the following proof term :

$\lambda f : nat{\rightarrow}nat{\rightarrow}nat.$
$\lambda f\_assoc\_l : (n0, m0, p0 : nat)(f\ n0\ (f\ m0\ p0)) = (f\ (f\ n0\ m0)\ p0).$
$\quad \lambda f\_sym : (n0, m0 : nat)(f\ n0\ m0) = (f\ m0\ n0).$
$\quad \lambda n, m, p : nat.$
$\quad\quad (eq\_ind\_r\ nat\ (f\ (f\ m\ n)\ p)\ \lambda n0 : nat.(f\ n\ (f\ m\ p)) = n0$
$\quad\quad\quad (eq\_ind\_r\ nat\ (f\ n\ m)\ \lambda n0 : nat.(f\ n\ (f\ m\ p)) = (f\ n0\ p)$
$\quad\quad\quad\quad (f\_assoc\_l\ n\ m\ p)\ (f\ m\ n)\ (f\_sym\ m\ n))$
$\quad\quad\quad (f\ m\ (f\ n\ p))$
$\quad\quad\quad (f\_assoc\_l\ m\ n\ p))$

2.2   Instantiations

We now check that we can instantiate *f_permute*, with addition and the proof of its associativity and commutativity, to obtain a more specific instance of the theorem. We get:

$Lemma\ plus\_permute$:
$\quad \forall n, m, p : nat.\ ((plus\ n\ (plus\ m\ p)) = (plus\ m\ (plus\ n\ p))).$
$\quad Proof\ (f\_permute\ plus\ plus\_assoc\_l\ plus\_sym).$

Before going further, the following section gives the broad outline of an interactive tool helping the user to carry out such generalizations.

---

[3] In our example these two identifiers also appear in the script, however this is not always the case.

3   A Tool to Assist the Generalization

3.1   Basic Principle

Given a statement S and a proof term P, the user provides the name of the function that he wishes to abstract in S (to simplify the following explanation we suppose that he selects only one function identifier denoted by f). In a graphical environment such as *CtCoq* [Ber97] or *Proof-General* [Asp99], this can be done by a single mouse click. The following operations are then done automatically:

1. Find the type t associated with the identifier f.
2. Recover the list of all the free identifiers which appear in the proof term P and do not already appear in the statement S.
3. Find the types associated with these identifiers.
4. Among these identifiers, select all those whose type refers to the identifier f that has been abstracted. We will denote by f_{P_i} the selected identifiers and t_{P_i} their types (which express the properties of f which have been used).
5. Build the statement of the theorem generalized by abstracting f and the f_{P_i} in S. We obtain a generalized statement of the form:

$$\forall f : t.\forall f_{P_1} : t_{P_1}.\ldots.\forall f_{P_i} : t_{P_i}.\ldots.S$$

6. Build the proof term associated with this statement by abstracting on P. We get a term of the form :

$$\lambda f : t.\lambda f_{P_1} : t_{P_1}.\ldots.\lambda f_{P_i} : t_{P_i}.\ldots.P$$

Remark. Actually, steps 1 to 4 can be done once for all by computing the dependency graph of the development. Each node is annotated by an identifier, a type and a proof term. After that when abstracting an identifier I corresponding to the node N_i in a statement S corresponding to the node N_s, we also abstract all the identifiers associated with a node that appears in a path between N_i and N_s.

3.2   Choosing the Names of the Abstracted Identifiers

As we can notice in the statement of the lemma $f\_permute$, discussed in the previous section, the names given to the abstract identifiers are not randomly selected. We propose a small algorithm to name the properties of the abstracted terms.

It is usual to express a property of an operator by the name of this operator followed by the name of the property (often separated by a character *underscore*). To express that it corresponds to an unspecified function, we will use f (then f_i if we abstract more than one operator) to denote the abstract operator. Thus, it is also necessary to substitute f to all the occurrences of the abstracted operator in the statement.

Once we have recovered the list of identifiers resulting from step 4 in our algorithm, we use their names to build new names by substituting f to the name of the operator in the old name of the property; thus *mult_assoc* becomes *f_assoc* which remains meaningful. Then we substitute f for all the occurrences of the abstracted operator in the types of the identifiers recovered at step 3. Finally we substitute f and the names of properties formed from f in the proof term.

4   More Abstractions

So far, we have limited the abstractions to function identifiers. Let us consider again the example of section 2.1. The permutation property remains true for any function $f : E \rightarrow E \rightarrow E$ which is associative and commutative, whatever the set E. Let us try to continue our generalization by abstracting the type of the function arguments. A new statement and a new proof are obtained:

$$
\begin{aligned}
&Lemma\ f2\_permute\ : \\
&\forall E : Set. \\
&\quad \forall f : E \rightarrow E \rightarrow E. \\
&\quad \forall f\_assoc1\ : \forall n, m, p : E.(f\ n\ (f\ m\ p)) = (f\ (f\ n\ m)\ p). \\
&\quad \forall f\_sym\ : \forall n, m : E.(f\ n\ m) = (f\ m\ n)). \\
&\quad \forall n, m, p : E.((f\ n\ (f\ m\ p)) = (f\ m\ (f\ n\ p))). \\
&Proof \\
&\lambda E : Set. \\
&\quad \lambda f : E \rightarrow E \rightarrow E. \\
&\quad \lambda f\_assoc1 : \forall n0, m0, p0 : E.(f\ n0\ (f\ m0\ p0)) = (f\ (f\ n0\ m0)\ p0). \\
&\quad \lambda f\_sym : \forall n0, m0 : E.(f\ n0\ m0) = (f\ m0\ n0). \\
&\quad\quad \lambda n, m, p : E. \\
&\quad\quad (eq\_ind\_r\ E\ (f\ (f\ m\ n)\ p)\ \lambda n0 : E.(f\ n\ (f\ m\ p)) = n0 \\
&\quad\quad\quad (eq\_ind\_r\ E\ (f\ n\ m)\ \lambda.n0 : E.(f\ n\ (f\ m\ p)) = (f\ n0\ p) \\
&\quad\quad\quad (f\_assoc1\ n\ m\ p)\ (f\ m\ n)\ (f\_sym\ m\ n))\ (f\ m\ (f\ n\ p)) \\
&\quad\quad (f\_assoc1\ m\ n\ p))
\end{aligned}
$$

4.1   Instantiations

Trivially, this generalization can be instantiated with the multiplication on natural numbers but we can also instantiate the theorem with other functions such as the addition on multi-variable monomials as defined by Pottier[4]. We get :

$$
\begin{aligned}
&Lemma\ mon\_permute\ : \\
&\forall k : nat.\forall n, m, p : (mon\ k). \\
&\quad ((mult\_mon\ k\ n\ (mult\_mon\ k\ m\ p)) = (mult\_mon\ k\ m\ (mult\_mon\ k\ \ n\ p))). \\
&\\
&Proof\ \lambda k : nat.(f2\_permute\ (mon\ k) \\
&\quad (\lambda i : (mon\ k).\lambda j : (mon\ k).(mult\_mon\ k\ i\ j\ )) \\
&\quad (\lambda i : (mon\ k).\lambda j : (mon\ k).\lambda l : (mon\ k).(mult\_mon\_assoc\ k\ i\ j\ l)) \\
&\quad (\lambda i : (mon\ k).\lambda j : (mon\ k).(mult\_mon\_com\ k\ i\ j)))
\end{aligned}
$$

---

[4] (http://www-sop.inria.fr/croap/CFC/buch/Monomials.html).

5   Abstracting Theories

Generalization can be represented in a more structured way. We have said that the permutation property remains true for any binary operator f, which is associative and commutative. Any set equipped with such an operator is a commutative semi-group. Such an algebraic structure may be represented by a record.

Record semi_group : *Type* :=
  {*sg_s* : *Set*;
   sg_law : *sg_s* → *sg_s* → *sg_s*;
   sg_assoc : ∀x,y,z : *sg_s*.(*sg_law* x (*sg_law* y z)) = (*sg_law* (*sg_law* x y) z)}.

Record semi_group_com : *Type* :=
  {*sg* :> *semi_group*;
   sg_com : ∀x,y : (*sg_s* *sg*).(*sg_law* *sg* x y) = (*sg_law* *sg* y x)}.

The notation :> is a coercion facility that allows us to see a commutative semi-group as a semi-group.

Now we can build a lemma expressing that the internal law of a commutative semi-group satisfies the permutation property:

Lemma sg_permute : ∀*sg*1 : *semi_group_com*.∀n,m,p : (*sg_s* *sg*1).
  ((*sg_law* *sg*1 n (*sg_law* *sg*1 m p)) = (*sg_law* *sg*1 m (*sg_law* *sg*1 n p))).

Proof
λ*sg*1 : *semi_group_com*.
λn,m,p : (*sg_s* *sg*1).
 (*eq_ind_r* (*sg_s* *sg*1) (*sg_law* *sg*1 (*sg_law* *sg*1 m n) p)
    λ$n0$: (*sg_s* *sg*1).(*sg_law* *sg*1 n (*sg_law* *sg*1 m p)) = $n0$
    (*eq_ind_r* (*sg_s* *sg*1) (*sg_law* *sg*1 n m)
       λ$n0$: (*sg_s* *sg*1).
         (*sg_law* *sg*1 n (*sg_law* *sg*1 m p)) = (*sg_law* *sg*1 $n0$ p)
         (*sg_assoc* *sg*1 n m p) (*sg_law* *sg*1 m n) (*sg_com* *sg*1 m n))
 (*sg_law* *sg*1 m (*sg_law* *sg*1 n p)) (*sg_assoc* *sg*1 m n p)).

In this generalization of our initial *permute* lemma, all abstracted identifiers (*i.e.* the set, the operator and the properties) have been grouped in one record that denotes a commutative semi-group. The internal law and its properties have been substituted for the operator and the operator properties respectively.

As usual, we check that this generalization can be specialized as shown below for addition on natural numbers. We first define the commutative semi group *sg_com_nat* :

$Definition\ sg\_com\_nat\ :=$
$(Build\_semi\_group\_com$
  $(Build\_semi\_group\quad nat\ plus\ plus\_assoc\_l)$
  $plus\_sym).$

$Lemma\ plus\_permute\ :$
$\forall\ n,m,p:nat.((plus\ n\ (plus\ m\ p))=(plus\ m\ (plus\ n\ p))).$
$Exact\ (sg\_permute\ sg\_com\_nat).$
$Save.$

6   Correctness Problems

Unfortunately all these transformations do not always produce a type-checkable term. Actually our generalization mechanism may produce a badly typed term because proof objects are not completely explicit.

In fact for efficiency and clarity reasons most systems add to the traditional deduction rules some computational rules. The application of such a computational rule does not appear in the proof term.

A classical example of such rule is the iota-reduction. Let us consider the definition of the function *plus* using *Fixpoint* and *Cases*

$Fixpoint\ plus\ \lambda n:nat.\ :\ nat\rightarrow nat\ :=$
  $\lambda m:nat.$
  $Cases\ n\ of$
    $O\quad \leadsto\ m$ (rule r1)
    $|\ (S\ p)\ \leadsto\ (S\ (plus\ p\ m))$ (rule r2)
$end.$

The *iota*-reduction let us identify $(plus\ 2\ 2)$ and 4 without any demonstration, just using the computational rules r1 and r2 to reduce $(plus\ 2\ 2)$.

Now let us show when such a reduction step may interfere with our generalization mechanism. We consider the right associativity of *plus* :

$Lemma\ plus\_assoc\_r\ :$
  $\forall n,m,p:nat.((plus\ (plus\ n\ m)\ p)=(plus\ n\ (plus\ m\ p))).$

We will see two ways to build the proof. The first one produces a term in which *plus* may be abstracted safely. The second one produces a term for which the abstraction of *plus* brings out a badly typed term. The last subsection proposes a solution for the bad case.

6.1   The Correct Case

If we prove the statements *plus_assoc_r* by the small script below, using **explicitly** the proof of the left associativity:

```
Intros;Apply   sym_eq;  Apply   plus_assoc_l
```

we get the proof term

$\lambda n, m, p : nat.$
$(sym\_eq \ nat \ (plus \ n \ (plus \ m \ p)) \ (plus \ (plus \ n \ m) \ p)$
$\qquad (plus\_assoc\_l \ n \ m \ p))$

in which *plus* can be abstracted as previously to get a lemma *f_assoc_r*.

6.2   The Problematic Case

Now, if we use the script below to build the proof term of *plus_assoc_r*

`Intros;Apply` *sym_eq*`;Elim` *n*`; Simpl;Auto with *.`

This will produce a different proof term that does not refers to the left associativity[5].

$\lambda n, m, p : nat.$
$(sym\_eq \ nat \ (plus \ n \ (plus \ m \ p)) \ (plus \ (plus \ n \ m) \ p)$
$\quad (nat\_ind \ \lambda n0 : nat.(plus \ n0 \ (plus \ m \ p)) = (plus \ (plus \ n0 \ m) \ p)$
$\qquad (refl\_equal \ nat \ (plus \ m \ p))$
$\qquad \lambda n0 : nat.\lambda H : ((plus \ n0 \ (plus \ m \ p)) = (plus \ (plus \ n0 \ m) \ p)).$
$\qquad\quad (f\_equal \ nat \ nat \ S \ (plus \ n0 \ (plus \ m \ p)) \ (plus \ (plus \ n0 \ m) \ p) \ H) \ n))$

Abstracting *plus* in this term we can build a term *f_assoc_r*.

$Definition \ f\_assoc\_r \ :=$
$\lambda f : nat{\rightarrow}nat{\rightarrow}nat.\lambda n, m, p : nat.$
$(sym\_eq \ nat \ (f \ n \ (f \ m \ p)) \ (f \ (f \ n \ m) \ p)$
$\quad (nat\_ind \ \lambda n0 : nat.(f \ n0 \ (f \ m \ p)) = (f \ (f \ n0 \ m) \ p)$
$\qquad (refl\_equal \ nat \ (f \ m \ p))$
$\qquad \lambda n0 : nat.\lambda H : ((f \ n0 \ (f \ m \ p)) = (f \ (f \ n0 \ m) \ p)).$
$\qquad\quad (f\_equal \ nat \ nat \ S \ (f \ n0 \ (f \ m \ p)) \ (f \ (f \ n0 \ m) \ p) \ H) \ n))$

But this term is badly typed and a type checker will make a complaint like :

`In environment` $[n \ : \ nat; m \ : \ nat; p \ : \ nat; f \ : \ nat{\rightarrow}nat{\rightarrow}nat]$
`The term` *nat_ind* `of type` :
$\quad \forall P : (nat{\rightarrow}Prop).(P \ O){\rightarrow}(\forall n : nat.(P \ n){\rightarrow}(P \ (S \ n))){\rightarrow}\forall n : nat.(P \ n)$
`Cannot be applied to the terms`
$\lambda \ n0 : nat.(f \ n0 \ (f \ m \ p)) = (f \ (f \ n0 \ m) \ p) \ : \ nat{\rightarrow}Prop$
$(refl\_equal \ nat \ (f \ m \ p)) \ : \ (f \ m \ p) = (f \ m \ p)$
$\lambda n0 : nat.\lambda \ H : ((f \ n0 \ (f \ m \ p)) = (f \ (f \ n0 \ m) \ p)).$
$(f\_equal \ nat \ nat \ S \ (f \ n0 \ (f \ m \ p)) \ (f \ (f \ n0 \ m) \ p) \ H)$
$\quad : \ \forall n0 : nat.$
$\qquad (f \ n0 \ (f \ m \ p)) = (f \ (f \ n0 \ m) \ p)$
$\qquad {\rightarrow}(S \ (f \ n0 \ (f \ m \ p))) = (S \ (f \ (f \ n0 \ m) \ p))$
$n \ : \ nat$

In such a case, the transformation is considered illegal and is rejected by our basic generalization mechanism.

---

[5] Because it does not contains *plus_assoc_l* as free identifier.

6.3   How to Enlarge the Set of Legal Transformations?

To get a correct transformation when the proof term produced by the basic abstraction mechanism does not type-check, we will try to replace the badly typed subterm by a variable with the right type which is then abstracted.

To do this, subterms are annotated with their type and during the transformation substitutions are also applied to the annotations to get new type annotations. Then each time the type of a translated subterm (the proposed one) differs from the translated type (the expected one), we replace the subterm by a variable with the expected type.

For example the badly typed version of f_assoc_r of the previous subsection becomes :

```
Definition f_assoc_r :=
 λf : (nat→nat→nat).
λn, m, p : nat.
 λp1 : (λn0 : nat.(f n0 (f m p)) = (f (f n0 m) p) n).
 (sym_eq nat (f n (f m p)) (f (f n m) p) p1)
: ∀f : (nat→nat→nat). ∀n, m, p : nat.
     (f n (f m p)) = (f (f n m) p) n)→(f (f n m) p) = (f n (f m p))
```

This simply expresses that on nat left associativity implies right associativity!

Remark. Managing such "bad cases" not always produces useful results. The trivial case where the transformation is stupid is when the result is a tautology (of the form $\forall c : tc \ldots \forall c_n : tc_n.tc_n$). We will reject such terms.

7   Limitations

We could be tempted to push the abstraction mechanism further. Indeed, why limit the statement to the equality relation eq. The only property of this relation which is used is the rewriting principle :

$$eq\_ind\_r : \forall A : Set.\forall x : A.\forall P : A→Prop.(Px)→\forall y : A.y = x→(Py)$$

The statement thus remains true for any relation R verifying this principle. Thus it is possible to get the following generalization:

```
Lemma   f3_permute :
 ∀E : Set.∀f : E→E→E).
 ∀R : ∀A : Set.A→A→Prop.
 ∀R_ind_r : ∀A : Set.∀x : A.∀P : A→Prop.(P x)→∀y : A.(R A y x)→(P y))
  ∀f_assoc_l : ∀n, m, p : E.(R E (f n (f m p)) (f (f n m) p)).
   ∀f_sym  : ∀n, m : E.(R E (f n m) (f m n)).
    ∀n, m, p : E.(R E(f n (f m p)) (f m (f n p)))).
```

which can be proved by the term below:

$\lambda E : Set.$
  $\lambda f : E{\rightarrow}E{\rightarrow}E.$
    $\lambda R : (A : Set)A{\rightarrow}A{\rightarrow}Prop.$
    $\lambda R\_ind\_r : \forall A : Set.\forall x : A.\forall P : A{\rightarrow}Prop.(P\ x){\rightarrow}\forall y : A.(R\ A\ y\ x){\rightarrow}(P\ y).$
    $\lambda f\_assoc\_1 : \forall n0, m0, p0 : E.(R\ E\ (f\ n0\ (f\ m0\ p0))\ (f\ (f\ n0\ m0)\ p0)).$
    $\lambda f\_sym : \forall n0, m0 : E.(R\ E\ (f\ n0\ m0)\ (f\ m0\ n0)).$
      $\lambda n, m, p : E.$
      $(R\_ind\_r\ E\ (f\ (f\ m\ n)\ p)\lambda n0 : E.(R\ E\ (f\ n\ (f\ m\ p))\ n0)$
        $(R\_ind\_r\ E\ (f\ n\ m)\ \lambda n0 : E.(R\ E\ (f\ n\ (f\ m\ p))\ (f\ n0\ p))$
          $(f\_assoc\_1\ n\ m\ p)\ (f\ m\ n)\ (f\_sym\ m\ n))\ (f\ m\ (f\ n\ p))$
          $(f\_assoc\_1\ m\ n\ p)).$

7.1   Instantiations

We still check this generalized theorem by instantiating it with the multiplication on *nat*:

$Definition\ mult\_permute\ :=$
  $\lambda n, m, p : nat.(f3\_permute$
        $nat\quad plus\ eq\ eq\_ind\_r\ plus\_assoc\_1\ plus\_sym) :$
  $\forall n, m, p : nat.((plus\ n\ (plus\ m\ p)) = (plus\ m\ (plus\ n\ p))).$

But we can not always reuse this theorem to proof the same permute property on any type provided with an "equality" relation and a commutative and associative operator.

Let us consider for example the polynomials such as they were defined by Pottier and Théry during their certification of the Buchberger's algorithm [Thé98]. They are represented by the table of their coefficients. The addition (*plusP*) of two polynomials (of type P) is obtained by summing their coefficients of the same degree. An equality relation *eqP* is defined to identify for example 0 and $0 + (0 * x)$ which are different for the equality *eq*.

Thus, we tried to prove the *permute* property for the addition of these polynomials, that is to say the following lemma:

$Lemma\ plusP\_permute\ :$
  $\forall n, m, p : P.(eqP\ (plusP\ n\ (plusP\ m\ p))\ (plusP\ m\ (plusP\ n\ p))).$

In order to reuse the general theorem $f3\_permute$ we have to be able to give a witness of of the property R_ind_r.

We could try to establish a generalization of eq_ind_r such as for example in the case of the polynomials with integer coefficients.

$Lemma\ eq\_ind\_r\_int :$
$\forall x : (P\ nat).\forall\ P1 : (P\ nat){\rightarrow}Prop.$
  $(P1\ x){\rightarrow}\forall y : (P\ nat).(eqP\ nat\ (eq\ nat)\ \lambda x : nat.x = O\quad y\ x){\rightarrow}(P1\ y).$

But this result is not true, otherwise by taking $(\lambda z : (Pnat).x = z)$ as $P1$ we could prove that eqP implies eq, which is not true.

We can nevertheless complete the proof of *plusP_permute* using the appropriate script but the corresponding proof term, has nothing to do with the one we previously tried to instantiate.

This example shows that when abstracting "too specific property", the result can not alway be reinstanciated by other objects that the abstracted one. Here, *eq_ind_r* characterizes the Leibniz' equality, and we will not be able to reuse our lemma with other definite equalities.

The problem is not particular to equality, which abstraction may be useful in many cases, but comes from the use of a "too specific property" of equality. Those "too specific" objects relies on the abstraction of inductively defined objects[6] and is discussed in the next section.

8   Remarks on the Abstraction of Inductively Defined Objects

In the previous sections we have discussed several examples of abstraction where inductively defined objects are abstracted. In fact we can distinguish two cases:

In the first case, the identifiers associated with the constructors of the abstracted objects do not appear in the term to be generalized. We say that this is a "silent object" in the term. In such cases, we just have to abstract this identifier as we have done to build *f2_permute* in section 4.

In the second case, the constructors identifiers appear in the term to be generalized. Here we can distinguish two sub-cases.

The first subcase is when the identifiers do not appear in a *case*-expression. Then they have just to be abstracted. The example below shows how to abstract *nat* in the dependent type Tab $(where(Tab$ s $n)$ is the type of the s lists of length n). In $TabAbs$ the type of the dependent parameter, nat, and its two constructors O and S have been abstracted.

```
Inductive Tab λs : Set.  :   nat→Set  :=
    Tnil  :  (Tab s O)
   | Tcons  :  ∀n : nat.s→(Tab s n)→(Tab s (S n)).

Inductive TabAbs λs : Set.λs1 : Set.λ.OAbs : s1λ.SAbs : s1→s1.  :  s1→Set  :=
    TnilAbs  :  (TabAbs s s1 OAbs SAbs OAbs)
   | TconsAbs  :  (n : s1)s→(TabAbs s s1 OAbs SAbs n)→
                         (TabAbs s  s1  OAbs SAbs (SAbs n)).
```

The second subcase occurs when the identifiers appear in a *case expression*. This happens for example when abstracting inductively defined types in function definitions. For example let us see what happens if we try to abstract *nat* in the definition of *plus* defined with a recursor.

---

[6] The equality *eq* is an inductive type defined by:
   Inductive eq λA : Set.λx : A. : A → Prop := refl_equal : (eqAxx).

$Definition\ \ \ plusl\ :=$
$\lambda H : nat.$
$(nat\_rec\ \lambda\_ : nat.nat{\rightarrow}nat\ \lambda m : nat.m$
$\quad\ \lambda\_ : nat.\lambda H0 : (nat{\rightarrow}nat).\lambda H1 : nat.(S\ (H0\ H1))\ H).$

As the before the generalization is syntactic, we just abstract the type identifier, the constructors and the recursor to get:

$Definition\ \ \ plusAbs\ :=$
$\lambda s1.\lambda Set.\lambda OAbs : s1.\lambda SAbs : s1{\rightarrow}s1.$
$\lambda s1\_rec : (\forall P : (s1{\rightarrow}Set).$
$\qquad\qquad\qquad (P\ (OAbs)){\rightarrow}((n : s1)(P\ n){\rightarrow}(P\ (SAbs\ n))){\rightarrow}(n : s1)(P\ n)).$
$\quad\lambda\ H : s1.$
$\qquad (s1\_rec\ \lambda\_ : s1.s1{\rightarrow}s1\ \lambda m : s1.m$
$\qquad\quad\ \lambda\_ : s1.\lambda H0 : (s1{\rightarrow}s1).\lambda H1 : s1.(SAbs\ (H0\ H1))\ H).$

Remark that here $OAbs$, and $SAbs$ may not be constructors of the $s1$, so that a *case analysis* may not be possible on element of type $s1$. For this reason we have to work with a version of *plus* written with a recursor and not with *Fixpoint* and *Case*. More Generally if we want to generalize a function written with *Fixpoint* and *Case* we have first to translate it to get a definition using a recursor.

But as highlighted by the example of the inductive type eq and its recursor eq_ind_r in section 7 such generalization is not always useful for reuse.

Nevertheless, there is one situation where such a generalization is very useful: to switch from one data representation to another one. In this case we know that the abstracted type and the one used in the reinstantiation are isomorphic so that the property expressed by the recursor or the first type is provable for the second one.

For example, this allows us to switch from the usual unary representation of natural number to the dyadic one defined by[7] :
$Inductive\ bin : Set := BH : bin|BO : bin{-}>bin|BI : bin{-}>bin.$

Now, in order to instantiate $TabAbs$, $plusl$ or another abstraction, we have first to define the successor

$Definition\ Sbin\ \ \ : bin{\rightarrow}bin\ :=$
$\lambda b : bin.Cases\ b\ of$
$\qquad BH\ \ =>(BO\ BH)$
$\qquad |(BO\ x)\ =>(BI\ x)$
$\qquad |(BI\ x)=>(BO\ (Sbin\ x))$
$end.$

Next to define the unary recursor for bin proving the following lemma :

$Lemma\ bin\_rec\ \ : \forall P : (bin{-}>Set).$
$\quad(P\ (BH)){\rightarrow}(\forall n : bin.(P\ n){\rightarrow}(P\ (Sbin\ n))){\rightarrow}\forall n : bin.(P\ n).$

we get

---

[7] Here, intuitively $BH \equiv O$, $BO\ y \equiv 2y + 1$ and $BI\ y \equiv 2y + 2$.

> $Definition\ TabBin\ :=\lambda s:Set.(TabAbs\ s\ bin\ BH\ Sbin).$
> $Definition\ binplus\ :=\lambda n:bin.(plusAbs\ bin\ zero\ S\ bin\_rec\ n).$

9   Related Work

Proof reuse by generalization and reasoning by analogy have been intensively studied in Artificial Intelligence and in automatic proof systems (in which they are used to generate intermediate lemmas in order to avoid diverging proof search), (see e.g. [DFGS99]), but to our knowledge their is little work in the framework of interactive proof systems based on type theory.

The idea of analogy[Cur95,Mel95,Mel98,KW94] is that when proving a statement S, the proof of a "similar" problem may help. Therefore to prove S we first search for a similar problem S_2 and a mapping between S and S_2 and then we try to use this mapping to build a proof P of S from the proof P_2 of S_2. This relies on higher-order pattern matching, higher-order unification and proof generalization.

Melis [Mel95,Mel98] works in analogy in a context of proof planning and proposes a technique based on a replay of *proof-plan*.

Reif and Stenzel focus on program verification. In [RS93] they address the reuse of (possibly failed) proofs after certain changes of the program. When the program is changed they compute an optimal presentation of the new program as a combination of parts of the old program and new parts. When constructing the new proof, the corresponding part of the old proof is copied to the new proof.

Reconstruction of partial proofs as part of recovering from errors in definitions or in proof strategies is also studied by Felty in [FH94]. In her system written in lambda-prolog a tactic proof is a tree-structured sequent proof where steps may be justified by tactic programs. Her mechanism is based on a proof generalization step that is done after a tactic is applied.

Kolbe and Walther [KW94,KW95] work (in a context of a simple equational calculus) were our first inspiration source. Their idea is to analyze the proof to find its relevant features yielding to a *proof catch* (basically the set of result used in the proof). Next the goal and the catch are generalized by abstracting symbol names as far as possible. This produces a so called *proof shell*. When presented with a new goal, the generalized old goal is mapped onto the new one and the result of the generalized catch are mapped to get a new set of lemma needed for the new proof. The resulting proof obligation may be managed in a same way or proved by a traditional method. Remark that the proof is not replayed but analysis and mapping guarantees that the new goal is proved if only the new lemma can be proved. Their technique also manages the storage and efficient retrieval of applicable generalized proofs.

As seen in section 8, the problem of changing from on particular representation of data to another is also related to this work. It is discussed in details in [MB01]. The basic idea of their work is very close to idea of section 8 but they also propose several heuristics to make the transformation transparent and automatic. In [BP01], we have also proposed another approach involving a modification of the type system. The result is an extension of the CIC, in which we give a computational interpretation of type isomorphism. We show that enhancing the usability of generic frameworks and allowing to switch between equivalent representations of mathematical theories or data structures, type isomorphisms can be used to good effect for proof reuse.

The idea of structuring "a posteriori" as sketched in section 5, also take inspiration from Siff and Reps's works [SR96] in the field of programming languages, when they try to generate generic $C++$ code starting from C code.

10   Conclusion and Perspectives

We have suggested a practical generalization mechanism and detailed a number of situations in which such a mechanism may be useful. We also highlighted some problems and difficulties that appear in practical use and propose pragmatic solutions. We ensure the correctness of the transformations by rejecting terms that do not typecheck after transformation.

In a proof developments many proofs use only some properties of the "objects" on which they work. The relations between objects may be highlighted once and for all by computing a dependency graph [PBR98] allowing a more abstract theory to be built.

The certification of mathematical algorithms for computer algebra must be based on a certified implementation of the basic mathematical structures (groups, rings, algebra etc). In the long run, generalization should make it possible to recover for free all the theorems in existing developments which are still valid for new structures. For example, in the case of groups, proofs which have been done for an instance of these theories such as integers or polynomials may be recovered. In relation to the introduction of the concept of modules in proof assistants, this could make it possible to develop parameterized theories starting from specific instances.

When proving result about mathematic we have an intuition of the similarity between result that help to chose the useful generalization. It is not so clear how such a generalization mechanism could be used when studying programming language semantics and much work remain to be done. As experiment, using the works of Boite [Boi01], we are studying how to use a generalization mechanism to allow reuse of type soundness proof when extending a small functional language.

References

[Asp99] D. Aspinall, et al.   Proof   general   3.0,   December   1999. http://www.proofgeneral.org.

[Bar99] B. Barras, et al. *The Coq Proof Assistant Reference Manual.* INRIA, December 1999. Version 6.3.1.

[Ber97] J. Bertot, et al. User guide to the CtCoq proof environment. Rapport technique RT0210, INRIA, 1997.

[Boi01] O. Boite. Proving type soundness of simply typed ml-like language with refer-
         ences. In *Supplementary Proceedings of the 14th International Conference on
         Theorem Proving in Higher Order Logics (TPHOLs'2001)*, September 2001.
[BP01]   G. Barthe and O. Pons. Type Isomorphisms and Proof Reuse in Dependent
         Type Theory. In F. Honsell and M. Miculan, editors, *Proceedings of FOSSACS
         2001*, LNCS 2030, pages 57–71. Springer, 2001.
[Con86] R. Constable, et al. *Implementing mathematics with the Nuprl proof develop-
         ment system*. Prentice-Hall, 1986.
[Cur95] Régis Curien. *Outils pour la preuve par analogie*. Thèse de doctorat, Univer-
         siteé Henri Poincaré, Nancy I, 1995.
[DFGS99] J. Denzinger, M. Fuchs, C. Goller, and S. Schul. Learning from previ-
         ous proof experience: a survey. Technical report, Technischen Universität
         München, 1999.
[FH94]   A. Felty and D. Howe. Generalization and reuse of tactic proofs. LNCS 822,
         1–15, 1994.
[HR92]   R. Hasker and U. Reddy. Generalization at higher types. In D. Miller, editor,
         Proceedings of the Workshop on the λProlog Programming Language, pages
         257–271, Philadelphia, Pennsylvania, July 1992. University of Pennsylvania.
         Available as Technical Report MS-CIS-92-86.
[KW94]   T. Kolbe and C. Walther. Reusing proofs. In *Proc. of the 11th ECAI*, pages
         80–84, Amsterdam, The Netherlands, 1994.
[KW95]   T. Kolbe and C. Walther. Patching proofs for reuse. In N. Lavrac and
         S. Wrobel, editors, *Proceeding of the European Conference on Machine Learn-
         ing (ECML-95)*, Springer LNAI 912, pages 303–306. Heraklion, Greece, 1995.
[LEG]   The LEGO World Wide Web page. http://www.dcs.ed.ac.uk/home/lego.
[MB01]   N. Magaud and Y. Bertot. Changement de représentation des structures de
         données en Coq: le cas des entiers naturels. In *JFLA 2001*, 2001.
[Mel95] E. Melis. a model of analogy-driven proof-plan construction. In Frank Pfen-
         ning, editor, *Proceedings of the 14th International Joint Conference on Artifi-
         cial Intelligence*, pages 182–189, Montreal, 1995.
[Mel98] E. Melis. AI-techniques in proof planning. In *European Conference on Arti-
         ficial Intelligence*, pages 494–498, 1998.
[PBR98] O. Pons, Y. Bertot, and L. Rideau. Notions of dependency in proof assistants.
         In *Electronic Proceedings of "User Interfaces for Theorem Provers 1998 "*,
         Sophia-Antipolis, France, 1998.
[Pfe91] F. Pfenning. Unification and anti-unification in the Calculus of Constructions.
         In *Sixth Annual IEEE Symposium on Logic in Computer Science*, pages 74–85,
         Amsterdam, The Netherlands, July 1991.
[Plo69] G. D. Plotkin. A note on inductive generalization. In B. Meltzer and D. Michie,
         editors, *Machine Intelligence 5*, pages 153–163, Edinburgh, 1969. Edinburgh
         University Press.
[RS93]   W. Reif and K. Stenzel. Reuse of Proofs in Software Verification. In R. Shyama-
         sundar, editor, *Foundation of Software Technology and Theoretical Computer
         Science. Proceedings*, Springer LNCS 761, pages 284–293. Bombay, India, 1993.
[SR96]   M. Siff and T. Reps. Program generalization for software reuse: From C to
         C++. In *Proceedings of the Fourth ACM SIGSOFT Symposium on the Foun-
         dations of Software Engineering*, volume 21.6 of *ACM Software Engineering
         Notes*, pages 135–146, New York, October 16–18 1996. ACM Press.
[Thé98] L. Théry. A certified version of Buchberger's algorithm. In *Automated Deduc-
         tion (CADE-15)*, LNAI 1421, Springer-Verlag, July 1998.

An Inductive Version of Nash-Williams' Minimal-Bad-Sequence Argument for Higman's Lemma

Monika Seisenberger

[1] Mathematisches Institut der Universität München* * *
[2] Department of Computer Science, University of Wales Swansea[†]

Abstract. Higman's lemma has a very elegant, non-constructive proof due to Nash-Williams [NW63] using the so-called minimal-bad-sequence argument. The objective of the present paper is to give a proof that uses the same combinatorial idea, but is constructive. For a two letter alphabet this was done by Coquand and Fridlender [CF94]. Here we present a proof in a theory of inductive definitions that works for arbitrary decidable well quasiorders.

1 Introduction

This paper is concerned with Higman's lemma [Hig52], usually formulated in terms of well quasi orders.

If $(A, \leq_A)$ is a well quasiorder, then so is the set A^* of finite sequences in A, together with the embeddability relation $\leq_{A^*}$,

where a sequence $[a_1, \ldots, a_n]$ is embeddable in $[b_1, \ldots, b_m]$ if there is a strictly increasing map $f \colon \{1, \ldots, n\} \to \{1, \ldots, m\}$ such that $a_i \leq_A b_{f(i)}$ for all $i \in \{1, \ldots n\}$.

Among the first proofs of Higman's lemma which all were non-constructive the proof of Nash-Williams using the so-called minimal-bad-sequence argument is considered most elegant. A variant of this proof was translated by Murthy via Friedman's A-translation into a constructive proof [Mur91], however resulting in a huge proof whose computational content couldn't yet be discovered. More direct constructive proofs were given by Schütte/Simpson [SS85], Murthy/Russell [MR90], and Richman/Stolzenberg [RS93]. The Schütte/Simpson proof uses ordinal notations up to ϵ_0 and is related to an earlier proof by Schmidt [Sch79], the other proofs are carried out in a (proof theoretically stronger) theory of inductive definitions. However, their computational content is essentially the same, but does not correspond to that one of Nash-Williams' proof. (The proof theoretic strength of the general minimal-bad-sequence argument is $\Pi_1^1-\mathsf{CA}_0$, as

---

* * * Research supported by the DFG Graduiertenkolleg "Logik in der Informatik".
† Research supported by the British EPSRC.

P. Callaghan et al. (Eds.): TYPES 2000, LNCS 2277, pp. 233–242, 2002.

was shown by Marcone, however it is open whether the special form used for Higman's lemma has the same strength [Mar96].)

The objective of this paper is to present a constructive proof that captures the combinatorial idea behind Nash-Williams' proof. For an alphabet A consisting of two letters this was done by Coquand and Fridlender [CF94]. Their proof can quite easily be extended to a finite alphabet. To obtain a proof for arbitrary decidable well quasiorders, more effort is necessary, as we will describe in section 3.

A proof of Higman's lemma which in contrast to all proofs mentioned above does not require decidability of the given relation $\leq_A$ was given by Fridlender [Fri97]. His proof is based on a proof by Veldman that can be found in [Vel00]. In our formulation of Higman's lemma we will also use an accessibility notion, as it was done in Fridlender's proof.

2   Basic Definitions and an Inductive Characterization of Well Quasiorders

In the whole paper we assume $(A, \leq_A)$ to be a set with a reflexive and transitive, decidable relation.[1]

Definition 1. We use

$a, b, \ldots$ for letters, i.e., elements of a A,

$as, bs, \ldots$ for finite sequences of letters, i.e. elements of A^*,

$v, w, \ldots$ for words, i.e., elements of A^*, [2]

$us, ws, \ldots$ for finite sequences of words, i.e., elements of A^{**} .

By $as * a$ we denote the sequence obtained from the sequence as by appending the element a. $ws * w$ is defined similarly. At some places we add brackets to keep the expressions legible. However, unary function application will be written without brackets, in general.

For a finite sequence ws of non-empty words let $\mathsf{lasts}\, ws$ denote the finite sequence consisting of the end-letters of the words of ws, that is,[3]

$$\mathsf{lasts}\,[w_1 * a_1, \ldots, w_n * a_n] = [a_1, \ldots, a_n], \ n \geq 0.$$

---

[1] Whereas transitivity is only required for historical reasons, but is not used in our proof, decidability plays an essential role.

[2] Although of the same kind we distinguish between finite sequences (of letters) and words, because they will play different rolls, as is illustrated in the picture on the right.

[3] In our picture we have $\mathsf{lasts}\,[w_1, \ldots, w_5] = as$.

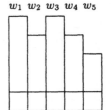

$w_1 \ w_2 \ w_3 \ w_4 \ w_5$

as

Definition 2 (Higman Embedding). The embedding relation on A^* can be inductively described by the following rules:

$$\frac{}{[] \leq_{A^*} []} \qquad \frac{v \leq_{A^*} w}{v \leq_{A^*} w*a} \qquad \frac{v \leq_{A^*} w,\ a \leq_A b}{v*a \leq_{A^*} w*b}.$$

Definition 3 (Good/Bad). A finite sequence $[a_1, \ldots, a_n]$ (respectively an infinite sequence $a_1, a_2, \ldots$) of elements in A is good if there exist $i < j \leq n$ ($i < j < \omega$) such that $a_i \leq_A a_j$; otherwise it is called bad.

Furthermore, we use the notion $\mathsf{good}(\mathit{as}, a)$ if there is an element in as, say the i-th one, such that $(\mathit{as})_i \leq_A a$. $\mathsf{bad}(\mathit{as}, a)$ stands for $\neg\mathsf{good}(\mathit{as}, a)$.

Finally, $\mathsf{badsubseq}(\mathit{as})$ determines the first occurring bad subsequence in as:

$$\mathsf{badsubseq}([]) \quad = []$$
$$\mathsf{badsubseq}(\mathit{as}*a) = \begin{cases} \mathsf{badsubseq}(\mathit{as})*a & \text{if } \mathsf{bad}(\mathsf{badsubseq}(\mathit{as}), a), \\ \mathsf{badsubseq}(\mathit{as}) & \text{otherwise.} \end{cases}$$

Definition 4 (Well Quasiorder). $(A, \leq_A)$ is a well quasiorder (wqo) if every infinite sequence of elements in A is good.

Definition 5 (The Relation $\ll_A$ and Its Accessible Part). The relation $\ll_A \subseteq A^* \times A^*$ is defined by

$$\mathit{bs} \ll_A \mathit{as} :\leftrightarrow \mathit{bs} = \mathit{as}*a \text{ for some } a \in A \text{ s.t. } \mathsf{bad}(\mathit{as}, a).$$

The accessible part (also called the well-founded part) of the relation $\ll_A$ is inductively given by the rule

$$\frac{\forall \mathit{bs} \ll_A \mathit{as} \quad \mathsf{acc}_{\ll_A} \mathit{bs}}{\mathsf{acc}_{\ll_A} \mathit{as}}$$

and provides the following induction principle[4] for any formula ϕ:

$$\frac{\forall \mathit{as}.\, \forall \mathit{bs} \ll_A \mathit{as}\ \phi(\mathit{bs}) \to \phi(\mathit{as})}{\forall \mathit{as}.\, \mathsf{acc}_{\ll_A} \mathit{as} \to \phi(\mathit{as})}.$$

Definition 6 (Well Quasiorder, Inductive Characterization[5]). $(A, \leq_A)$ is a well quasiorder if $\mathsf{acc}_{\ll_A} []$ holds.

Definitions 3 to 6 should be understood for arbitrary (reflexive and transitive) relations, not only for our fixed $(A, \leq_A)$. We will use them also for $(A^*, \leq_{A^*})$. Moreover, the operation acc will also be applied to the relations $\ll_{A^*}$ and $\prec$, still to be defined.

---

[4] At some places we use a seemingly stronger induction principle where the premise is of the form $\forall \mathit{as}.\mathsf{acc}_{\ll_A} \mathit{as} \to \forall \mathit{bs} \ll_A \mathit{as}\ \phi(\mathit{bs}) \to \phi(\mathit{as})$. This principle can be easily derived from the above one.

[5] In our paper we only deal with this second definition of a well quasiorder since it is very suitable for a constructive proof. For sake of completeness we give an argument for the equivalence of definition 4 and definition 6. To prove that definition 6 implies definition 4 one shows more generally that for all as such that $\mathsf{acc}_{\ll_A} \mathit{as}$ holds every infinite sequence, starting with as, is good. The reverse direction is an instance of Brouwer's axiom of bar induction.

3   Towards a Constructive Proof

In order to motivate further definitions we first want to give the idea behind the constructive proof. This is best done by showing the connection between the classical and the constructive proof. To this end we shortly recall Nash-Williams' minimal-bad-sequence proof and show how the main steps are captured by the inductive proof. We also include an informal idea of the latter.

The steps of the Nash-Williams' proof:

1. In order to show "wqo(A) implies wqo(A^*)", assume for contraction that there is a bad sequence of words.
2. Among all infinite bad sequences we choose (using classical dependent choice) a minimal bad sequence, i.e., a sequence, say $(w_i)_{i<\omega}$, which is minimal with respect to a lexicographical order on infinite sequences of words (where w_1 is less or equal w_2, if w_1 is an initial segment of w_2).
3. Since $w_i \neq [\,]$, let $w_i = v_i * a_i$ for all i. Using Ramsey's theorem and the fact that our alphabet A is a well quasiorder, we know that there exists an infinite subsequence $a_{\kappa_1} \leq_A a_{\kappa_2} \leq_A \cdots$ of the sequence $(a_i)_{i<\omega}$. This also determines a corresponding sequence $w_1, \ldots, w_{\kappa_1-1}, v_{\kappa_1}, v_{\kappa_2}, \ldots$.
4. The sequence $w_1, \ldots, w_{\kappa_1-1}, v_{\kappa_1}, v_{\kappa_2}, \ldots$ must be bad (otherwise $(w_i)_{i<\omega}$ would be good), but this contradicts the minimality in 2.

In the constructive proof this steps correspond to

1. Prove inductively "$\mathrm{acc}_{\ll_A} [\,] \to \mathrm{acc}_{\ll_{A^*}} [\,]$".
2. The minimality argument will be replaced by structural induction on words.
3. Given a bad sequence $ws = [w_1, \ldots, w_n]$ s.t. $w_i = v_i * a_i$, we are interested in all subsequences $a_{\kappa_1} \leq_A \cdots \leq_A a_{\kappa_l}$ of maximal length[6] and their corresponding sequences $w_1, \ldots, w_{\kappa_1-1}, v_{\kappa_1}, \ldots, v_{\kappa_l}$. In the proof these sequences will be computed by the procedure forest which takes ws as input and yields a forest labeled by pairs in $A^{**} \times A$. In the produced forest the right-hand components of each path form a weakly ascending subsequence of $[a_1, \ldots, a_n]$ and the corresponding sequence of form $w_1, \ldots, w_{\kappa_1-1}, v_{\kappa_1}, \ldots, v_{\kappa_l}$ could be read off in the left-hand component of the endnode of such a path. If we extend the sequence ws badly by a word $v*a$, then in the existing forest either new nodes, possibly at several places, are inserted, or a new singleton tree with node $\langle ws*v, a \rangle$ is added. Now the informal idea of the inductive proof is: if in forest ws not infinitely often new nodes could be inserted and if also not infinitely often new trees could be added, then ws could not be extended badly infinitely often. Formally this will be captured by the statement: $\forall ws.\ \mathrm{acc}_{\ll_A} \mathrm{badsubseq}(\mathrm{lasts}\ ws) \to \mathrm{acc}_{\prec} \mathrm{forest}\ ws \to \mathrm{acc}_{\ll_{A^*}} ws$.
4. The first part of item 4 corresponds to lemma 1.

---

[6] By maximal length we mean that we only look at those subsequences which are ascending, but not contained in other ones, for instance our chosen subsequences of $[1, 4, 3, 0, 3]$ are $[1, 4]$, $[1, 3, 3]$ and $[0, 3]$.

We proceed with the formal definition of forest and the relation $\prec$ on forests.

Definition 7. We use

t for elements in $T(A^{**} \times A)$, i.e., trees labeled by pairs in $A^{**} \times A$,

f, ts for elements in $(T(A^{**} \times A))^*$, i.e., forests.

The tree with root $\langle ws, a \rangle$ and finite sequence of immediate subtrees ts is written $\langle ws, a \rangle ts$. We use the destructors left and right for pairs and the destructors root and subtrees for trees, hence $\text{root}\,\langle ws, a \rangle ts = \langle ws, a \rangle$ and $\text{subtrees}\,\langle ws, a \rangle ts = ts$. For better readability we set:

$$
\begin{aligned}
\text{newtree}\,\langle ws, a \rangle &:= \langle ws, a \rangle [\,], \\
\text{roots}\,[t_1, \ldots, t_n] &:= [\text{root}\,t_1, \ldots, \text{root}\,t_n], \\
\text{lefts}\,[\langle ws_1, a_1 \rangle, \ldots, \langle ws_n, a_n \rangle] &:= [ws_1, \ldots, ws_n], \\
\text{rights}\,[\langle ws_1, a_1 \rangle, \ldots, \langle ws_n, a_n \rangle] &:= [a_1, \ldots, a_n].
\end{aligned}
$$

Definition 8. Let $ws \in A^{**}$ be a sequence of non-empty words. Then forest $ws \in T((A^{**} \times A))^*$ is defined recursively by: [7] [8]

$$
\begin{aligned}
\text{forest}\,[\,] \quad &= [\,], \\
\text{forest}\,ws{*}(w{*}a) &= \begin{cases} \text{insertforest}(\text{forest}\,ws, w, a) & \text{if good}(\text{badsubseq}(\text{lasts}\,ws), a) \\ (\text{forest}\,ws) * \text{newtree}\,\langle ws{*}w, a \rangle & \text{otherwise,} \end{cases}
\end{aligned}
$$

where

$$
\text{insertforest}(f, w, a) = \text{map}\left(\lambda t \begin{bmatrix} \text{if} & \text{right}\,(\text{root}\,t) \leq_A a \\ & \text{inserttree}(t, w, a) \\ & t \end{bmatrix} \right) f
$$

and

$$
\begin{aligned}
&\text{inserttree}(\langle ws, a' \rangle ts, w, a) = \\
&\quad \begin{cases} \langle ws, a' \rangle \text{insertforest}(ts, w, a) & \text{if good}(\text{rights}\,(\text{roots}\,ts), a), \\ \langle ws, a' \rangle (ts * \text{newtree}\,\langle ws{*}w, a \rangle) & \text{otherwise.} \end{cases}
\end{aligned}
$$

---

[7] For sake of simplicity we define insertforest by a map operation, i.e., we insert a new node at every possible place. However, it would already suffice to insert at least once and it even could be arbitrarily chosen where to insert.

[8] In case of a finite alphabet forest ws has only non-branching trees where the right-hand components of such a tree are constant. So, if in the notion of [CF94] we have $T_0(ws, ws)$ or $R_0(ws, ws)$, in our setting the sequence ws could be read off as the left-hand component in the endnode of the tree whose right-hand components are 0.

Definition 9. Let f and f' be forests in $T((A^{**} \times A))^*$. Then

$$f' \prec f :\leftrightarrow \begin{cases} f' = \text{insertforest}(f, w, a) \text{ for some } w \in A^*, a \in A \\ \text{such that } f' \neq f \text{ and the left-hand component of} \\ \text{each label in } f' \text{ is a bad sequence in } A^{**}. \end{cases}$$

Lemma 1. Let ws be a bad sequence of non-empty words. Then in every label of forest ws the left-hand component is a bad sequence.

Proof. IND(structure of ws). 1. $ws = []$. Clear.

2. Assume that every left-hand component of a label in forest ws is bad and look at the nodes in forest $ws*(w*a)$ where $ws*(w*a)$ is assumed to be bad.

Case 1: bad(badsubseq(lasts ws), a). Then in forest $ws*(w*a)$ only one node was added, i.e., the node with label $\langle ws*w, a \rangle$ where by assumption $ws*w$ is bad.

Case 2: good(badsubseq(lasts ws), a). In this case[9] some nodes of the form $\langle vs*w, a \rangle$ were inserted in forest ws where vs is a left-hand component of a node in forest ws which by assumption is bad. Assume good(vs, w), that is, $\exists i (vs)_i \leq_{A^*} w$ and show $\bot$.

Case 2.1: $(vs)_i$ is a word in ws. Then, by the Higman embedding we obtain $(vs)_i \leq_{A^*} w*a$ – contradicting the badness of $ws*(w*a)$.

Case 2.2: $(vs)_i$ is a word in ws cut by an end letter a_0 and by the construction of the forests it holds $a_0 \leq_A a$.[10] Then, again by the Higman embedding it follows $(vs)_i*a_0 \leq_{A^*} w*a$. Contradiction.   □

Lemma 2. i) $\text{acc}_\prec []$. ii) $\text{acc}_\prec f \wedge \text{acc}_\prec [t] \rightarrow \text{acc}_\prec f * t$.

Proof. i) $\text{acc}_\prec []$ holds by definition, since there is no tree in which new nodes could be inserted. ii) Clear, since insertforest is defined by a map-operation.   □

Lemma 3. Assume $\text{acc}_{\ll_A} []$. Then $\forall ws. \text{acc}_{\ll_{A^*}} ws \rightarrow \forall a. \text{acc}_\prec [\text{newtree} \langle ws, a \rangle]$.

Proof. $\text{IND}_1(\text{acc}_{\ll_A})$: $\text{IH}_1: \forall vs \ll_{A^*} ws, \forall a. \text{acc}_\prec [\text{newtree} \langle vs, a \rangle]$. Let $a \in A$. Instead of proving $\text{acc}_\prec [\text{newtree} \langle ws, a \rangle]$ we show more generally that this assertion holds for all t with root $t = \langle ws, a \rangle$ such that

(a) the subtrees of t form a forest in $\text{acc}_\prec$ and

(b) rights (roots (subtrees t)) is sequence in $\text{acc}_{\ll_A}$.[11]

We do this by main induction on (b) and side induction on (a), i.e., formally we prove

$\forall as. \text{acc}_{\ll_A} as \rightarrow$
$\forall ts. \text{acc}_\prec ts \rightarrow$
$\forall t. \text{root } t = (ws, a) \wedge \text{subtrees } t = ts \wedge as = \text{rights (roots (subtrees } t)) \rightarrow$
$\quad \text{acc}_\prec [t]$.

---

[9] Here, we only sketch the combinatorial part of the proof; a formal proof involves an induction on the tree structure.

[10] Note that transitivity is not required.

[11] It's intended that [t] lies in the image of the partial function forest, however we don't need this restriction in the formulation of the lemma.

$\mathsf{IND}_2(\mathsf{acc}_{\ll_A})$. Assume that we have an $\textit{as}$ such that $\mathsf{acc}_{\ll_A}\,\textit{as}$.

$\mathsf{IND}_3(\mathsf{acc}_\prec)$. Let $\textit{ts}$ be such that $\mathsf{acc}_\prec\,\textit{ts}$ and fix t such that $\mathsf{root}\,t = \langle \textit{ws}, a\rangle$, subtrees $t = \textit{ts}$, and $\textit{as} = \mathsf{rights}\,(\mathsf{roots}\,(\mathsf{subtrees}\,t))$.

We have to prove $\mathsf{acc}_\prec\,[t]$. By the definition of $\mathsf{acc}_\prec$ and $\prec$ if suffices to show $\forall t'.\,[t'] \prec [t] \to \mathsf{acc}_\prec\,[t']$ where $t' = \mathsf{inserttree}(t, w, a') \neq t$ for some $w \in A^*, a' \in A$ and all left-hand components of nodes in t' are required to be bad. We prove the assertion by case distinction on the definition of $\mathsf{inserttree}$.

Case 1: $t' = \langle \textit{ws}, a\rangle\,(\textit{ts} * \mathsf{newtree}\,\langle \textit{ws}*w, a'\rangle)$ for some w and a' such that $\mathsf{bad}(\textit{as}, a')$. Then we have

$$\begin{aligned}\mathsf{root}\,t' &= \langle \textit{ws}, a\rangle, \\ \mathsf{subtrees}\,t' &= \textit{ts} * \mathsf{newtree}\,\langle \textit{ws}*w, a'\rangle, \\ \textit{as}*a' &= \mathsf{rights}\,(\mathsf{roots}\,(\textit{ts} * \mathsf{newtree}\,\langle \textit{ws}*w, a'\rangle)).\end{aligned}$$

Since all left-hand components in t' are supposed to be bad, in particular, we have that $\textit{ws}*w$ is bad, i.e., $\textit{ws}*w \ll_{A^*} \textit{ws}$. By IH_1 we obtain $\mathsf{acc}_\prec\,[\mathsf{newtree}\,\langle \textit{ws}*w, a'\rangle]$, and hence by lemma 2

$$\mathsf{acc}_\prec\,\textit{ts} * \mathsf{newtree}\,\langle \textit{ws}*w, a'\rangle.$$

Now, since $\textit{as}*a' \ll_A \textit{as}$, we may apply IH_2 to $\textit{as}*a', \textit{ts} * \mathsf{newtree}\,\langle \textit{ws}*w, a'\rangle$ and t' and conclude $\mathsf{acc}_\prec\,[t']$.

Case 2: $t' = \langle \textit{ws}, a\rangle\,\mathsf{insertforest}(\textit{ts}, w, a')$ where a' such that $\mathsf{good}(\textit{as}, a')$. In this case we have

$$\begin{aligned}\mathsf{root}\,t' &= \langle \textit{ws}, a\rangle, \\ \mathsf{subtrees}\,t' &= \mathsf{insertforest}(\textit{ts}, w, a'), \\ \textit{as} &= \mathsf{rights}\,(\mathsf{roots}\,(\mathsf{subtrees}\,t')).\end{aligned}$$

Moreover $[t'] \prec [t]$ implies subtrees $t' \prec$ subtrees $t = \textit{ts}$, and by IH_3, applied to subtrees t' and t', we end up with $\mathsf{acc}_\prec\,[t']$.

Now, the proof of the general assertion is completed, and we may put $\textit{as} = []$, $f = []$ and $t = \mathsf{newtree}\,\langle \textit{ws}, a\rangle$. Since we have $\mathsf{acc}_{\ll_A}\,[]$ by assumption and $\mathsf{acc}_\prec\,[]$ by lemma 2, we obtain $\mathsf{acc}_{\ll_{A^*}}\,\textit{ws} \to \mathsf{acc}_\prec\,[\mathsf{newtree}\,\langle \textit{ws}, a\rangle]$. $\qquad\square$

4   The Proof of Higman's Lemma

Proposition 1 (Higman's Lemma). $\mathsf{acc}_{\ll_A}\,[] \to \mathsf{acc}_{\ll_{A^*}}\,[]$.

Proof. Assume $\mathsf{acc}_{\ll_A}\,[]$. We show more generally

$$\begin{aligned}&\forall \textit{as}.\ \mathsf{acc}_{\ll_A}\,\textit{as} \to \\ &\forall f.\ \mathsf{acc}_\prec\,f \to \\ &\forall \textit{ws}.\ \textit{as} = \mathsf{badsubseq}(\mathsf{lasts}\,\textit{ws}) \wedge f = \mathsf{forest}\,\textit{ws} \to \mathsf{acc}_{\ll_{A^*}}\,\textit{ws}.\end{aligned}$$

$\mathsf{IND}_1(\mathsf{acc}_{\ll_A})$. Let $\textit{as}$ be such that $\mathsf{acc}_{\ll_A}\,\textit{as}$ and $\mathsf{IH}_1: \forall \textit{bs} \ll_A \textit{as}, \forall f.\,\mathsf{acc}_\prec\,f \to \forall \textit{ws}.\,\textit{bs} = \mathsf{badsubseq}(\mathsf{lasts}\,\textit{ws}) \wedge f = \mathsf{forest}\,\textit{ws} \to \mathsf{acc}_{\ll_{A^*}}\,\textit{ws}$.

$\text{IND}_2(\text{acc}_\prec)$. Let f be s.t $\text{acc}_\prec f$ and IH_2: $\forall f' \prec f$, $\forall ws$. $\text{badsubseq}(\text{lasts } ws) = as \wedge f' = \text{forest } ws \rightarrow \text{acc}_{\ll_A}. ws$ and assume that we have ws such that $as = \text{badsubseq}(\text{lasts } ws)$ and $f = \text{forest } ws$. In order to prove $\text{acc}_{\ll_A}. ws$ we fix w s.t. $ws*w$ is bad and show $\text{acc}_{\ll_A}. ws*w$ by induction on the structure of w:

$\text{IND}_3(w)$. 1. $\text{acc}_{\ll_A}. ws[]$ holds by definition of $\text{acc}_{\ll_A}.$.

2. Now, assume that we have a word of form $w*a$. We show $\text{acc}_{\ll_A}. ws*(w*a)$ by case analysis on whether or not $\text{bad}(as, a)$.

Case 2.1: $\text{bad}(as, a)$. Then we have

$$as*a = \text{badsubseq}(\text{lasts } (ws*(w*a))),$$
$$f * \text{newtree } \langle ws*w, a \rangle = \text{forest } (ws*(w*a)).$$

First, we show $\text{acc}_\prec f * \text{newtree } \langle ws*w, a \rangle$. By assumption we already have $\text{acc}_\prec f$ and by IH_3 $\text{acc}_{\ll_A}. ws*w$. Hence, by lemma 3 we obtain $\text{acc}_\prec [\text{newtree } \langle ws*w, a \rangle]$ and by lemma 2 we may conclude

$$\text{acc}_\prec f * \text{newtree } \langle ws*w, a \rangle.$$

Now we are able to apply IH_1 (to $as*a$, $f * \text{newtree } \langle ws*w, a \rangle$ and $ws*(w*a)$) and end up with $\text{acc}_{\ll_A}. ws*(w*a)$.

Case 2.2: $\text{good}(as, a)$. In this case it follows

$$as = \text{badsubseq}(\text{lasts } (ws*(w*a))),$$
$$\text{insertforest}(f, w, a) = \text{forest } (ws*(w*a)).$$

By lemma 1 all left-hand components of nodes in $\text{insertforest}(f, w, a)$ are bad. Moreover, $\text{insertforest}(f, w, a) \neq f$ since $\text{good}(as, a)$ and $\text{badsubseq}(\text{lasts } ws) = as = \text{rights}(\text{roots}(\text{forest } ws))$ imply that indeed at least one node was inserted. Hence, we obtain

$$\text{insertforest}(f, w, a) \prec f$$

and we may apply IH_2 (to $\text{insertforest}(f, w, a)$ and $ws*(w*a)$) and conclude $\text{acc}_{\ll_A}. ws*(w*a)$.

This completes the proof of the general assertion. Now, by putting $as = [], f = []$ and $ws = []$ and the fact that $\text{acc}_\prec []$ holds by definition we obtain $\text{acc}_{\ll_A} [] \rightarrow \text{acc}_{\ll_A}. []$. $\qquad\square$

5 Conclusion

We presented a new constructive proof of Higman's lemma for arbitrary decidable well quasiorders in a theory of inductive definitions. We hope not only that this proof gives more insight in the interplay of classical proofs using a minimal bad sequence argument and constructive proofs, but also that this strategy is extendible to other non-constructive theorems, for instance Kruskal's tree theorem and the so-called extended Kruskal theorem, also known as Kruskal's theorem with gap condition. Both have proofs using a minimal-bad-sequence argument (see [NW63] resp. [Sim85]), however no constructive proof at all is known for the latter. Kruskal's theorem was proved constructively (see [RW93] for a proof using ordinal notations or [Sei01] for an inductive reformulation of this proof, and [Vel00] for a proof not requiring decidability). These proofs, however, are quite involved in comparison with the minimal-bad-sequence proof.

We do not claim that our proof of Higman's lemma is 'better' than the other constructive proofs mentioned in the introduction, but, as already stated, it uses a different combinatorial idea, hence results in another algorithm. An analysis of these different algorithms is still missing and could give rise to an interesting case study in machine supported theorem proving.

Acknowledgment

I am grateful to Ulrich Berger, Thierry Coquand, and the referees for their helpful comments and suggestions on this article, and particularly I would like to thank Daniel Fridlender for valuable discussions during his Munich visit in June 2000. The idea to use a tree structure is due to him.

References

[CF94]   Thierry Coquand and Daniel Fridlender. A proof of Higman's lemma by structural induction, 1994.
ftp://ftp.cs.chalmers.se/pub/users/coquand/open1.ps.Z

[Fri97]   Daniel Fridlender. *Higman's Lemma in Type Theory*. PhD thesis, Chalmers University of Technology and University of Göteburg, Sweden, Oktober 1997.

[Hig52]   Graham Higman. Ordering by divisibility in abstract algebras. *Proc. London Math. Soc.*, 2:326–336, 1952.

[Mar96]   Alberto Marcone. On the logical strength of Nash-Williams' theorem on transfinite sequences. In: Logic: from Foundations to Applications; European logic colloquium, pp. 327–351, 1996.

[MR90]   Chetan R. Murthy and James R. Russell. A Constructive proof of Higman's Lemma. In Proc. Fifth Symp. on Logic in Comp. Science, pp. 257–267, 1990.

[Mur91]   Chetan R. Murthy. An Evaluation Semantics for Classical Proofs. In Proc. Sixth Symp. on Logic in Computer Science, pp. 96–109, 1991.

[NW63]   Crispin St. J. A. Nash-Williams. On well-quasi-ordering finite trees. *Proc. Cambridge Phil. Soc.*, 59:833–835, 1963.

[RW93]   Michael Rathjen and Andreas Weiermann. Proof–theoretic investigations on Kruskal's theorem. *Annals of Pure and Applied Logic*, 60:49–88, 1993.

[RS93]   Fred Richman and Gabriel Stolzenberg. Well Quasi-Ordered Sets. *Advances in Math.*, 97:145–153, 1993.

[Sei01]   Monika Seisenberger. Kruskal's tree theorem in a constructive theory of inductive definitions In: Reuniting the Antipodes - Constructive and Nonstandard Views of the Continuum. Synthese Library, Kluwer, Dordrecht, forthcoming.

[SS85]   Kurt Schütte and Stephen G. Simpson. Ein in der reinen Zahlentheorie unbeweisbarer Satz über endliche Folgen von natürlichen Zahlen. *Archiv für Mathematische Logik und Grundlagenforschung*, 25:75–89, 1985.

[Sim85] Stephen G. Simpson. Nonprovability of certain combinatorial properties of finite trees. In L.A. Harrington, et al., eds., *Harvey Friedman's Research on the Foundations of Mathematics*, pp. 87–117. North–Holland, Amsterdam, 1985.

[Sch79] Diana Schmidt. Well-orderings and their maximal order types, 1979. Habilitationsschrift, Mathematisches Institut der Universität Heidelberg.

[Vel00] Wim Veldman. An intuitionistic proof of Kruskal's Theorem. Report no. 0017, Department of Mathematics, University of Nijmegen, 2000.

Author Index

Lecture Notes in Computer Science

For information about Vols. 1–2194
please contact your bookseller or Springer-Verlag